Guilford L. Spencer

A Hand-Book for Chemists of Beet-Sugar Houses and Seed-Culture Farms

Containing selected methods of analysis, sugar-house control, reference tables, etc.

First Edition

Guilford L. Spencer

A Hand-Book for Chemists of Beet-Sugar Houses and Seed-Culture Farms
Containing selected methods of analysis, sugar-house control, reference tables, etc. First Edition

ISBN/EAN: 9783337139155

Printed in Europe, USA, Canada, Australia, Japan

Cover: Foto ©berggeist007 / pixelio.de

More available books at **www.hansebooks.com**

A HANDBOOK

FOR

CHEMISTS OF BEET-SUGAR HOUSES

AND

SEED-CULTURE FARMS.

CONTAINING

SELECTED METHODS OF ANALYSIS, SUGAR-HOUSE CONTROL, REFERENCE TABLES, ETC., ETC.

BY

GUILFORD L. SPENCER, D.Sc.,

OF THE U. S. DEPARTMENT OF AGRICULTURE,

Author of "A Handbook for Sugar Manufacturers."

FIRST EDITION.

FIRST THOUSAND.

NEW YORK:

JOHN WILEY & SONS.

LONDON: CHAPMAN & HALL, LIMITED.

1897.

ROBERT DRUMMOND, ELECTROTYPER AND PRINTER, NEW YORK.

PREFACE.

AT the time the writer's "Handbook for Sugar Manufacturers" was published, 1889, the sugar industry of the United States was confined almost exclusively to the cane sections of the South. Sorghum was attracting attention in the North, with some prospect of success; the beet industry was represented by two factories in California and dismantled factories in several other States. The conditions at this time are quite different. The beet-sugar industry bids fair to attain enormous proportions, and sorghum, for the present, at least, has given up the struggle.

Under these changed conditions there appears to be an opening for a book devoted exclusively to the sugar-beet, hence this work.

In the preparation of this book it is assumed that the reader is familiar with many of the ordinary chemical manipulations, but the fact is recognized that on account of the short manufacturing season many factories are compelled to employ assistants whose chemical knowledge is somewhat limited.

In order to avoid repetition, methods of sampling are described in a special chapter.

It is appropriate to mention here some of the men through whose efforts the sugar-beet has been successfully introduced into the United States. Among these are Dr. William McMurtrie, who visited the beet-sugar districts of Europe in 1880 and published a very complete report on the industry. Dr. H. W. Wiley, Chemist of the U. S. Department of Agriculture, has labored incessantly for the promotion of sugar-manufacture in this country, and has

iii

published many able and exhaustive reports upon the subject. Mr. E. H. Dyer, after repeated disappointments
which would have discouraged the bravest advocates of the
sugar-beet, succeeded in establishing the Alvarado factory
in California, the pioneer of the successful American beet-
sugar houses. Mr. Claus Spreckels, through his large in-
vestments in the Watsonville, Cal., works, and the prestige
of his renown as a successful sugar-manufacturer, has
given the advocates of the industry great encouragement.
The work of Mr. Henry T. Oxnard gave renewed impetus
to beet-sugar manufacture, and has been of material value
in demonstrating its financial success when backed by
thoroughly scientific and systematic preparations. Many
others have done much to encourage the culture of the
sugar-beet. Among these may be mentioned Mr. Lewis S.
Ware, of Philadelphia, who has for several years published a journal devoted to the sugar-beet without other
compensation than the satisfaction of encouraging a new
and promising industry.

I take this opportunity of acknowledging many refer-
ences to methods and suggestions given me by Mr. Ervin
E. Ewell, Assistant Chemist of the U. S. Department of
Agriculture, and of thanking him for many courtesies.

G. L. Spencer.

Washington, D. C., 1897.

TABLE OF CONTENTS.

References are to pages.

ANALYSIS OF LIME.

ANALYSIS OF SULPHUR.

ANALYSIS OF COKE.

LUBRICATING OILS.

ANALYSIS AND PURIFICATION OF WATER.

SEED-SELECTION.

SEED-TESTING.

MISCELLANEOUS NOTES.

SUGAR-HOUSE NOTES.

SPECIAL REAGENTS.

LIST OF ILLUSTRATIONS.

HANDBOOK

FOR

SUGAR-HOUSE CHEMISTS.

SUGAR-HOUSE CONTROL.

1. General Remarks.—The control of sugar-house work requires the analysis of the various products at each stage of the manufacture, and the tabulation of the results. From the data supplied by the analyses, the weights and measures of the raw material and the products, the chemist endeavors to trace the losses. The sugar received by the factory, in the beets, is charged on one side of the account, and that in the products and known losses is credited on the other side. The two sides of this account never balance owing to small unavoidable inaccuracies in methods, and to losses which cannot be located or measured.

The question of the detection, location, and estimation of the losses of sugar in the processes of the manufacture is often very complicated, and its solution requires the highest degree of skill on the part of the chemist. As the processes become more complicated through efforts to extract the uttermost grain of sugar from the beet, the difficulties which beset the chemist increase.

In many houses it is impossible to trace the losses quantitatively, through lack of tank-room, etc.

The slightest analytical error will sometimes result in figures of negative value and necessitate their rejection. The so-called "losses from unknown sources," "undeterminable losses," and "mechanical losses," are probably in many cases the result of unavoidable errors in weights and measures or in sampling and analysis.

If an apparent loss be too large to be attributable to a reasonable allowance for error, it is well to view its existence with doubt, until it is verified by repeated observations.

The work of the chemist is further complicated in sugar-houses which treat the molasses by a saccharate process, especially a lime process in which the saccharate is used in liming the juice.

The adjustment of the analytical instruments should be frequently verified. The calibration of graduated ware should be checked. (*See* pages 231 and 250.)

The chemical control of a sugar-house does not end with the tracing and location of losses; it is also necessary to control the processes of manufacture. Each product should be studied, and the influence of each of the processes on the yield of the sugar noted. Slight modifications in the treatment of the material at various stages of the manufacture are often suggested by the work of the chemist, and result in an increased yield of sugar.

Analytical data should be promptly obtained and tabulated, also all manufacturing data. Blank forms are given in pages 302 *et seq.* for permanent records for the chemist's use. The comparison of the data obtained in one period with those of another will always raise the questions, ''Why is the yield of sugar smaller in one period than in the other?'' and '' Why are the losses greater or less this week than last?''

The writer has always made it a practice, in the control of sugar-house work, to divide the season into periods of one week each, and estimate the yield and losses, so far as practicable, in each. (*See* **14.**)

2. The Basis of Sugar-house Control.—It is evident that sugar-house control must begin at a stage where the amount of sugar entering the factory can be accurately determined. In order to include the diffusion it must begin with the weight of the beets. The weight of the beets cannot be deduced with accuracy from the average volume of a definite weight of cuttings as measured in the diffusers.

The objections to the use of the net weight as determined by the deduction of the estimated tare from the gross weight are (1) the element of uncertainty due to an estimate, and (2) that portions of the beet, for which a deduction is made in the tare, reach the diffusion-battery.

In those countries where the clean beets are weighed as they enter the cutters, by the government officials, the control should begin with the cuttings. This affords the only strictly reliable method of checking the work of the diffusion-battery, since the losses at this stage must be the difference between the weight of sucrose in the beets, as determined by analysis of the cuttings, and that in the diffusion-juice.

In the absence of the weights of the beets as indicated above, the control of the general work of the factory must begin with the weight of the diffusion-juice.

It is very probable that the so-called "losses from unknown sources," "mechanical losses," and "undetermined losses" are largely due to errors in weights and measures, and inaccuracies in sampling and analysis, rather than to actual losses.

This suggests that all instruments and graduated ware be carefully checked, and that weights of the raw material be adopted, instead of gauging, where practicable.

Claassen,[1] a prominent German authority, recommends the automatic scale constructed by Reuther & Reisert, Hennef, Germany, for weighing the beets immediately before they are sliced. He states that this scale is prefectly reliable.

The eminent French sugar engineer Charles Gallois has devised an apparatus which insures accurate weights. This apparatus is so arranged that the small car in which the roots are weighed cannot leave the scale unless it contain the correct weight of beets.

WEIGHTS AND MEASURES.

3. System of Weights.—In view of the fact that all chemists employ the metric system in their analytical work,

---

[1] *Zeit. Rübenzucker-Industrie*, 1895, 1084.

and that manufacturers in this country still adhere to the
English, it is necessary in a work of this kind to use both
systems of weights and measures.

4. Net Weight of the Beets. — The beets as received at the factory have been topped with more or less
care, and have variable quantities of earth and pebbles adhering to them. These conditions necessitate the careful
determination of an allowance for tare.

As nearly an average sample of the roots as is practicable
is selected. This sample should consist of as many beets
as can be conveniently taken, the larger the number the

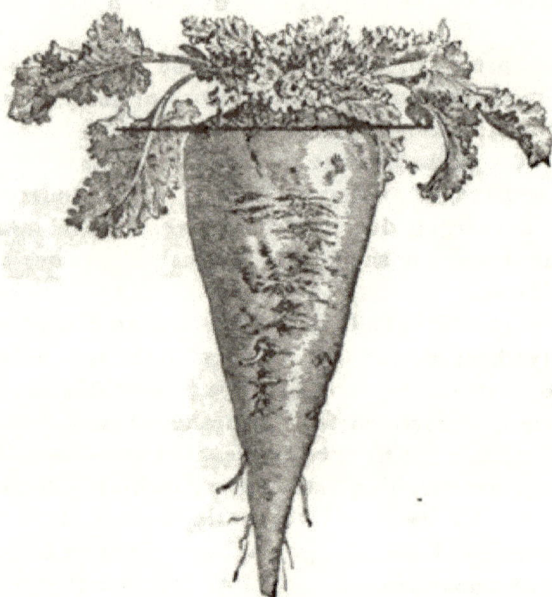

FIG. 1.

better. This number may afterwards be reduced by subsampling by the method of "quartering."

The roots are weighed, then thoroughly washed, using a
brush to remove adhering soil and rootlets, and are then
dried. A cloth may be used for drying them, but where
many samples are to be examined it is usually more convenient to dry the roots by exposure to a free circulation of
the air for a short time.

The next operation is the removal of the neck or crown, *i.e.*, that portion of the beet from just below the lowest leaf-bud. The cut should be made at the line shown in Fig. 1.

The roots are again weighed, the difference between this weight and the first being recorded as the tare. The number of beets included in the sample and their average weight should also be recorded.

The beets, which have been employed in determining the deduction for tare, conveniently serve as a sample for analysis when the roots are purchased upon a basis of their sugar content. These roots, however, would not be a satisfactory average for calculating the sugar entering the factory.

5. Measurement of the Juice.—At the present time, the diffusion process has replaced all others in the extraction of the juice from the beet. This process requires that definite volumes of juice be drawn from the battery for definite quantities of beets.

The juice is drawn into a measuring-tank which is alternately filled and emptied. If this measurement be made with accuracy and reliable samples of the juice be drawn, a basis is supplied for subsequent control work. Unfortunately this measurement as usually made is only an approximation. Errors are introduced through variations in the temperature of the juice and the difficulty of closing the inlet-valve at the proper instant. Hence special apparatus is essential to accurate measurement. This apparatus should be so arranged that it is wholly or partly automatic in its functions.

Whatever the system of tank measurements, it is essential that the measuring-tank be carefully calibrated by means of a known volume of water rather than by calculation. A slight error in the calibration is multiplied many times before the end of the manufacturing season.

6. Measurement of the Juice—Automatic Recording Apparatus.—The errors mentioned above may be reduced to a minimum by a careful supervision of the battery temperatures, the use of automatic recording apparatus, and overflow pipes.

The apparatus illustrated in Fig. 2, the invention of

9. Automatic Determination of the Weight of the Juice.—It is preferable to determine the weight of the juice by actual weighing when practicable. The automatic scale shown in Fig. 3 and in the diagrams (1, 2,

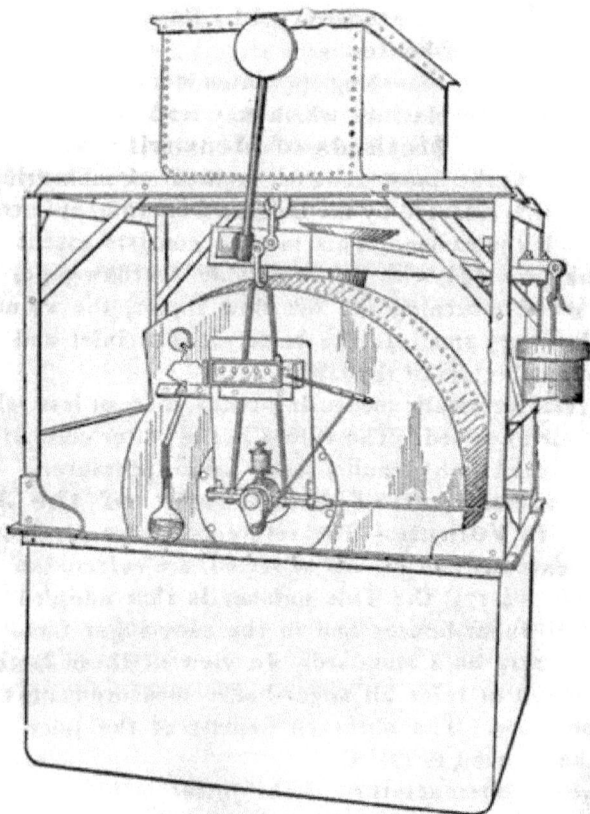

FIG. 3.

3, and 4), Fig. 4, is the invention of John Paul Baldwin, and was devised especially for sugar-house purposes.

The machine consists essentially of a revolving drum mounted upon a suitable scale. The liquid enters through the central pipe and flows into one of the compartments of the drum. When the weight of liquid for which the scale is set has entered the compartment, the liquid is automatically diverted to the second compartment, the lead in

which soon revolves the drum so that the weighed liquid
runs into the receiver beneath.   The drum continues to revolve until it assumes its original position.

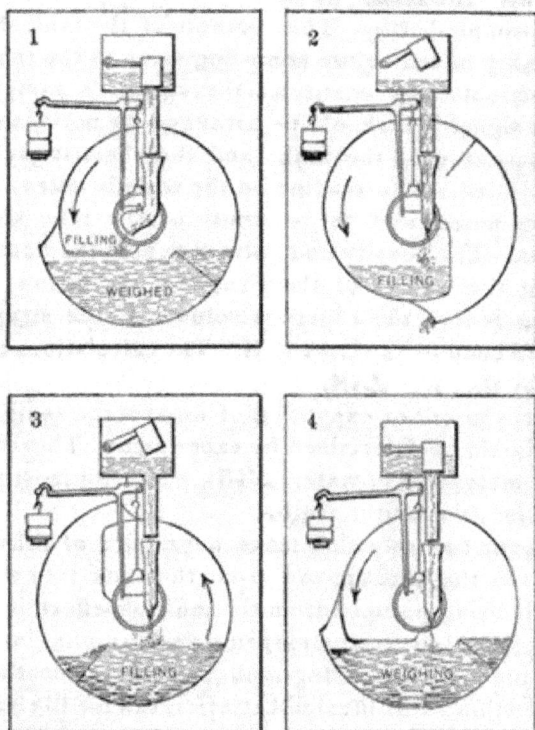

FIG. 4.

A counter records the number of weighings.   A cup
removes a small sample of the liquid from each load and
stores it in a bottle, as shown in Fig. 3.

10. Measurement and Weight of the Sirup.—
The sirup is pumped from the multiple-effect evaporator to
storage-tanks.   It is not always easy to obtain accurate
measurements of the sirup in these tanks.   Rectangular
tanks should be thoroughly stayed with rods.   In case the
tanks are bulged or uneven, it may be necessary to calibrate
them by running in a measured volume of water.   If the
tanks are of uniform sectional area from top to bottom,
they may be fitted with gauge-glasses similar to the water-

9. Automatic Determination of the Weight of the Juice.—It is preferable to determine the weight of the juice by actual weighing when practicable. The automatic scale shown in Fig. 3 and in the diagrams (1, 2,

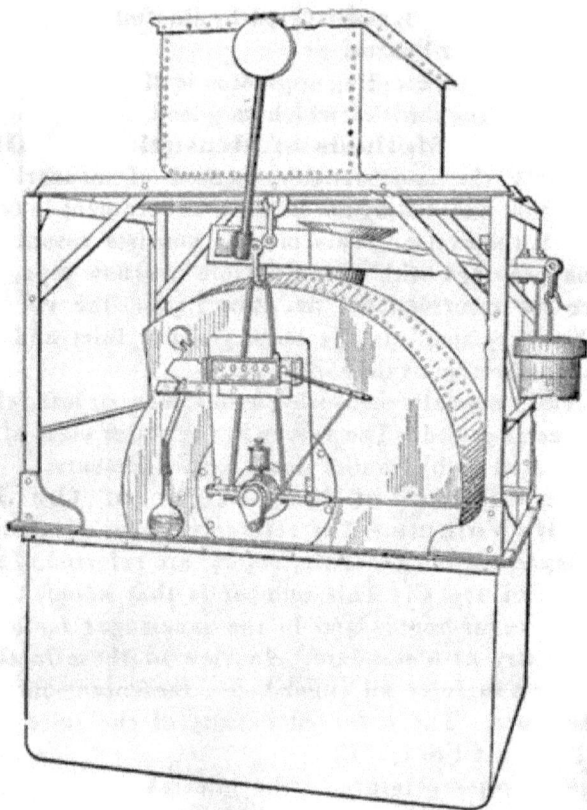

FIG. 3.

3. and 4), Fig. 4, is the invention of John Paul Baldwin, and was devised especially for sugar-house purposes.

The machine consists essentially of a revolving drum mounted upon a suitable scale. The liquid enters through the central pipe and flows into one of the compartments of the drum. When the weight of liquid for which the scale is set has entered the compartment, the liquid is automatically diverted to the second compartment, the lead in which soon revolves the drum so that the weighed liquid runs into the receiver beneath. The drum continues to revolve until it assumes its original position.

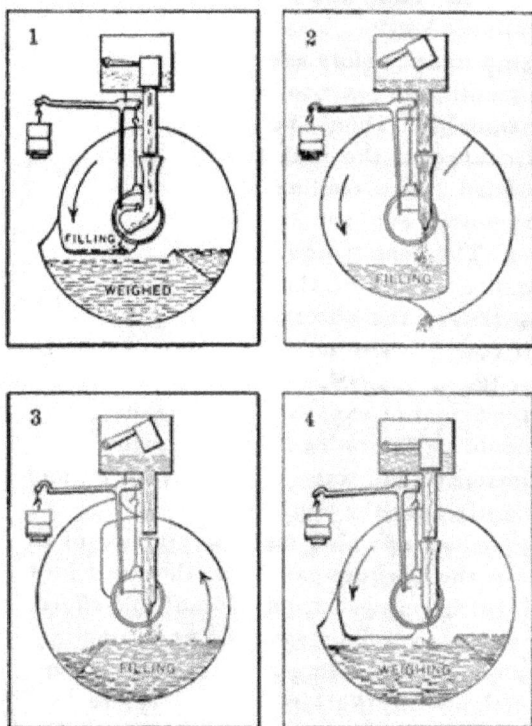

FIG. 4.

A counter records the number of weighings. A cup removes a small sample of the liquid from each load and stores it in a bottle, as shown in Fig. 3.

10. Measurement and Weight of the Sirup.— The sirup is pumped from the multiple-effect evaporator to storage-tanks. It is not always easy to obtain accurate measurements of the sirup in these tanks. Rectangular tanks should be thoroughly stayed with rods. In case the tanks are bulged or uneven, it may be necessary to calibrate them by running in a measured volume of water. If the tanks are of uniform sectional area from top to bottom, they may be fitted with gauge-glasses similar to the water-

instant the water reaches the tube, it rises some distance by capillarity, and affords prompt means of ascertaining when the vessel has been filled to a certain point. If water slightly colored with phenolphthalein be used, the rise of the water may be observed with ease. The difference between the volume of the cylinder to the capillary tube and the volume of water added is the required volume of the massecuite.

It is evident that this method can only be used in a building free from vibrations. Under proper conditions, a measurement to within two or three tenths of a cubic centimetre can be made by this method [1] in a large cylinder.

A convenient-sized cylinder is 8 centimetres in diameter by 25 centimetres in depth, holding approximately 1500 grams of massecuite.

In houses where the massecuite is run into large rectangular tanks or into small portable tanks, the volume may be roughly approximated by the above method; but where the various forms of "crystallizers with movement" are used, or the massecuite is run directly into the mixer, the weight can only be calculated from the analysis and the volume of the lower products. In order to estimate approximately the loss of sucrose at this stage when "boiling in" is practised, the analysis and volume of the molasses used must be known. It is not possible to do more than closely approximate the loss without knowing the actual weight of the massecuite.

12. Measurement and Weight of the Second Massecuite, etc.—With modern methods of boiling first-sugar, *i.e.*, "boiling in" molasses on first-sugar, there is comparatively little of the lower grades of massecuite made. Such massecuite is boiled to "string-proof," and may be easily sampled as it flows to the crystallizing-tanks. The weight of a definite volume may be determined as in **11**. The measurement may be made in the tank after the massecuite attains approximately the temperature of the hot-room. A correction for expansion should be made, or

---

[1] This method of measurement is given in Mohr's *Titremethode*. The author used it several years, supposing it to be original with himself.

the weight of a measured volume at the hot-room temperature should be determined.

13. Sugar-weights.—The sugar-weights should be reported to the chemist for tabulation and for his use in calculating the yield and losses.

ESTIMATION OF LOSSES AND THE DIVISION OF THE MANUFACTURING SEASON INTO PERIODS.

14. Division of the Season into Periods.—In factories which suspend manufacturing operations every Sunday, it is a simple matter to divide the season into periods of one week each, but in other factories it requires a systematic scheme of estimates to do this.

The following plan has given excellent results in the hands of the author, and is suggested : Sunday is a convenient time for beginning a period; for example, let each period begin at 6 A.M. that day. At six o'clock the chemist and his assistants pass through the sugar-house and measure and estimate the quantities of material in the different stages of manufacture. The cossettes in the diffusion-battery are estimated as equivalent to a certain weight of beets, and the juice in the tanks is also calculated to terms of the beet. The sum of these estimates is subtracted from the weight of the beets as reported by the weigher, to obtain the weight of the beets worked. An estimate of the quantity of sirup in the multiple-effect and a measurement of that in the storage-tanks are made. The weight of beets equivalent to the cossettes in the battery and the quantity of sirup in the multiple-effect may be considered constant quantities, and the same numbers be used from week to week. The volume of the sirup in the tanks should be calculated to a volume of a standard density (see table, page 242), and from the average yield of sugar per 100 units of volume, as indicated by experience, its sugar value may be determined, or the sirup may be calculated back to juice, then to terms of the weight of beets, and this weight deducted from the beets charged to the factory.

If a practice be made of allowing the vacuum-pan-man a certain volume of sirup per strike, *i.e.*, equivalent to a certain volume of a standard density (**225**), it is easy to calculate the quantity of massecuite in the pan by difference and its equivalent in sugar or beets. If it be the practice to "boil in" molasses on first-sugar, the quantity of this product must be determined and its sugar value estimated. The practice of "boiling in" molasses complicates the calculation. The lower grades of sugar are estimated from the quantity of massecuite, and the first-sugar is actually weighed. It is difficult, if at all possible, to carry the separation of the work into periods beyond the first-sugar.

15. Loss of Sucrose in the Exhausted Cossettes (Pulp).—In the analysis of the exhausted cossettes, the percentage of sucrose is expressed in terms of the cossettes. In order to calculate the loss of sucrose, it is necessary to know the weight of exhausted cossettes per 100 pounds of beets. This number can only be accurately determined by actually weighing the cossettes from a definite weight of beets. This is manifestly impracticable, hence the chemist must necessarily base his calculations upon the average of a few weighings made each season.

It is also evident that different diffusion-battery conditions result in differences in the percentage of exhausted cossettes. The depth of the diffuser, the working temperature, the condition of the beets, the thickness of the cossette, and the use of water-pressure only or water-pressure and compressed air, all have their influence upon the weight of exhausted cossettes produced.

In general, it is usually considered that 100 pounds of beets, when working by water-pressure only, produce approximately 90 to 100 pounds of well-drained exhausted cossettes, and working with compressed air, 100 pounds of beets produce approximately 80 to 85 pounds of exhausted cossettes.

16. Loss of Sucrose in the Waste Water.—It is not practicable to measure the waste water in the diffusion process. In order to figure the loss of sucrose at this stage of the manufacture it is necessary that this quantity be known; hence, being unable to ascertain it by actual measurement, it must be determined approximately by calculation.

The total volume of the diffuser and its connections must be known, also the weight and specific gravity of the exhausted cossettes.

It is more convenient to use the metric system in these calculations.

Calculation.

Let x = the required volume of waste water in hectolitres;

D = specific gravity of the exhausted cossettes;

W = the weight of the exhausted cossettes per diffuser in kilograms;

V = the net volume of the diffuser in hectolitres, *i.e.*, the volume between the upper and lower strainers;

$$x = V - \frac{W}{100D} = \text{the waste water in the net diffuser}$$

in hectolitres.

To obtain the total volume of the waste water, add the calculated volume of the "dead space," *i.e.*, the space above and below the strainers and of the parts of the pipes which drain into the diffuser.

Example.

(A diffusion-battery using water-pressure only.)

Volume of the diffuser (net), hectolitres.......... 30

Weight of the exhausted cossettes per diffuser, kilograms................................... 1300

Weight of fresh cossettes per diffuser, kilograms.. 1530

Specific gravity of the exhausted cossettes........ 0.984

Per cent sucrose in the waste water.............. .05

Volume of the "dead space," hectolitres......... 2.5

$$x = V - \frac{W}{100D} = 30 - \frac{1300}{98.4} = 30 - 13.2 = 16.8 \text{ hectolitres,}$$

and $16.8 + 2.5 = 19.3$ hectolitres total waste water. This water contains so little solid matter in solution that its specific gravity may be considered to be 1, hence 19.3 hectolitres of the waste water weigh 1930 kilograms or $\frac{1930}{1530} \times 100$

$= 126$ kilograms per 100 kilograms of beets. $126 \times .05 \div 100$

certain volume of a standard density (**225**), it is easy to calculate the quantity of massecuite in the pan by difference and its equivalent in sugar or beets. If it be the practice to " boil in" molasses on first-sugar, the quantity of this product must be determined and its sugar value estimated. The practice of "boiling in" molasses complicates the calculation. The lower grades of sugar are estimated from the quantity of massecuite, and the first-sugar is actually weighed. It is difficult, if at all possible, to carry the separation of the work into periods beyond the first-sugar.

15. Loss of Sucrose in the Exhausted Cossettes (Pulp).—In the analysis of the exhausted cossettes, the percentage of sucrose is expressed in terms of the cossettes. In order to calculate the loss of sucrose, it is necessary to know the weight of exhausted cossettes per 100 pounds of beets. This number can only be accurately determined by actually weighing the cossettes from a definite weight of beets. This is manifestly impracticable, hence the chemist must necessarily base his calculations upon the average of a few weighings made each season.

It is also evident that different diffusion-battery conditions result in differences in the percentage of exhausted cossettes. The depth of the diffuser, the working temperature, the condition of the beets, the thickness of the cossette, and the use of water-pressure only or water-pressure and compressed air, all have their influence upon the weight of exhausted cossettes produced.

In general, it is usually considered that 100 pounds of beets, when working by water-pressure only, produce approximately 90 to 100 pounds of well-drained exhausted cossettes, and working with compressed air, 100 pounds of beets produce approximately 80 to 85 pounds of exhausted cossettes.

16. Loss of Sucrose in the Waste Water.—It is not practicable to measure the waste water in the diffusion process. In order to figure the loss of sucrose at this stage of the manufacture it is necessary that this quantity be known; hence, being unable to ascertain it by actual measurement, it must be determined approximately by calculation.

The total volume of the diffuser and its connections must be known, also the weight and specific gravity of the exhausted cossettes.

It is more convenient to use the metric system in these calculations.

Calculation.

Let $x =$ the required volume of waste water in hectolitres;

$D =$ specific gravity of the exhausted cossettes;

$W =$ the weight of the exhausted cossettes per diffuser in kilograms;

$V =$ the net volume of the diffuser in hectolitres, *i.e.*, the volume between the upper and lower strainers;

$x = V - \dfrac{W}{100D} =$ the waste water in the net diffuser in hectolitres.

To obtain the total volume of the waste water, add the calculated volume of the "dead space," *i.e.*, the space above and below the strainers and of the parts of the pipes which drain into the diffuser.

Example.

(A diffusion-battery using water-pressure only.)

Volume of the diffuser (net), hectolitres.......... 30

Weight of the exhausted cossettes per diffuser, kilograms................................... 1300

Weight of fresh cossettes per diffuser, kilograms.. 1530

Specific gravity of the exhausted cossettes........ 0.984

Per cent sucrose in the waste water.............. .05

Volume of the "dead space," hectolitres......... 2.5

$x = V - \dfrac{W}{100D} = 30 - \dfrac{1300}{98.4} = 30 - 13.2 = 16.8$ hectolitres,

and $16.8 + 2.5 = 19.3$ hectolitres total waste water. This water contains so little solid matter in solution that its specific gravity may be considered to be 1, hence 19.3 hectolitres of the waste water weigh 1930 kilograms or $\dfrac{1930}{1530} \times 100$

$= 126$ kilograms per 100 kilograms of beets. $126 \times .05 \div 100$

= .063 kilogram of sucrose lost per 100 kilograms of beets or .063 pound of sucrose per 100 pounds of beets.

The quantity of sucrose in the waste water is so small that a very considerable error in figuring the volume of the waste water has but little influence.

With a battery employing compressed air, the volume of the waste water is very small, and is determined by deducting the volume of diffusion-juice drawn from the volume of the waste water as calculated above.

In the above example, assuming a "draw" of 115 litres of diffusion-juice per 100 kilograms of beets, using compressed air, the volume of the waste water would be calculated as follows :

$15.3 \times 115 = 1759.5$ litres = 17.595 hectolitres of juice drawn and $19.3 - 17.595 = 1.705$ hectolitres of waste water = 170.5 kilograms, or 11.1 kilograms per 100 kilograms of beets. The loss of sucrose would be $11.1 \times .05 \div 100 = .0056$ kilogram per 100 kilograms of beets or .0056 pound per 100 pounds of beets.

17. Estimation of the Losses of Sucrose in the Diffusion by Difference. — If it were always practicable to ascertain the exact weight of the beets entering the diffusers, the simplest method of estimating the loss of sucrose in the diffusion would be by deducting the sucrose obtained in the diffusion-juice from that present in the beets, as ascertained by direct analysis. There are several probable sources of error in this method when not based upon the actual net weight of the beets. The tare (3) includes that part of the neck of the beet which should be removed in the field, but which has been left through careless topping ; this passes into the diffusion-battery and contributes its sugar to the juice. This sugar increases the quantity in the juice without being charged to the beet supplying it.

In brief, except in houses where the beets are weighed immediately before they are sliced, the only method of determining the losses in the diffusion is by direct gauging and analysis of the waste products. It is always advisable to make these analyses.

Many chemists consider that there is usually some loss

through decomposition of sucrose in the battery. Such loss has not been clearly proven.

There is probably not often an appreciable inversion of sucrose in the diffusion of beets, except when there are long delays.

In the event of inversion the loss may be calculated by the formulæ used in cane-sugar-houses, which were first proposed by Dr. Stubbs of Louisiana (**263**).

18. Loss of Sucrose in the Filter Press-cake.— The weight of the press-cake per ton of beets ✕ per cent sucrose in the press-cake ÷ 100 = pounds of sucrose lost per ton of beets. In sugar-houses in which it is not convenient to weigh the press-cake the approximate weight may be estimated by the following method : Weigh several entire press-cakes and figure the average weight; multiply the average by the number of cakes per press. A record must be kept of the number of presses emptied. The average weight should occasionally be verified.

19. Loss of Sucrose in the Evaporation to Sirup.—An examination of the ammoniacal waters from the multiple-effect apparatus will sometimes reveal the presence of sucrose. It is practically impossible to estimate this loss from the analyses of these waters, since the weight of the water is unknown and the percentage of sucrose small. The quantity of sucrose lost is best determined by the difference between the weight of sucrose in the purified juice and that in the sirup. To obtain the weight of sucrose in the purified juice otherwise than by direct analysis, the loss in the filter press-cakes and at the mechanical filters must be deducted from the weight of sucrose entering the house in the diffusion-juices.

The following are some of the sources of loss of sucrose in the evaporation : Priming, *i.e.*, juice entrained with the vapors ; caramelization and decomposition of the sugar. The liquors should always be alkaline, hence there is no loss from inversion.

20. Loss of Sucrose in the Vacuum-pan.—The estimation of the loss in the granulation of the sugar in the vacuum-pan is difficult. The sources of loss are the same as those in the multiple-effect. If the weight of the massecuite can be accurately ascertained (**11**), the loss can be determined with certainty, as the weight of sucrose in the sirup should balance that in the massecuite. The "boiling in" of molasses with first-sugar complicates the determination in so far as it requires that the quantity and analysis of such molasses be known. The weight of sucrose in the massecuite can be ascertained indirectly when boiling "straight strikes" from the weight of sugar obtained and the volume of molasses produced, the weight of sucrose in the "wash" used in the centrifugals being deducted. In the event of its not being convenient to gauge the molasses, the measurement may be made after concentration to second massecuite, the loss indicated being that of the two boilings.

SUGAR ANALYSIS. OPTICAL METHODS.

APPARATUS AND MANIPULATION.

21. The Polariscope.—The instrument employed in the optical methods of determining cane-sugar and other sugars is termed a polariscope, or saccharimeter. This instrument depends in theory and construction upon the action of sugar upon the plane of polarization of light.

Polariscopes may be divided into two general classes, viz., shadow and transition-tint instruments. The shadow instruments may be subdivided into polariscopes employing white light, as from an ordinary kerosene lamp, and those employing monochromatic light, supplied by a sodium lamp.

The principal instruments in use are the half-shadow, triple-field and the transition-tint polariscopes. The shadow instruments are constructed for use with white light and with the yellow monochromatic light. The former are usually employed in commercial work, and the latter in scientific investigations.

The transition-tint instruments are being rapidly displaced by the shadow polariscopes, since these latter leave little to be desired in the matter of accuracy and convenience.

The reader is referred to the manuals of Wiley and others for the theory and construction of polariscopes.

A brief description of the polariscopes in general use will suffice for the purposes of this book.

22. Half-shadow Polariscope (Schmidt and Haensch).—The optical parts of this instrument are indicated in Fig. 6. At O there is a slightly modified Jellet-Corny Nicol prism, at G is a plate of dextrogyratory quartz, at E is a quartz wedge, movable by means of the screw M, and at F is a quartz wedge, fixed in position, to which is attached the vernier. The scale is attached to the

movable wedge.   These quartz wedges are of lævogyratory
quartz.   The parts G, E, and F constitute the compensating
apparatus, *i.e.*, the apparatus which compensates for the
deviation of the plane of polarization due to the influence
of the solution of the optically active body placed in the

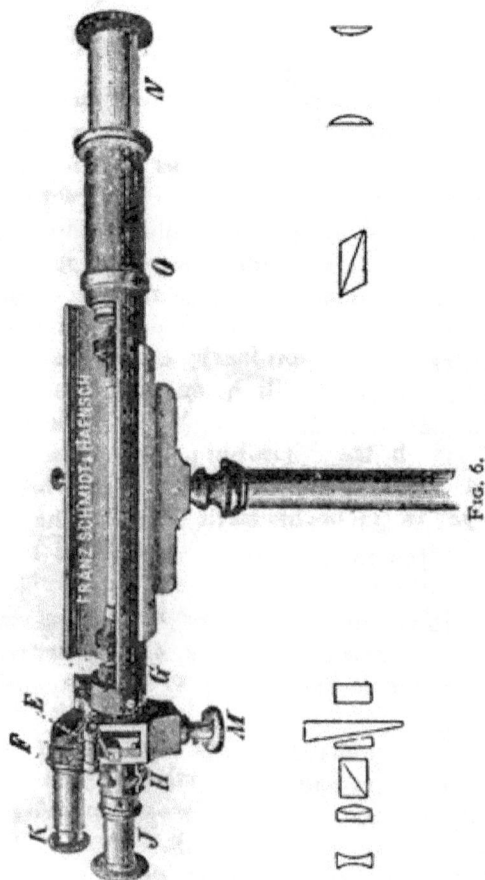

Fig. 6.

observation-tube as shown in the figure.   At H is the analyzer, a Nicol prism.   At J is the telescope used in making
the observation, and K is the telescope and reflector for
reading the scale.   The two lenses, shown in the diagram
at the extreme right, are for concentrating the rays of

light from the lamp and transmitting them in parallel lines
to the polarizing Nicol prism.

The instrument above described is of the single compensating type.

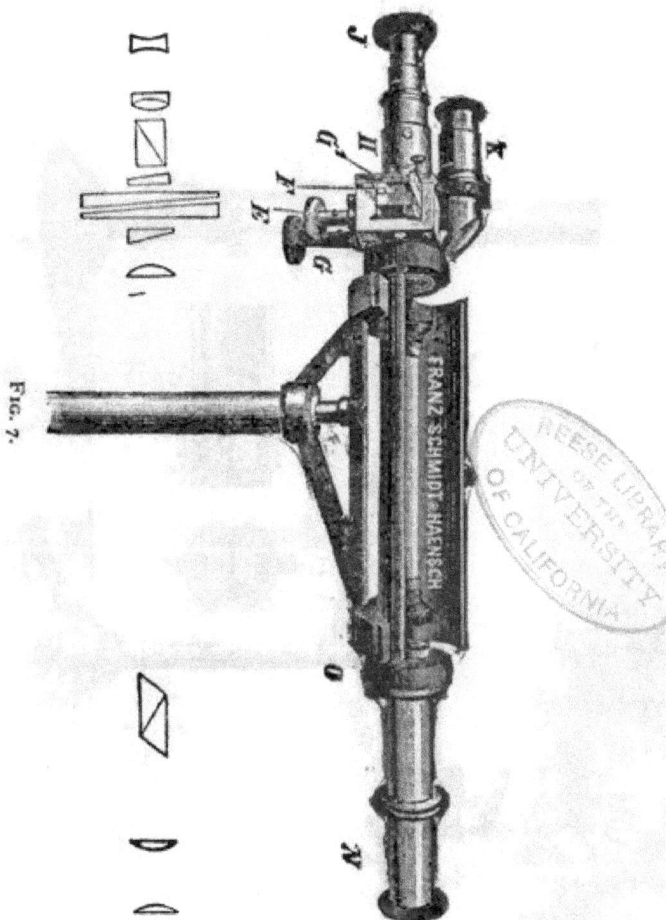

Fig. 7.

The double compensating instrument is shown in Fig. 7.
This polariscope differs from the single compensating instrument in having two sets of quartz wedges of opposite
optical properties and two scales and verniers.

The field of vision of the above instruments when set at

the neutral point is a uniformly shaded disk.   If the milled screw controlling the compensating wedge be slightly turned to the right or left, one half the disk will be shaded and the other light.   It is from this half-shaded disk that this type of instruments takes its name.

23. Triple - Field Polariscope (Schmidt and Haensch).—This instrument differs from the preceding in

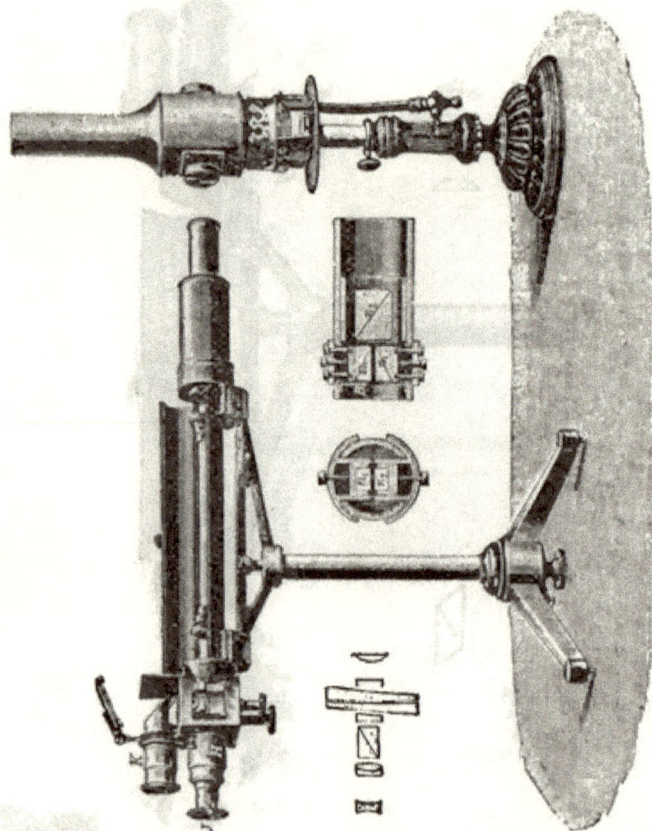

Fig. 8.

having two small Nicol prisms placed in front of the polarizer, as shown in Fig. 8.   The field of the instrument is divided into three parts, 1, 2, and 3 of the diagram, Fig. 9. This figure shows the arrangement of the Nicol prisms (I, II, III) and a diagram of the field of observation.

When the scale is set at the zero point, no optically active body being interposed, the field is uniformly shaded; in other positions 1 is shaded and 2 and 3 are light, or *vice versa*. This arrangement permits a very high degree of accuracy in the adjustment of the field in polariscopic observations. According to the experiments of Wiley[1] this instrument is extremely sensitive and is capable of results but little inferior to those with the Landolt-Lippich apparatus. It is probably the superior, in point of accuracy, to other instruments designed for industrial work.

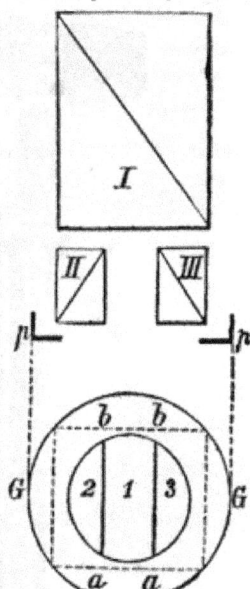

FIG. 9.

24. Laurent Polariscope.— The Laurent polariscope (Fig. 10) is a half-shadow instrument. It was originally designed for use with a monochromatic flame, but these instruments, as now made, are provided also with compensating apparatus for use with white light.

In the Laurent polariscope the analyzer is revolved by means of a milled screw, to compensate for the deflection of the plane of polarization by the sugar solution. The angular rotation is measured by means of a scale and vernier. This instrument is also provided with a second scale, termed the cane-sugar scale, on which the per cents may be read directly.

As stated above, the Laurent polariscope is also often provided with a compensating apparatus (Fig. 11), which permits the use of white light.

A distinctive feature of the Laurent instrument is the adjustable polarizer. This Nicol prism may be rotated through a small angle, thus permitting the sensitiveness of the instrument to be varied.

The polarized light is passed through a disk of glass,

---

[1] Agricultural Analysis, 3, 91.

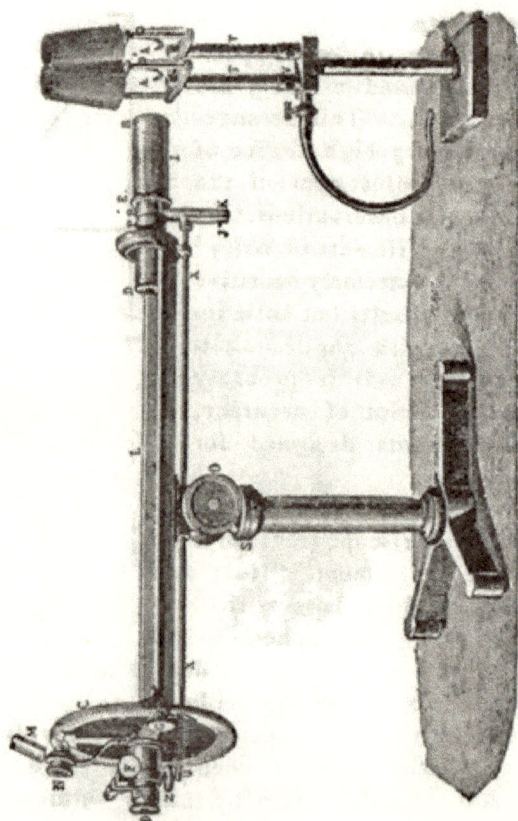

FIG. 10.

FIG. 11.

one half of which is covered with a thin plate of quartz, thus producing the half-shadow feature of the instrument.

25. The Transition-tint Polariscope. Soleil-Ventzke-Scheibler. — The tint polariscope, Fig. 12,

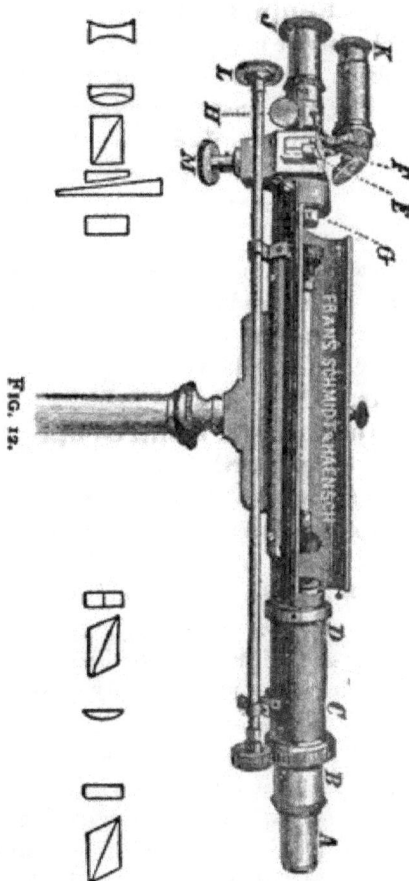

FIG. 12.

resembles in appearance the half-shadow instrument of Schmidt and Haensch. It differs from this in being provided with an additional Nicol prism at *A* and a quartz plate *B*, which produce the color. The tint is varied by means of a spur-wheel and pinion, revolved by a rod with a milled head, *L*. The optical parts at the front end of the instrument are the same as in the Schmidt and Haensch half-shadow polariscope.

The field is colored, and when the instrument is set at the neutral point the tint is uniform. The sensitive tint for most eyes is a rose-violet.

26. General Remarks upon Polariscopes. — The Laurent polariscope is very extensively used in France, and to but a limited extent in other parts of Europe and in this country. The tint instruments were formerly used almost exclusively, but have been largely replaced by the various forms of half-shadow polariscopes. Tint instruments, obviously, cannot be used by persons who are color-blind.

All polariscopes are made to receive observation-tubes of various lengths. The standard length is 200 millimetres.

There are several forms of polariscopes in addition to those described, but for industrial work it is unnecessary to mention others.

27. Manipulation of a Polariscope. —Having dissolved the normal weight (**28**) of the material under examination in water, clarify the solution as described in **30**. Fill the observation-tube with a portion of the clarified solution, and pass the light from a suitable lamp into the instrument. The observer, with his eye at the small telescope *J* of the Schmidt and Haensch instruments, Figs. 6, 7, 8, and 12, or the corresponding part of the Laurent, will notice that one half the disk is shaded or more deeply colored, according to the kind of polariscope, provided the instrument is not set at the neutral point. The vertical line dividing the half-disks should be sharply defined; if not, the ocular should be slipped backwards or forwards until a sharp focus is obtained. Turn the milled screw until the field appears uniformly shaded or tinted on both sides of the vertical line, and then read the scale (**29**). A little practice will enable the observer to detect very slight differences in the depth of the shadow or color and to attain great accuracy in this manipulation.

The manipulation of the triple-field polariscope is as described above except as to the position of the shadows (**23**).

In polarizing very clear solutions with the shadow polariscope, the light may dazzle or fatigue the eye, making it somewhat difficult to obtain concordant readings. In such cases substitute the eyepiece, fitted with a plate of bichromate of potash, for the ordinary ocular of the observation-telescope. It is also well to check the adjustment of the zero point after changing the eyepiece.

The Laurent instrument is fitted with a device for varying its sensitiveness. This is convenient in polarizing dark-colored solutions, since a slight change in the position of the lever which rotates the polarizer will increase the intensity of the light, though at the same time decreasing the sensitiveness of the instrument; and *vice versa* in polarizing very clear light-colored solutions, the rotation of the polarizer in the opposite direction, through a small angle, increases the sensitiveness.

The double compensating Schmidt and Haensch polariscopes are provided with two scales, one graduated in black and the other usually in red. The black scale is operated by a black milled screw, and the red scale by a brass screw. For ordinary work, set the red scale at zero and equalize the field with the black screw. To check the readings, remove the observation-tube and equalize the field with the brass screw. The readings on the two scales should agree. To make a reading with lævorotatory sugar, set the black scale at zero, and use the brass screw and red scale.

The manipulations of the tint instruments, as explained, are similar to those of the shadow polariscopes, except that a uniform tint must be obtained. The intensity of the tint varies with the position of the analyzer. The color is varied by turning the milled screw on the horizontal rod which revolves the regulator.

The Schmidt and Haensch shadow polariscopes, the Laurent with special attachment, and the tint instruments require a strong white light. A kerosene-lamp with duplex burner is usually employed. A gas-lamp such as shown in Fig. 8 is very convenient in many localities. The kerosene-lamp should be provided with a metal chimney.

Dr. Wiechmann uses the Welsbach light in his laboratory at the Havemeyer & Elder refinery, Brooklyn, and finds it very satisfactory. Dr. Wiley of the U. S. Department of Agriculture has investigated the use of the light from acetylene gas for polariscopic purposes, and states that the readings obtained are very accurate and that the light is especially convenient when polarizing very dark-colored solutions. This gas is readily and economically produced, in the small quantities required for polariscopic purposes in sugar-works, by the decomposition of calcium carbide in water, in a suitable gas-holder. A special gas-tip is required in burning this gas.

In the author's experience, the Wiley[1] constant-flame sodium gas-lamp is the best yet devised for those instruments requiring a monochromatic light. This lamp gives a sodium flame of constant intensity for hours at a time, if required. Other lamps for monochromatic light are the Laurent gas-sodium and the Landolt gas-sodium lamps, and the Laurent eolipyle, burning alcohol.

FIG. 13.

M. Dupont[2] has recently experimented with various sodium salts for use in monochromatic lamps. He finds that sodium chloride and tribasic phosphate of sodium, melted together in molecular proportions, give excellent results and are in every way superior to sodium chloride alone.

28. The Polariscopic Scale. The Normal Weight.—The scales of polariscopes for use in industrial work are usually so divided that if a certain weight of the

---

[1] Agricultural Analysis, H. W. Wiley, 3, 85.
[2] *Bulletin de l'Association des Chimistes de France*, 14, 1041.

substance be dissolved in water and the solution diluted
to 100 cc., and observed in a 20-centimetre tube, the reading will be in percentages of sucrose. This scale is termed
the "cane-sugar scale," and the weight of material required to give percentage readings is termed the "normal
weight," or sometimes the "factor of the instrument."

In commercial work the divisions of the scale are often
termed "degrees," especially in the polarization of sugars.

The normal weight for the German instruments is 26.048
grams, and for the Laurent 16.29 grams. The number
given for the Laurent polariscope is that adopted by the
2ᵉ Congrès International de Chimie Appliquée, 1896.

29. Reading the Polariscopic Scale.—Having
equalized the shadow or tint as directed in **27**, examine
the scale through the reading-glass. For example : Let
the scale and vernier have the positions shown in Fig. 14.

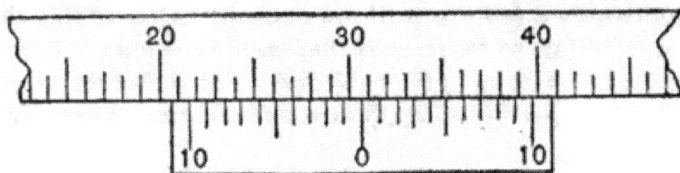

FIG. 14.

The zero of the vernier is between 30 and 31; record the
lower number; note the point to the right at which a line of
the vernier, the small scale, corresponds with a line of the
scale, in this case at 7; enter this number in the tenths
place. The completed reading is 30.7. The portions of the
scale and vernier to the left of the zeros are used in the
polarization of lævorotatory bodies. If the zero of the vernier correspond exactly with a division of the scale, the
reading is a whole number.

If the normal weight of the material have been dissolved
in a volume of 100 cc.[1] and a 20-centimetre observation-tube
have been used, the reading on the cane-sugar scale is the
percentage of sucrose in the substance, provided other optically active bodies than sucrose are absent. The read-

---

[1] The flasks should be graduated to hold 100 grams of distilled water at
17½° C., and not to true cubic centimetres, for the S. and H. instruments,
but to true cc. for the Laurent.

ings must be corrected for other weights of the substance than the normal, for other volumes than 100 cc., and for other tube lengths than 20 centimetres.

30. Preparation of Solutions for Polarization.—Dissolve the normal or other convenient weight of the material in water. Add sufficient subacetate of lead [1] to clarify the solution. It is difficult to specify the amount of the lead salt to use. If too little or too much be used, the solutions usually filter with difficulty and become turbid. With juice from immature beets the filtered solution will sometimes be perfectly clear and colorless when first obtained and in a few moments become too dark to polarize. In such cases the juice should be thoroughly mixed with the lead solution and stand some time before filtration. Usually 10 cc. of the dilute solution of subacetate of lead (**207**) or 2–3 cc. of the concentrated solution (**208**) will be sufficient for 100 cc. of beet juice. Sugars of high grade require only a few drops of the reagent. After adding the lead salt dilute to 100 cc., mix thoroughly and filter. Reject the first few drops of the filtrate. Fill the observation-tube with a portion of the filtrate, and polarize as described in **27**.

In sugar analysis the materials to be examined are most conveniently weighed in a nickel or German-silver capsule such as is shown in Fig. 15. A convenient filtering arrangement is illustrated in Fig. 16. *A* is a stemless funnel; *B* is a quarter-pint precipitating jar; *C* is a small cylinder. The stemless funnels may be made of tinplate or thin copper, planished. The latter, while more expensive, are preferable, as they are more durable. A plain cylin-

FIG. 15.

---

[1] The subnitrate and nitrate of lead have been used to some extent as clarifying agents. Their use would require the adoption of new constants in the inversion methods (**89** *et seq.*), and, so far as present experience indicates, without material advantage.

der is preferred by some chemists, as the funnel makes a close joint with the edge.

The advantage of the metal stemless funnels and the heavy glass precipitating jar or the lipped cylinder is the ease with which they may be washed and dried. The jar or cylinder is also a very convenient support for the funnel.

Stammer and Sickel advise the addition of at least four times the weight of the sucrose in the massecuite or mo-

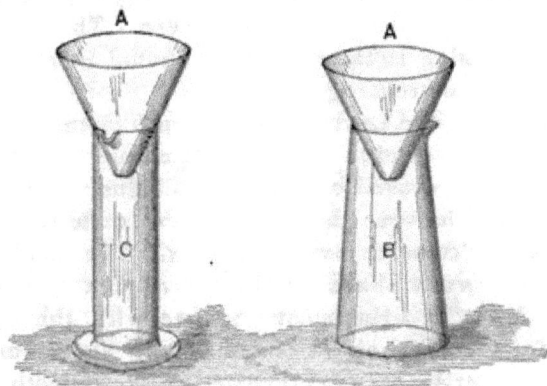

FIG. 16.

lasses, of strong alcohol in preparing solutions for polarization, and if the substance be alkaline to acidulate with acetic acid.[1] Herzfeld, as the result of his experiments, gives the same advice.[2] Other equally prominent chemists consider the use of alcohol unnecessary and liable to lead to error.

31. The Adjustment of the Polariscope.—The scale of the polariscope is the only part which is liable to get out of position. Fill an observation-tube with water and make an observation. If the scale be properly adjusted the reading should be zero.

The method of adjusting the instrument to read zero under the above conditions is the same with all the Schmidt

---

[1] *Revue Universelle de la Fabrication du Sucre*, 2d year, 578.
[2] *Deutsche Zuckerind.*, 1886, No. 24.

and Haensch polariscopes. A micrometer-screw, turned by means of a key, is arranged to move the vernier a short distance. The field is equalized as usual by manipulating the milled screw. The micrometer-screw is then turned until the zeros of the scale and vernier coincide. The scale is moved through several divisions and the field then equalized as before. If after several trials the zeros be found not to coincide, the adjustment must be repeated, turning the micrometer-screw very little. It usually requires several trials to set the instrument to read zero. This adjustment is very fatiguing to the eye, which should be rested a few seconds between readings.

In adjusting the Laurent polariscopes to read zero, the lever U, Figs. 10 and 11, is lifted to the upper limit; the ocular O is next focused on the vertical line which divides the field into halves; the zeros of the scale and vernier are made to coincide by the screw G. The field should then be uniformly shaded if the instrument is in adjustment; if not in adjustment, equalize with the screw F. This adjustment should be tested as with the Schmidt and Haensch polariscopes, and repeated until satisfactory. All parts of the instrument should be kept very clean, especially the exposed parts of the lenses. Chamois-skin is convenient for cleaning the metal parts and pieces of clean old linen for the lenses. All the crown-glass lenses should be occasionally removed from the instrument and cleaned with alcohol and wiped with old linen. The Nicol prisms should not be removed from the instrument or disturbed. The micrometer screw near H, Figs. 6, 7, 8, and 12, is for adjusting the analyzer, should the field be unevenly shaded. This adjustment should be left to an experienced workman. Should the prisms, etc., require adjustment owing to an accident to the instrument, it is advisable to send the polariscope to the dealer that he may have it repaired by an expert.

32. Notes on Polariscopic Work.—When solutions do not filter readily, the funnel employed should be covered with a glass plate to prevent evaporation.

The screw-caps of the observation-tubes should not bear heavily upon the cover-glasses, since glass is double-refracting under these conditions.   It is preferable to use caps held with a bayonet-catch rather than screw-caps.

In making an observation, the eye should be in the optical axis of the instrument, and should not be moved from side to side.

The cover-glasses should be of the best quality of glass, perfectly clean and with parallel surfaces.   A glass may be tested by holding it in front of a window and looking through it at the window-bars ; on turning the glass slowly, if the bars appear to move the surfaces of the cover are not parallel and the glass should be rejected.   Old glasses which have become slightly scratched by repeated wiping should not be used.

The planes of the ends of the observation-tube should be perpendicular to the axis of the tubes.   This may be tested by placing a tube, containing a sugar solution, in the instrument and making an observation.   On revolving the tube in the trough of the polariscope, should the readings in different positions vary, the ends of the tube have not been properly ground.

The manufacturers of polariscopes have attained such precision in their methods that errors in the adjustment of the instruments or accessories are rarely found.

The polariscope should be used in a well-ventilated room from which all light, except that from the polariscope-lamp, is excluded.   It is an excellent arrangement to have the lamp in an adjoining room and pass the light through a glass screen to the instrument.   Late models of the German polariscopes have mirrors arranged to reflect the light to the scale.   When the instrument is in a room adjoining the lamp-room, obviously the above arrangement cannot be used.   A small gas-jet or a candle should never be used for lighting the scale.   A convenient source of light is a half-candle-power incandescent electric lamp mounted near the scale and switched into the circuit by an ordinary push-button.   The lamp may be operated by a two-cell accumulator or in circuit with a 32-candle-power incandescent lamp, the latter being outside the polariscope-room.   Ordinary

Leclanché cells are cheap and will answer for several hundred polarizations.

The instrument should be occasionally tested with pure sugar (**206**), or more conveniently with standardized quartz plates, to be obtained of the makers.

Messrs. Schmidt and Haensch construct a control-tube, Fig. 17, with which all parts of the scale may be tested. The sugar solution is poured into the funnel T and flows into the tube as it is lengthened by turning the milled screw. The tube length is read on the scale N. The figure describes the tube sufficiently.

Errors which may occur in the polarization, but not through faulty manipulation of the instrument, are indicated in the following paragraphs.

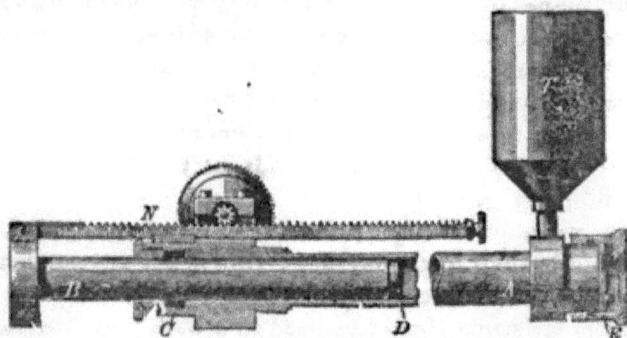

FIG. 17.

33. Error Due to the Volume of the Lead Precipitate.—The lead precipitate, formed in the clarification of the solutions, introduces errors in the polarization, some of which are probably offset by compensating errors, notably in the analysis of low-grade products.

An important error is that due to the volume of the lead precipitate. This question has been studied by a number of chemists, notably by Scheibler in Germany and Sachs in Belgium. Scheibler devised a simple method for the correction of this error, which is commonly termed the "method of double dilution." It was noticed by Rafey, Pellet, Commerson, and others that in low-grade products, the saline coefficient of which is large, there is apparently no error due to the volume of the precipitate, which is very large. They attributed this fact to an absorption of sucrose by the precipitate at the moment of its formation. Sachs[1] published an exhaustive paper on this question some years since, and demonstrated that there is no absorption of sucrose. He attributed the results with low products to the influence of acetates of potassium and sodium, formed with the acetic acid set free in the decomposition of the lead salt, upon the rotatory power of the sucrose. This view is strengthened by the fact that there is a very perceptible error, in the polarization of juices, due to the precipitate. The precipitate in juices contains but little of the acetates of potassium and sodium, whereas these salts are formed in considerable quantities in molasses and low products.

Sachs' experiments were made by increasing the concentration of the solutions instead of by dilution as practised by Scheibler. Sachs dissolved x grams of molasses in water, added sufficient subacetate of lead for clarification, completed the volume to 100 cc. and polarized as usual. The quantity x increases from experiment to experiment by practically equal increments. Since the quantity of molasses is increased with each experiment, the volume of the precipitate must increase in the same ratio. An increase in the volume of the precipitate, if this were the only disturbing influence, should increase the polarization, since the volume of the solution is decreased and the concentration is increased.

Letting $x =$ the weight of molasses, and $y =$ the polariscopic reading, the ratio $\dfrac{y}{x}$ should increase with each increment of molasses if there be an error due to the volume of the precipitate, not compensated for by other influences. Sachs employed quantities of molasses ranging from 5 to 35 grams in 100 cc., and substituting the values of x and y in the ratio and reducing, obtained the following figures:

1st series :   1.906,   1.900,   1.900,   1.906,   1.896;
2d series :   2.14,   2.13,   2.14,   2.14

---

[1] *Revue Universelle de la Fabrication du Sucre*, **1**, 451.

The practically constant value of $\frac{y}{x}$ shows that a minus error or errors have fully compensated for that due to the volume of the precipitate. Sachs' deductions are given above.

A similar experiment with beet-juices gave the following values of $\frac{y}{x}$:

1st series :	0.5446,	0.5474,	0.5480,	0.5497;
2d series :	0.5800,	0.5830,	0.5842,	0.5860.

It is thus shown that there is an increase in the ratio and an error due to the volume of the precipitate in the analysis of juices. The volume of the lead precipitate from 100 cc. of normal juice is approximately 1 cc.

It is not improbable that at least to some extent the so-called "losses from unknown sources" in sugar-house practice are due to errors in analysis which, with our present information, are unavoidable.

34. Error Due to the Volume of the Lead Precipitate — Scheibler's Method of Double Dilution.—The error due to the volume of the lead precipitate may usually be determined by Scheibler's[1] method.

To 100 cc. of the juice add the requisite quantity of subacetate of lead, complete the volume to 100 cc. and polarize as usual; a second portion of 100 cc. of the juice is treated with lead as above, diluted to 220 cc. and polarized.

Calculation.—Multiply the second reading by 2, subtract the product from the first reading, multiply the remainder by 2.2, and deduct this product from the first reading. The remainder is the required per cent sucrose.

---

[1] *Zeit., Rübenzucker-Industrie,* 25, 1054.

Example.

Degree Brix of the juice.................... 18.
First polariscopic reading (110 cc.)......... 57.6
Second polariscopic reading (220 cc.)........ 28.7

$2 \times 28.7 = 57.4$; $57.6 - 57.4 = .2$; $2.2 \times .2 = .44$; $57.6 - .44 = 57.16$, = the corrected reading. By Schmitz' table, as described on page 76, we have

$$15.18$$
$$.03$$
$$.02$$
$$\overline{}$$
$$15.23 = \text{required per cent.}$$

In the application of this method to other products using the normal or multiple-normal weight, calculate as follows:

1st volume, 100 cc.;   2d volume, 200 cc.

Multiply the second polariscopic reading by 2 and subtract the product from the first reading; multiply the remainder by 2 and subtract the product from the first reading. This remainder is the required per cent sucrose.

It is evident that this method requires extreme care in the polarization, since an error is multiplied.

35. Sachs'[1] Method of Determining the Volume of the Lead Precipitate.—Clarify 100 cc. of juice with subacetate of lead as usual, using a tall cylinder instead of a sugar-flask. Wash the precipitate by decantation, first using water and finally hot water. Continue the washing until all the sucrose is removed. Transfer the precipitate to a 100-cc. sugar-flask and add the one-half normal weight (13.024 grams) of pure sugar, dissolve and dilute to 100 cc., mix, filter, and polarize, using a 400-mm. tube.

---

[1] *Op. cit.*, **1**, 451.

Calculation.

Let P = the per cent of sucrose in the sugar;

$\quad P'$ = the polarization of the solution in the presence of the lead precipitate;

$\quad x$ = volume of the lead precipitate.

Then
$$x = \frac{100P' - 100P}{P}.$$

Example.

Let P = 99.9;

$\quad P'$ = 100.77.

Then $x = \dfrac{100 \times 100.77 - 100 \times 99.9}{100.77}$

$\quad\quad$ = .86 cc., the volume of the lead precipitate.

36. Influence of Subacetate of Lead and Other Substances upon the Sugars and Optically Active Non-sugars[1] in Beet Products.—*Sucrose.*—The

rotatory power of sucrose in aqueous solution is not modified by subacetate of lead under the conditions which usually obtain in analysis. In the presence of a very large excess of the lead salt there is a slight diminution in the rotatory power; there is a decided diminution in alcoholic solution in the presence of the lead salt.

Farnsteiner[2] made the following observations relative to the influence of certain inorganic salts:

" With a constant relation of sugar to water, the chlorides of barium, strontium, and calcium cause a decrease in the rotation which continues to decrease as the salt is increased; calcium chloride causes a decrease, but when the salt reaches a maximum further addition causes an

---

[1] The beet and beet products contain other substances which are optically active in addition to those given here, but the quantities present are exceedingly small and would not appreciably influence the analytical results. The following optically active substances are also present : tartaric acid, leucine, coniferine, and cholesterine.

[2] *Berichte deut. chem. Gesel.*, **23**, 3570; *Journ. Chem. Soc.*, **60**, 283.

increase which finally exceeds that of the pure sugar solution.

"If the relation of the sugar to that of the salt be kept constant, it is found that the addition of water causes in all cases an increase in the specific rotatory power, *i.e.*, the action of the salts is lessened. The specific rotatory power is almost unaffected by varying the quantity of sugar with a constant relation between the salt and water. The chlorides of lithium, sodium, and potassium behave in a similar manner.

"An examination of the action of the same quantities of different salts shows that in the case of strontium, calcium, and magnesium the depression varies inversely with the molecular weight, and that the product of the two quantities is approximately a constant. Barium chloride does not act in the same manner, but the chlorides of the alkalis show a similar relation. The relation, however, only holds good within each group of chlorides and not for two salts belonging to different groups."

The rotatory power of sucrose in water or alcohol solution is not modified by the presence of nitrates of sodium and potassium even when the quantity of the nitrate amounts to as much as 50 per cent of the sucrose (E. Gravier).

In investigating the influence of the lead precipitate (33), Sachs found that the presence of acetate of potassium very perceptibly diminished the rotation. The diminution was also noticeable with the sulphates of potassium and lead, but was not so marked with the corresponding sodium salts. Sachs also states that he has demonstrated that citrate of potassium, carbonate of sodium, and several other salts have an influence analogous to that of the acetates. The presence of free acetic acid reduces this influence in part. Sachs, in the same paper, urges that the use of tannic acid in decolorizing solutions is very objectionable, on account of the volume of the precipitate formed with the lead.

Dextrose.—The rotatory power of dextrose is not modified, or, if at all, but very slightly, under ordinary analytical conditions by either the subacetate or the neutral acetate of lead. See also *Invert-sugar.*

Levulose.—The rotatory power of levulose is very greatly diminished by the presence of subacetate of lead. Under certain conditions, a levulosate of lead is probably formed. This levulosate is precipitated in the presence of certain chlorides, in quantities more or less considerable according to the relative proportions of the salts, lead, and levulose. There is no precipitation by the normal acetate of lead (Pellet).

Invert-sugar. Dextrose and Levulose.—In the presence of the salts formed in the decomposition of the subacetate of lead, dextrose and levulose are precipitated in part (Pellet, Edson). The influence of the basic lead salt on the rotatory power of levulose (*see* above), or the formation of a levulosate of lead of little optical activity, gives undue prominence to the dextrose and results in a plus error. In 1885 the author recommended the acidulation of solutions containing invert-sugar with acetic acid. This restores the normal or nearly the normal rotatory power to the levulose.

Acetic acid slightly lowers the rotatory power of invert-sugar; hydrochloric acid has an opposite effect. Sodic acetate and sodic chloride increase the rotation (H. A. Weber and Wm. McPherson). Sulphuric and hydrochloric acids increase the rotation; oxalic acid has no effect. The rotation increases as the quantity of mineral acid is increased.[1]

Raffinose.—The rotatory power of raffinose is greatly diminished in concentrated solution by subacetate of lead in large quantity, and not at all in dilute solution, especially in the presence of sucrose. The normal rotation is restored by slight acidulation with acetic acid (Pellet). Raffinose is precipitated by highly basic subacetate of lead as readily as with ammoniacal acetate of lead solution (Svoboda).

Asparagine.—Not precipitable by subacetate of lead, but is rendered dextrorotatory, instead of lævorotatory, by the

---

[1] Gubbe, *Bulletin Assoc. Chimistes de France*, 3, 131.

lead salt. Asparagine is insoluble in alcohol, and in the presence of acetic acid is inactive (Pellet). In neutral and alkaline solution, lævorotatory; in presence of a mineral acid, dextrorotatory; in the presence of acetic acid the rotation is diminished and with 10 molecules of the acid becomes 0°, and with additional acid dextrorotatory (Degener).

Aspartic Acid.—From asparagine by the action of lime; the lime salt is soluble. In alkaline solutions aspartates are lævorotatory, and acid solutions dextrorotatory; aspartic acid is precipitated by subacetate of lead.

Glutamic acid is dextrorotatory, and in the presence of subacetate of lead it becomes lævorotatory. Not precipitated by lead acetate except in the presence of alcohol.

Malic acid is lævorotatory. The artificial malic acid is optically inactive. Malic acid is precipitated by subacetate of lead.

Pectine and *parapectine* are dextrorotatory and are both precipitated by subacetate of lead, and the latter by normal acetate of lead.

CHEMICAL METHODS.

37. Determination of Sucrose by Alkaline Copper Solution.—Dissolve a weighed quantity of the material in water and dilute to 50 cc. Invert by means of hydrochloric acid as described in **89.** Transfer to a litre flask, cool, neutralize with caustic soda, and dilute to 1000 cc. The quantity of material to be used depends upon the method of further procedure selected.

It is, however, convenient to use 5 grams or a multiple of 5 grams and to dilute to a multiple of 100 cc. in order that the table of reciprocals on page 294 may be used for the calculations if a volumetric method be selected.

Determine percentage of invert-sugar by one of the methods in **72** or **73.** Multiply the per cent invert-sugar by .95, since sucrose on inversion yields invert-sugar in the ratio 100 : 95.

38. Determination of Sucrose in the Presence of Reducing Sugars.—Determine the reducing sugar before inversion and after, as indicated in **37**.

Calculation.—Per cent reducing sugar after inversion — per cent reducing sugar before inversion $\times$ .95 = the required per cent sucrose.

GENERAL ANALYTICAL WORK.

SAMPLING AND AVERAGING.

39. General Remarks on Sampling and Averaging.—Accurate sampling is essential to successful chemical control. The samples must be strictly representative of the average composition of the substance or subsequent analytical work will be wasted.

The method of sampling should be by aliquot parts. This consists in drawing a definite quantity from each lot of the material, which must be the same aliquot part in each case.

Example.—Given four lots of sirup, *A*, *B*, *C*, and *D*, from which an average sample is to be drawn. Let *A* = 1000, *B* = 800, *C* = 500, and *D* = 200. Each of these lots differs in analysis. Manifestly a mixture of equal parts of *A*, *B*, *C*, and *D* would not be a true average sample, but a mixture of 10 parts of *A*, 8 parts of *B*, 5 parts of *C*, and 2 parts of *D* would be a representative sample.

In calculating an average analysis from a large number of analyses the same principle must be applied.

Example.—Given the following per cents of sucrose, representing the analyses of the beets each day for a week: 15%; 14%; 13%; 14.5%; 15%; 15.5%; 16%. The following numbers of diffusers of beets were worked each day: 168; 144; 140; 150; 165; 160; 145—a total of 1072. Required the mean percentage of sucrose in the beets.

Multiply each analysis by the number of diffusers of beets it represents, and divide the sum of the products by the total number of diffusers worked.

$$15 \times 168 = 2520$$
$$14 \times 144 = 2016$$
$$13 \times 140 = 1820$$
$$14.5 \times 150 = 2175$$
$$15 \times 165 = 2475$$
$$15.5 \times 160 = 2480$$
$$16 \times 145 = 2320$$

$$\overline{1,072 \quad 15,806}$$

$$\frac{15,806}{1,072} = 14.74 =$$

the mean per cent
sucrose for the week.

It is obvious that the weight of juice obtained per day, or the weight of the beets worked, may be used as a factor in the above calculation and strictly accurate averages secured.

Usually, however, if the diffusers receive practically uniform charges of beets, the average analysis, as calculated above, will approximate the true mean very closely.

40. Sampling Beets in the Field.—Beets growing side by side may differ greatly in sugar content; the same is true of beets grown within a few feet of one another as well as from widely different parts of the field. This indicates the difficulty if not impossibility of selecting a strictly representative sample. In point of fact, samples selected in the field only approximately represent the general average.

A convenient plan for sampling in the field is as follows: When drawing the beets to the factory, take a definite number at random from each load until all the beets have been hauled; unite the subsamples and proceed as indicated farther on.

If the sample is to be taken after the beets have been lifted and placed in piles, select a number of beets from each pile as above, or from every second or third pile, etc., and unite the subsamples. If the roots be still in the ground, lift a beet at definite intervals in the row, from every second, third, or fourth row as may be deemed best, and unite the subsamples as before.

The importance of the sample and the size of the field must determine the number of beets to be drawn, but this number should in any case be as large as practicable.

Having selected the beets, they should be sorted into three or four classes according to size and ranged in rows in a convenient place, protected from the rays of the sun. The number of beets is now reduced by subsampling, taking from each row in proportion to the number of beets in the row. For example, take every fifth or every tenth beet in the row. If the number of beets drawn in this way be too large, the subsample should be rearranged in rows and again subsampled.

41. Subsampling of Beets for Analysis in Fixing the Purchase Price.—As will be shown (p. 177), the sucrose is not uniformly distributed throughout the beet, and further the juice obtained by pressure from the same sample varies in composition with the pressure exerted and the state of division of the pulp. The more finely divided the pulp, and the heavier the pressure, the nearer the juice obtained approaches the mean juice in composition.

The proportion of juice in the beet varies from sample to sample, and often materially from the average (95 %); hence the practice of employing a coefficient, *e.g.* .95, to calculate the percentage of sucrose in the beet from the analysis of the juice, should be discouraged. In the course of an entire season this may be just to the manufacturer, but undoubtedly is an injustice in many cases to the producer of the beets.

If the indirect method of analysis be employed, the same models of rasp and press should be used by the chemists of the buyer and the seller. Further, the conditions of sampling and analysis should be the same in both laboratories.

The beets should be divided longitudinally into quarters or eighths, and an entire segment should be rasped. This insures the reduction of a portion of the beet in proportion to its size.

In many factories it is the custom to remove a small plug or cylinder from each beet for the analysis. Owing to the unequal distribution of the sugar in the beet this method cannot be depended upon to give a strictly representative sample, but experience has shown that the variations from the true average sample are not great, provided the cylinder be taken in the proper direction. The method and direction of removing the cylinder are indicated in Fig. 18. The boring-rasp (Keil and Dolle) is well adapted for removing a sample of pulp from each of a number of beets.

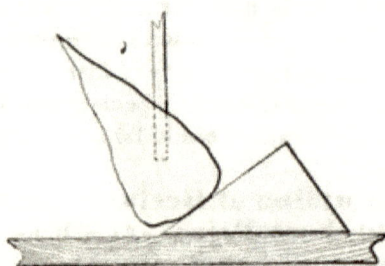

FIG. 18.

This machine, which is shown in Fig. 19, may be used in Pellet's instantaneous diffusion method (62).

FIG 19.

The beet is pressed carefully against the rasping-tool, which revolves at the rate of 2000 revolutions per minute.

FIG. 20.

An opening in the rasp, which is shewn in detail in Fig. 20, permits the pulp to pass into the tool, which is hollow, and thence to the box shown in the figure. Practice is necessary in using this machine in order to produce a suitable pulp. The pulp from the first perforation should be rejected.

It is evident that this machine does not remove a portion of pulp bearing a fixed relation to the size of the beet. This is essential in order that the analysis may represent the mean composition of the roots. The following method of sampling has been proposed by Kaiser[1] to obviate this difficulty.

The form of the beet is a cone the height of which is approximately three times the radius of the base, hence its volume is calculated by the formula $\pi r^3 =$ volume; in other words, the volume of the beet increases as the cube of its largest radius. For example, we have three beets whose radii are $4:5:6$; their volumes are then in the ratio $4^3:5^3:6^3$, or $64:125:216$. The beet whose radius is 4 should be perforated once; the second, whose radius is 5, should be perforated ($\frac{125}{64}$) 2 times, and the third, having a radius of 6, should be perforated ($\frac{216}{64}$) 3 times, and so on. Kaiser uses a scale which indicates the number of perforations to be made in each beet. Such a scale may easily be made which will show at a glance the number of times each beet should be perforated.

This method of sampling gives approximately correct results, even if the relation between the greatest radius of the beet and its length be different from Kaiser's numbers.

If the number of beets in the sample be small enough to permit, it is advisable to divide the beets longitudinally and reduce an entire segment of each to a pulp suitable for a direct method of analysis.

After the sample of washed beets is received in the laboratory its weight should be noted, that a correction may be made for the loss of weight by drying prior to the analysis.

42. Sampling Beets at the Diffusion-battery. —Samples of beets can be drawn at the battery with moderate certainty of obtaining a fair average. In the various

---

[1] *Deutsche Zuckerindustrie*, Nov. 1896.

manipulations from the field to the factory, including the transport and washing, the beets are pretty thoroughly mixed; hence if a beet be taken at random at regular and frequent intervals, the united subsamples so drawn will be very nearly of the mean composition of the beets entering the sugar-house. It is not usually necessary to sample beets in this way, since the method given in the following paragraph is simpler and the sample drawn is more satisfactory.

43. Sampling the Fresh Cossettes at the Diffusion-battery.—The proper time to sample the beets is after they have been sliced. A handful of the cossettes should be taken from the elevator or drag at regular intervals and stored in a covered receptacle. Large granite- or agate-ware pails are very convenient for the purpose, as they can be easily inspected as to their cleanliness. It is not advisable to use a mechanical device to divert a part of the cossettes to the pail, since the sample so obtained is not usually a fair average.

The samples should be drawn at very frequent intervals, if practicable every two or three minutes. In practice it is more convenient to take a small portion of the cuttings shortly after they begin to fall into the diffuser, a second when the diffuser is half filled, and a third before directing the cuttings into the next diffuser. The sample obtained in the manner described should be taken to the laboratory for immediate treatment. It is perfectly reliable, and if the beets be weighed immediately before entering the cutter, it may enter into the chemical control of the diffusion. It is necessary to keep the sample-pails scrupulously clean, using boiling water in washing them; they should be large enough to contain the subsamples from two or three hours' work.

44. Sampling the Exhausted Cossettes.—The exhausted cossettes should be sampled in a similar manner to the fresh cuttings. This sample may be taken from the elevator leading to the pulp-presses, and should be stored in a covered galvanized-iron pail having the bottom perforated for drainage.

45. Sampling Waste Waters.—A definite volume of the waste water should be drawn from each diffuser and these subsamples stored in a loosely stoppered bottle, with corrosive sublimate as a preservative.

46. Sampling Diffusion-juice, etc.—In sampling diffusion-juice, a definite volume should be drawn from each measuring tankful. This volume once decided upon should not be changed during the sampling period except there be a change in the volume of juice drawn into the tank, and then the sample should be changed in a like proportion. This is not easily accomplished, except by the use of an automatic sampler.

In sampling purified juices the same method should be observed.

47. Sampling Filter Press-cake.—In sampling the press-cake, small portions should be taken systematically from different parts of the press, bearing in mind that parts of the cake contain more moisture than others, according to the kind of press. The number of presses filled should be recorded for use in estimating the weight of the press-cake and in averaging the analyses.

A very simple and satisfactory instrument for sampling filter press-cake is made from a small brass tube with a cutting edge at one end. Several cork-borers of the same diameter are more convenient than a single brass tube for this purpose.

In using this instrument small cylinders of the press-cake are cut out in precisely the same manner as one would bore a hole through a cork. The subsamples are left in the tubes until a sufficient quantity of material has been collected. Each subsample pushes its predecessor farther into the tube.

48. Sampling Sirups.—A method is recommended in **10** for the measurement of sirups. In this method gauge-tubes, similar to the water-gauges on steam-boilers, are used. The sirup should be thoroughly mixed in the tanks before admitting it to the tubes. If several tanks be used, a volume of sirup should be drawn from each included in the analytical period, as advised in **10**.

49. The Preservation of Samples.—The sample of diffusion-juice is effectually preserved from fermentation by the addition of subacetate of lead. It may be preserved in this way several weeks or even months without perceptible change in the sucrose content. The most convenient preservative is mercuric chloride, 1 part to 10,000 parts of juice. It is not advisable to store juices treated with mercuric chloride for a longer period than 24 hours. The advantage of the mercuric chloride is that it permits the usual determinations, viz., sucrose, total solids, ash, etc., to be made with the same sample, thus obviating the necessity of drawing a second sample as is usual when subacetate of lead is used.

In many houses it is the practice to store the samples a week before analysis, uniting those drawn from day to day. In such cases it is advisable to determine the density, solids, and ash from day to day, and store a portion of the juice with subacetate of lead for the sucrose determination. The use of mercuric chloride simplifies the work, and as it is used in such minute quantities it does not perceptibly affect the accuracy of the results.

When subacetate of lead is employed as a preservative, it should be added in the proportions required for the clarification of the juice, i.e., about 2–3 cc. of the concentrated solution (**207**). It is convenient to use the concentrated lead solution, and, when preparing for the polarization, to measure the mixed juice and lead solution and add sufficient water to increase the volume to 110% that of the juice. The per cent sucrose is then readily calculated by the use of Schmitz' table(p.285)from the degree Brix of the juice and the polariscopic reading.

The preservative must be thoroughly mixed with juice as each portion is added. This is easily accomplished when an automatic sampler is employed by letting the delivery-tube dip to the bottom of the storage-bottle. The mouth of the bottle should be loosely plugged with cotton. When the sampling is by hand, it is advisable to use a wide-mouthed jar, provided with a cover, for the storage of the juice, and mix frequently. This facilitates the collection of the subsamples without the use of a funnel.

No preservative is required for sugar-house products other than the waste waters, juices, and sirups.

50. Automatic Sampling of Juices.—Automatic

samplers have for their object not only the relief of the
chemist from this duty, but the drawing of samples which
are probably more reliable than those obtained in any
other way.

This problem is not a simple one in the case of sampling
diffusion-juices at the measuring-tank. It is evident from
the method of conducting the diffusion, that the juice received into the measuring-tank is not of uniform composition. A sample drawn from the bottom of the tank will
differ slightly from one drawn at the centre or near the top.

Coombs' Automatic Sampler.—The apparatus shown in
Fig. 21 is the invention of Mr. F. E. Coombs, Chemist of

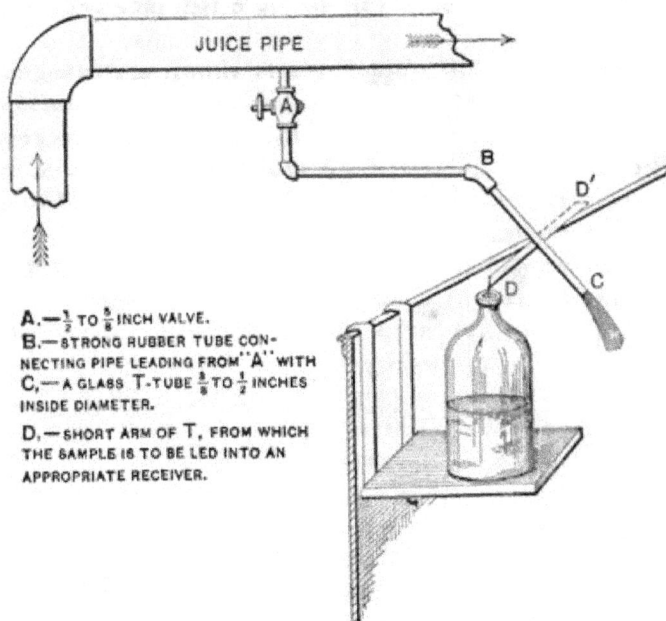

A.—$\frac{1}{2}$ TO $\frac{5}{8}$ INCH VALVE.
B.—STRONG RUBBER TUBE CON-
NECTING PIPE LEADING FROM "A" WITH
C,—A GLASS T-TUBE $\frac{3}{8}$ TO $\frac{1}{2}$ INCHES
INSIDE DIAMETER.
D,—SHORT ARM OF T, FROM WHICH
THE SAMPLE IS TO BE LED INTO AN
APPROPRIATE RECEIVER.

FIG. 21.

the Shadyside Plantation, Louisiana, and of the Esperanza
Estate, Trinidad, B. W. I., through whose courtesy this description and illustration were supplied the author.

This apparatus is applicable to the sampling of liquids
which are not too viscous to flow through small pipes. It
may be used in sampling juice and sirup, and has proved
quite reliable in practical work. It has the advantage of

being quickly set up wherever there is provision for returning a small quantity of overflow liquor to the tank.

Attempts to draw continuous samples of liquor from pipes by means of a small valve, depending upon the valve to regulate the flow of the sample, have usually failed, since the valve must be so nearly closed that fine pulp in the juice or, in the case of sirup, a mere change from a low to a high density clogs the opening and stops the sampling.

The flow must be sufficient to keep the valve free from obstruction. By the use of a T-tube, as shown in the figure, a strong current of liquor can be kept flowing through the pipe, and at the same time a small, continuous, easily regulated drip can be diverted into the sample-bottle.

In the figure, the apparatus is shown as arranged for drawing a sample of juice as it passes from the measuring-tank to the carbonatation. It is advisable to pass the juice through a distributing-tank in which the sampler is located, otherwise an arrangement must be provided for conducting the overflow to the carbonatation-tanks.

The sample-bottle at D rests upon a wooden shelf hung inside the tank by hangers of strap-iron which hook over the edge. It is apparent that when D, the short arm of the T-tube, is in its lowest position it will give its maximum discharge. By rotating the T-tube, which is of glass, in the strong rubber connecting-tube B to the position D', the drip will cease, all the liquor passing out at C. The position giving a sample of the required volume is readily ascertained by experiment. The sample, if juice, is preserved as indicated in **49**; sirups require no preservative.

With well-strained juice the drip is regular and there is rarely trouble from clogging.

It is evident, from the arrangement of the sampler, that the samples drawn, whether of juice or sirup, may be depended upon as being representative of the composition of the entire volume of the liquor.

It is necessary to connect the small pipe at the under side of the juice or sirup main, to insure a continuous flow, even when but little liquor is passing. The main should be tapped at its highest level, or on the discharge side of that level, to avoid drawing liquor left in the pipe when the flow is temporarily stopped. The valve on the sampling-pipe should be placed as close as possible to the point where the main is entered.

The valve *A* should always be opened as widely as possible to prevent clogging, but this must be regulated so that the current through the main arm of the T-tube shall not be too swift, since it will then act as an aspirator. For this reason it is advisable to avoid extending the discharge-tube *D* below the level of the sample in the bottle, otherwise the entire contents may be lost.

Horsin-Déon's Automatic Sampler.—This apparatus, shown in Fig. 22, consists of a three-way cock for connecting a small standpipe alternately with the measuring-tank and the sample-bottle, and is operated by a suitable float.

This sampler is placed inside the measuring-tank. It is so arranged that the volume of the sample drawn is proportionate to the quantity of juice in the tank. The discharge-pipe from the diffusion-battery should enter the measuring-tank at the bottom. The inlet to the sampler should be directly over the inlet from the battery, if practicable, projecting into the pipe. If this precaution be not ob-

Fig. 22.

served the sample drawn will not be a fair average.

It is obvious that this sampler is not applicable in sampling sirups.

51. Sampling Sugars.—Sugars are best sampled by means of a "trier" or sound (Fig. 23). This in-

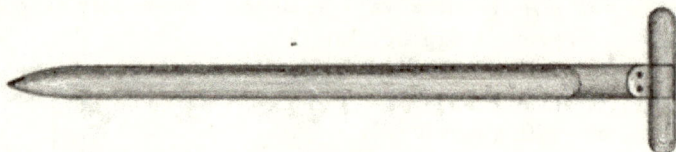

FIG. 23.

strument is so constructed that it may be plunged into a quantity of sugar and, on withdrawal, remove a sample representative of the sugar through which it has passed. The trier should be long enough to pass from end to end of the package of sugar, diagonally if necessary. The chemist must be guided largely by the grade of the sugar and the method of packing in drawing the sample. A portion should be drawn from every third, fifth, etc., package according to the size of the lot. The large sample should be well mixed, and all lumps broken, then subsampled by quartering.

heavier than water, and have the values .1, .01, .001, and .0001, respectively, when placed on the corresponding graduations of the beam, and for other graduations .300, .030, .003, .0003, etc. Each rider is provided with a hook

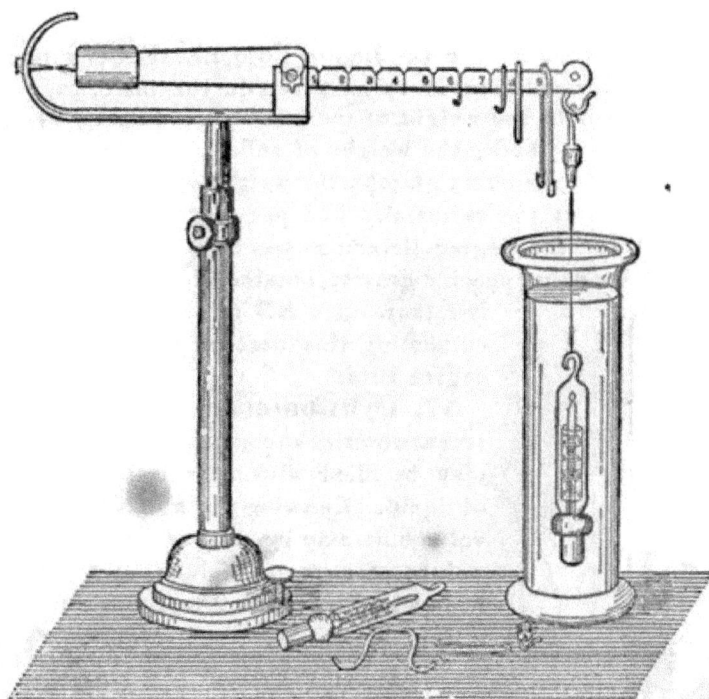

FIG. 27.

from which additional weights may be suspended in the case of more than one falling upon the same graduation.

The method of using the balance is as follows : Dissolve a weighed portion of the material in water and dilute to a measured volume at $17\frac{1}{2}°$ C.; for example, 25 grams to 100 cc. Suspend the bob of the balance, as described above, in this solution, and weight the beam with the riders until the balance is in equilibrium. Read off the specific gravity from the position of the weights on the beam. Example : 25 grams material dissolved and diluted to 100 cc. Position of the riders :

(1) at point of suspension of the bob $= 1.000$
(2) not on the beam.
(3) at 7 .... $= 0.07$
(4) at 9 .... $= 0.009$

Specific gravity..................... $= 1.079$

The degree Brix corresponding to 1.079, *i.e.*, the per cent solids in this solution, is 19, as given in the table, page 275. To obtain the weight of the solution, multiply 1.079 by 100 $= 107.9$; hence the weight of solids in the solution is $107.9 \times 19 \div 100 = 20.5$ grams $=$ the weight of solid matter in 25 grams of the material. The per cent solids in the material, *i.e.*, the degree Brix $= 20.5 \div 25 \times 100 = 82$, and the corresponding specific gravity, obtained from the table, is 1.4293. See **85** relative to the accuracy of this determination of the degree Brix.

57. Pyknometers. — Pyknometers are bottles so constructed that they may be filled with a definite volume of liquid. Knowing the weight of this volume, it may be compared with the weight of an equal volume of water, from which the density of the liquid is calculated. It is rarely necessary to use a pyknometer in the sugar industry, the more rapid density determination by the spindle being usually sufficiently accurate.

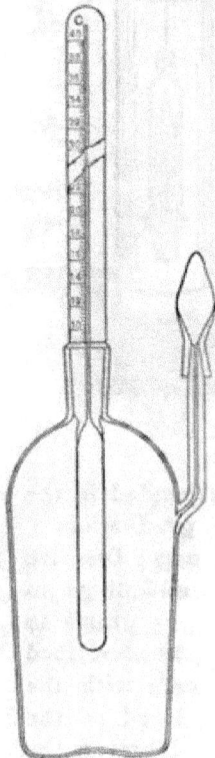

Fig. 28.

Pyknometers are made in a great variety of forms. One of the most convenient of these is shown in Fig. 28. The stopper is a fine thermometer ground into the neck of the bottle. The side tube provides an outlet for the excess of liquid when the stopper is put in place. The bottle should be filled at a somewhat lower temperature than that at which the density is to be determined. As the temperature gradually rises to the desired point, the excess of liquid is blotted off. At the required temperature, the cap is placed in position, and receives any further liquid, which may be expelled from the bottle, as the temperature rises to that of the room. There is a minute opening in the cap for the escape of the air.

In sugar work, the specific gravity should be determined at $17\frac{1}{2}°$ C. for reasons already stated. The weight of the corresponding volume of water may be determined at room temperature and a correction be made to reduce it to the standard temperature, the tables on page 251 being used for this purpose. It is customary to express specific gravities as follows: $\dfrac{17.5°}{17.5°}$, 1.0795; the numbers above and below the line being the temperatures at which the bottle was filled with water and the substance respectively.

Recently boiled and cooled distilled water should be used in density determinations.

To calculate the density of a liquid, divide the weight of a definite volume of it by the weight of an equal volume of water.

ANALYSIS OF THE BEET.

58. The Direct Analysis of the Beet.—The methods for the direct analysis of the beet may be divided into two general classes, according to the solvent used, viz.: (1) methods employing alcohol; (2) methods employing water. The alcoholic methods have found most favor in Germany, and the aqueous methods in France.

Certain modifications of the Scheibler alcoholic method and Pellet's aqueous methods, hot and cold, are the most important of their classes, and are the only ones which will be described in this book. It is probable, judging from the published statements of many chemists, that these methods are equally accurate if the instructions of their inventors be implicitly complied with. The alcoholic methods are usually considered the most scientific.

59. Scheibler's Alcoholic Method with Soxhlet's Extraction Apparatus.—Various modifications of Soxhlet's apparatus are used to such an extent in chemical laboratories that an illustration, Fig. 29, and a brief description of it will suffice. The apparatus is so arranged that the vapors of the solvent, which is boiled in the flask by means of a hot-water bath, pass up through the tube *B* to the reflux condenser, and the solvent falls back into the extractor in which the material is placed. When a sufficient quantity of the solvent accumulates in the extractor, it is siphoned into the flask by the tube shown at the right. The substance is thus extracted with successive portions of the solvent.

A very convenient and efficient modification of this apparatus is the siphon extraction-tube devised by A. E. Knorr, shown in Fig. 30. The connections with the flask and condenser are made with corks as in the Soxhlet apparatus. Knorr's apparatus, as arranged for general purposes, dispenses with corks, but requires a special flask, which is not convenient for sugar analysis.

The siphon-tube *S* is sealed into the bottom of the tube

A, and lies close to the wall so as to permit the insertion of the tube *B* containing the material. The lower end of *B* is closed with a perforated disk. A spiral of copper wire, *C*, prevents the tube *A* from closing the tube *D*.

This apparatus has the advantage of extracting the sucrose with a hot solvent.

Other convenient modifications of Soxhlet's apparatus are described by Wiley in his *Agricultural Analysis.*

In the direct analysis of the beet with the Soxhlet-Sickel apparatus, Fig. 29, proceed as follows for the extraction of the sucrose: Place a plug of absorbent cotton in the bottom of the tube, then introduce 26.048 grams of the pulped beet, or 2 × 16.29 grams, according to the polariscope in use, pressing the pulp lightly with a rod. Very small fragments of the beet may be used instead of pulp.

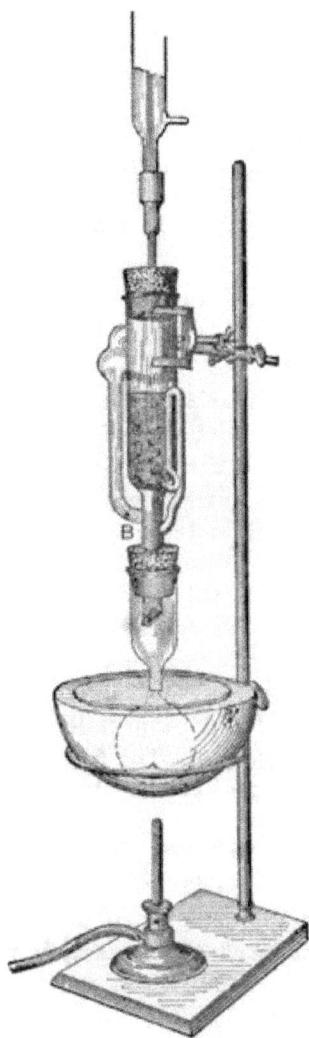

FIG. 29.

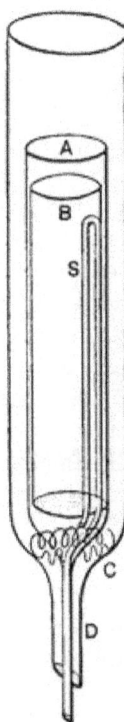

FIG. 30.

Connect the extractor with the reflux condenser as shown. Place 75 cc. of 95 per cent alcohol in the flask and connect with the extractor as indicated in the figure; heat the flask in the water-bath and continue the extraction from half an hour to two hours or more, according to the state of division of the sample. Use somewhat weaker alcohol if only 16.29 grams of pulp be taken. Cool and remove the flask, substituting a second containing 75 cc. of 75 to 80 per cent alcohol, and continue the extraction to ascertain whether the first extraction were complete.

Fill the first flask to the 100 cc. mark, after treating the sample with two or three drops of subacetate of lead solution. Mix the contents of the flask, filter, and polarize. Having extracted the normal weight of pulp, the polariscopic reading is the per cent of sucrose in the sample.

The extract in the second flask should also be polarized as a check upon the extraction.

Great care is essential in the polarization of alcoholic solutions. The least quantity of subacetate of lead, that will clarify the solution, should be used. The solution must be protected from evaporation during the filtration by a cover-glass. Avoid irregularities in the temperature of the solution in the observation-tube, due to the warmth of the hands; since the density of the solution in different parts of the tube will vary under such conditions, striæ will form, rendering an accurate reading impossible.

The Scheibler method, as above described, differs from the original only in a few minor details, especially in the arrangement of the extraction apparatus. The Soxhlet extraction apparatus is much more effective than Scheibler's original instrument.

60. Stammer's Alcoholic Digestion Method.[1] —This method differs from that of Pellet described in **62** in details of manipulation and in the use of alcohol instead of water. The pulp must be reduced to a cream, in fact should be as finely divided as is required in the Pellet method (**62**).

Wash 26.048 grams of pulp into a flask graduated at 100.55 cc. with 92 per cent alcohol, add subacetate of lead for clarification, and dilute to the mark with the alcohol. The least quantity of the subacetate that will effect clarification should be used. Acetic acid is not required. Mix thoroughly, and after allowing a few minutes for the digestion,

---

[1] *Zeit. Rübenzucker-Industrie,* **33**, 206.

filter and polarize, observing the precautions given in **59**
relative to the polarization of alcoholic solutions.

A method similar to this, Rapp-Degener, employs hot
digestion in a flask fitted with a reflux condenser.

61. Pellet's Aqueous Method. Hot Digestion.
—Any good rasp may be used in the preparation of the
pulp for this method. Pellet recommends the conical rasp

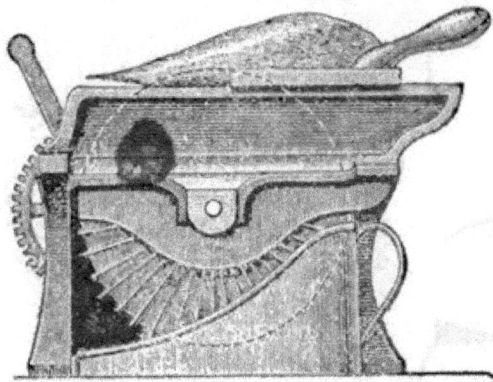

FIG. 31.

of Pellet and Lomont, as illustrated in Figs. 31, 32, and 33.
There is frequently a depression in the side of the beet, as
shown in section in Fig. 34. Since the segments OA and
OB are not of equal sugar con-
tent, two segments should be
reduced to pulp, or, if the sample include a large number of
beets, a single segment of each
may be pulped, taking care to
present alternately the large and
the small diameters of the beets
to the rasp.

The special flasks shown in
Fig. 35 are convenient for use in
this method. Transfer 26.048
grams of the pulp to the flask,

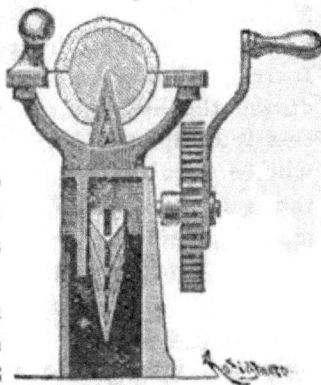

FIG. 32.

using a little water to wash the weighing capsule and
funnel, or, for the Laurent, employ 32.58 grams of pulp, *i.e.*,
2 × normal weight. The flasks are graduated to contain
201.35 cc. for the Schmidt and Haensch and 201.7 cc. for

the Laurent polariscopes, in order to compensate for the

FIG. 33.

volume of the marc and the lead precipitate. Add 5 to

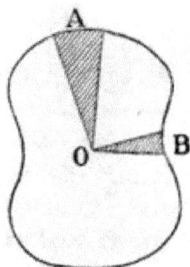

FIG. 34.

10 cc. subacetate of lead solution of 54.3° Brix (**207**) for the clarification. Approximately 6 to 7 cc. are required per 26 grams of beet-pulp. This reagent should be run into the flask in advance of the beet-pulp. Add a few drops of ether to beat down the foam, then sufficient water to increase the volume of the solution to about 190 cc. Heat to 80° C. in a water-bath and maintain this temperature about 30 minutes, occasionally giving the flask a circular movement to facilitate the escape of the air from the pulp. Increase the volume of the solution from time to time during the heating, so that when the operation is completed only a few drops of water will be required to complete the volume of the solution to the mark. After approximately 30 minutes' heating, cool the flask and contents and add strong acetic acid to the solution to acidity, dilute to the graduation, mix and filter. The state of division of the pulp will govern the time of heating. In polarizing the filtrate, use a 400-mm. observation-tube, thus directly obtaining the per cent sucrose in the beet with the Schmidt and Haensch polariscope, or double this percentage if the Laurent instrument be used.

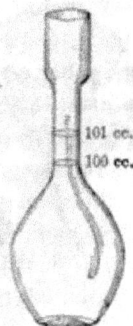

101 cc.
100 cc.

FIG. 35.

Pellet uses a special water-bath in this process that admits a considerable number of flasks at one time. The flasks are held in a rack and may all be removed from the bath at one time and plunged into cold water.

The solutions should be carefully protected from evaporation by covering the funnel during filtration.

There has been much controversy relative to this method, especially among the German chemists. Many claim that it gives results that are too high, and other chemists of equal prominence and experience contend that it gives correct results. Le Docte,[1] in a series of experiments, obtained percentages by hot digestion a few one-hundredths higher than by the cold diffusion method described below. The following method, using cold water, is usually preferred, provided a sufficiently fine pulp can be produced.

62. Pellet's Instantaneous Aqueous Diffusion Method.

—The method as described by Pellet will be given first, and then a few of the various modifications. The author prefers the Sachs-Le Docte modification given on page 181, which combines rapidity and accuracy.

Pellet's Original Method.—In following Pellet's original method the specifications as to the condition of the pulp and the quantity used must be strictly complied with in order to obtain satisfactory results.

For polariscopes whose normal weight is 26.048 grams wash this weight of pulp, with water, into a flask graduated to hold 201.35 cc., or 25.87 grams into a 200-cc. flask. Run 5 to 7 cc. of subacetate of lead solution of 54.3° Brix (207) into the flask before washing in the pulp, and then thoroughly mixed with the latter. Add several small portions of ether to beat down the foam. Rotate the flask to facilitate the escape of the air-bubbles. Add a few drops of acetic acid to acidulate the solution, complete the volume to the graduation, mix, filter, and polarize, using a 400-mm. observation-tube. The polariscopic reading is the per cent sucrose in the beet. With the Laurent instrument, use the normal weight of pulp and a flask graduated to hold 200.85 cc. The polariscopic reading,

---

[1] *Sucrerie Belge,* 25, 245, 273, 309.

using a 400-mm. observation-tube, is the per cent sucrose in the beet.

Success with this method demands (1) that the pulp shall be in a suitable state of division, neither too coarse nor too fine; (2) that no more pulp shall be used than indicated in the description of the method. If there be difficulty in removing the air occluded by the pulp, notwithstanding repeated additions of ether, the pulp is too fine. This may be remedied by altering the speed of the rasp. The occluded air is the source of error that requires greatest care to avoid.

Kaiser-Sachs Modification.—This method and the *Sachs-Le Docte* modification practically eliminate errors from the Pellet instantaneous diffusion method. Use flasks holding a little more than 200 cc. Also use the same quantities of pulp as indicated in the description of the original Pellet method. Run 5 cc. of subacetate of lead solution into the flask, then counterpoise the flask and contents on a balance. Wash the pulp into a flask and add sufficient water to make a total of 172 grams of water. Mix thoroughly, filter, and polarize the solution in a 400-mm. tube. The polariscopic reading is the per cent of sucrose in the beet. According to Pellet, acetic acid should always be added. This agrees with the author's experience.

Sachs-Le Docte.—This method, which is fully described on page 181, differs from the above in adding the water and subacetate of lead from an overflow or automatic pipette. This insures a very accurate measurement, with extreme rapidity.

The finest attainable pulp should be used with both the Sachs-Le Docte and the Kaiser-Sachs methods.

63. Determination of the Reducing Sugar in the Beet.—*Herzfeld's Modification of Claassen's Method.*— Digest 110 grams of finely divided pulp, or preferably creamed pulp, in a 500-cc. flask with 10 to 15 cc. of dilute subacetate of lead solution, 3 grams of precipitated carbonate of calcium, and sufficient water to nearly fill the flask. Digest 45 minutes at a temperature of 75° to 80° C. Cool and complete the volume to 500 cc., mix, and filter. If necessary, clarify 100 cc. of the filtrate with an additional portion of subacetate of lead ; add carbonate of sodium in small excess to precipitate the lead, dilute to 110 cc., and filter. Determine the reducing sugar in the filtrate by one of the methods given in **72** and **73**. The percentage of reducing substance in the beet is so small that no correction need be made for the volume of the marc.

64. Notes on the Direct Methods of Analysis. —With the exception of Scheibler's alcoholic method, it is necessary to make an arbitrary allowance for the volume of the marc in the direct analysis of the beet. Pellet has based this allowance upon the mean of a large number of marc determinations, made under practically the conditions which obtain in his cold diffusion method. The error introduced through an arbitrary allowance for marc is very small, and even in extreme cases may be neglected.

There should be no delay in the analysis of the pulp. As soon as it is obtained it should be thoroughly mixed and protected from the air.

65. Rasps and Mills for the Reduction of the Beet.—*The Cylindro-divider, Keil* (Gallois and Dupont, Paris).—This machine, Fig. 36, as indicated by its name, consists esssentially of two deeply grooved cylinders which revolve in opposite directions. Nearly all of the pulp adheres to the cylinders, but little dropping into the drawer. The particles which fall, if too large, should be returned to the mill and the grinding should be continued until the pulp is uniformly divided. The mill should be driven at 120 revolutions per minute, either by hand or power.

Should the beets be unripe or unsound, the juice may separate and collect in the drawer. In this event, the pulp, when fine enough, should be removed from the cylinders and thoroughly mixed with this juice. The pulp will absorb the juice, and may then be sampled as usual.

This mill is designed for grinding cossettes and fragments of beets, and produces a pulp which may be analyzed by Pellet's instantaneous method.

Pellet and Lomont's Conical Rasp.—This machine as illustrated in Figs. 31, 32, and 33 is fitted with saw-blades and is not applicable in the instantaneous diffusion method. The

machine is also constructed with a cast-steel disk, which may
be briefly described as a rotary file as cut for rasping wood.

FIG. 36

This form is applicable in the above-mentioned diffusion
method.

A little practice is necessary in the manipulation of this
and certain other rasps in order to produce a suitable pulp.

Neveu and Aubin's Rasp.[1]—This rasp may be used in the

---

[1] *Bulletin de l'Association des Chimistes*, **13**, 311.

reduction of beets, but not cossettes, to an extremely fine pulp for use in any of the direct methods of analysis or in the indirect method. The construction of the machine is shown in Fig. 37. It is driven by hand or power from 75 to 400 revolutions per minute.

FIG. 37.

Additional rasps, designed especially for use in seed selection, are described in **160** and **161**.

66. Indirect Analysis of the Beet.—The indirect analysis, *i.e.*, the analysis of the juice and calculation to terms of the weight of the beets, cannot be depended upon to supply data for the control of the factory. In order to calculate the analysis of the beet from that of the juice, it is necessary to assume that the juice extracted by the press is of the same composition as the average of all the juice contained in the beet. Experience has shown that this is not true, and that the juice obtained by moderate pressure differs materially from that obtained by heavy pressure. It also varies with the state of division of the pulp. There is also reason to believe that the beet contains water in which there is little, if any, sugar in solution. Further, in order to render an indirect method practicable, it is necessary to assume that the beet contains an average of a certain percentage of juice, and employ this percentage as a coefficient in reducing to terms of the beet. The fact

FIG. 38.

that the content of marc varies within rather wide limits is an argument against this method of analysis.

The indirect method is still employed in a large number of sugar-houses, hence is described in this book.

The following is the usual method of procedure:

The sample is finely rasped by a suitable machine, such as a special rasp or an efficient horseradish grater. The pulp is placed in a small cotton bag and the juice is expressed by means of a powerful press, such as that shown in Fig. 38. In operating the press as heavy pressure as possible is exerted by turning the upper wheel, then locking with the ratchet as shown in the figure, and completing the expression of the juice by means of the lower wheel. This press exerts a maximum pressure of nearly 2000 lbs. per square inch.

In order to closely approximate the true mean composition of the juice, it is essential that the pulp be very finely divided and that as much pressure be exerted in expressing the juice as is practicable.

The analysis is made as indicated in **67** *et seq.*

In the indirect analysis, it is customary to assume that the beet contains a mean of 95 per cent of juice ; therefore to calculate the percentage to terms of the weight of the beet, multiply the per cents on the weight of the juice by 95 and divide by 100.

This method permits an approximate determination of the coefficient of purity of the juice which is not possible with a direct method and which is often of value.

ANALYSIS OF THE JUICE.

67. Determination of the Density.—The density is usually determined by means of a Brix spindle. The degree Brix may be converted into terms of the specific gravity for use in calculating the weight of the juice by means of the table, page 275 ; or a Baumé spindle may be used and the readings converted into Brix and specific gravity by the above-mentioned table.

A cylinder is filled with a sample of the juice and is set aside for the escape of air-bubbles and to permit mechanical impurities to subside or rise to the surface. This time varies from a few minutes to half an hour. Care must be observed not to let the juice stand long enough for fermentation to set in. Those impurities which rise to the surface should be brushed off, and the spindle then floated in the juice. After allowing sufficient time for the spindle to reach the temperature of the juice, the scale is read as directed in **55** and illustrated in Fig. 26, and the temperature of the juice is noted.

To correct for temperatures above or below $17\frac{1}{2}°$ C., the standard temperature at which these instruments are usually graduated in Germany and the United States, consult the table on page 282. It is advisable that the temperature of the juice when spindling be as nearly $17\frac{1}{2}°$ C. as practicable.

It is necessary that the density be determined with great care, since the result obtained is employed in calculating the weight of the juice at an important stage of the control work.

Other methods of determining the density are indicated in pages 55 to 61.

68. Sucrose Determination . Special Pipette for Measurements.—The method of preserving the samples will, to some extent, influence the preliminary work of the analysis.

The method of analysis indicated in **69** is usually more convenient when subacetate of lead is used as a preservative. If, however, mercuric chloride be employed in the sampling, the special pipette devised by the author is convenient, since the polariscopic reading is a multiple of the percentage of sucrose.

This pipette is shown in Fig. 39. It is so graduated that one need simply note the degree Brix of the juice, then fill the pipette to the corresponding degree marked on its stem. The graduations indicate the volume of juice, of corresponding densities, which weighs 52.096 grams, *i.e.*, two times the normal weight.

The pipettes are usually graduated for ordinary work from 5 to 25 degrees Brix in tenths. It is recommended that, for control work, the pipettes be graduated with only a small range on each, and that there be an additional graduation as shown in the figure. The tubing should be of small internal diameter that the tenths may be the more easily read. The pipette as ordinarily made, without the additional mark near the outlet, should be graduated with a solution of approximately the viscosity of a sugar solution, of the mean degree Brix within the limits of the scale.

One should not blow into the pipette while emptying it, nor should the last portions of the juice be expelled in this way.

To calculate the percentage of sucrose, divide the polariscopic reading, with the German instruments, by 2. Pipettes for the Laurent instrument are graduated to deliver three times the normal weight (3 × 16.29 grams), hence the reading should be divided by 3. The juice should be measured at the temperature at which the degree Brix was determined.

69. Sucrose in the Juice . General Method.—The necessity of using a preservative in sampling, especially if the preservative be sub-

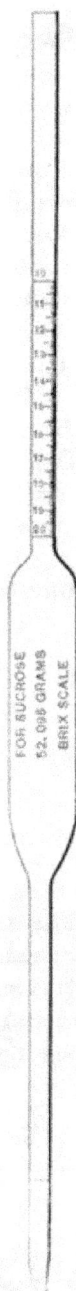

FIG. 39.

acetate of lead, complicates the weighing of the norma weight of the juice ; consequently measurements are used almost exclusively in this analysis. The sample may be weighed, however, when mercuric chloride is used, as the quantity of this substance employed is too small to appreciably influence the results.

When subacetate of lead is used, and this preservative is necessary when samples are accumulated for a period of several days, the following is a convenient method of procedure : Measure the entire sample and then add sufficient water to dilute to 110 per cent of the original volume of the juice; calculate the sucrose by Schmitz' table, page 285.

Example, showing methods of calculation :

Degree Brix of the juice as determined in duplicate samples = 12.2. Measure the day's sample, plus the lead subacetate solution, subtract the number of cubic centimetres of the lead solution, and calculate the water to be added as shown below:

> Volume of juice and lead solution.... 3750 cc.
> Volume of lead solution.............. 75 "
> _____
> Volume of juice...................... 3675 "

> Ten per cent of volume of juice
>   = one tenth of 3675 = 367.5 cc.
> Volume of lead solution   = 75   "
> _____
> Volume of water required   = 292.5 "

The total volume, *i.e.*, 3750 + 292.5 = 4042.5 cc. = 110 per cent of the volume of the juice (3675 cc.).

Having diluted the juice and lead solution to 4042.5 cc., mix and filter off a few cubic centimetres, and polarize in a 20-centimetre tube:

> Polariscopic reading = 38.3.

In Schmitz' table, in the column headed 12, the nearest degree Brix to the observed degree, and opposite 38, the integral part of the polariscopic reading, note the number 10.36; in the small table at the bottom of the page, opposite .3, the decimal part of the polariscopic reading, note the number .08, and add this to the number obtained above for the completed percentage: 10.36 + .08 = 10.44, the per cent sucrose in the juice.

Example including a storage period of seven days:

Each day's sample of juice is diluted as described in the preceding example, thoroughly mixed, and a volume proportionate to the day's work is measured into a storage-bottle. If a uniform volume of juice, per 100 lbs. of beets, have been drawn at the diffusion-battery, the amount of juice to store each day may be a certain number of cubic centimetres for each diffuser, as follows:

Day.	Number of Diffusers Drawn.	Cubic Centimetres of Juice, leaded and diluted to 110%, to be stored from day to day.
1	160	16.
2	155	15.5
3	148	14.8
4	135	13.5
5	165	16.5
6	163	16.3
7	158	15.8
		Total 108.4 cc.

If, as is the practice in some houses, the volume of juice drawn per diffuser be changed with the variations in the richness of the beets, the storage sample must be based on the volume of the juice and not upon the number of diffusers.

At the close of the sampling period, the united samples are thoroughly mixed, and a portion of the juice is filtered off and polarized, the calculations being made by Schmitz' table.

When mercuric chloride is employed as a preservative, samples of the juice obtained from day to day may be stored for a longer sampling period by the addition of 10 per cent, by volume, of diluted subacetate of lead solution (**207**). It is not advisable to depend upon preservation with mercuric chloride for a longer period than 24 hours.

70. Notes on the Clarification of Samples for Polarization.—Too little subacetate of lead solution or a decided excess in the clarification may result in cloudy filtrates, or solutions which filter too slowly. Experience will soon enable one to estimate the proper amount of the lead solution to use. Sufficient of the lead salt must be used, not only to produce a clear filtrate, but to precipitate all the matter precipitable by this reagent.   This is essential, since the beet contains other optically active bodies than sucrose (**36**).

71. Remarks on the Reducing Sugars in Beet Products.—Beet juices and products, under normal conditions, do not usually contain more than traces of reducing sugars.   There is a reducing substance present in small quantity, however, of which little is known.   It is usually termed " Bodenbender's substance," from the name of the chemist who first reported its presence.   There is little probability of inversion in the processes of manufacture, except at the diffusion-battery, since the liquors are always more or less alkaline.   There is probably rarely any inversion in the diffusion process, except during very irregular work or in treating unsound beets.   In view of these facts, the beet-sugar chemist is not often called upon to make reducing sugar determinations, except in the estimation of sucrose by the chemical inversion method.   The methods of estimating reducing sugars are given quite fully in the following pages, for use in any work in which chemical methods may be required.

72. Determination of Reducing Sugars (Glucose, etc.).   Gravimetric Methods.—In selecting a method for reducing sugars, the analyst should be guided by the probable composition of the material under examination.

Gravimetric Method for Material containing 1 *per cent or less of Invert-sugar* [1] *and a High Percentage of Sucrose.*—Dissolve 20 grams of the material in nearly 100 cc. of water.   If necessary, clarify with subacetate of lead (*see* **74**), precipitate the excess of lead by means of sodium carbonate in small excess, complete the volume to 100 cc., mix thoroughly and filter.   This clarification is usually advisable.   Place 50 cc. of Soxhlet's solution (**192**) in a beaker and add 50 cc. of the sugar solution.   Heat slowly, taking about four minutes to reach the boiling-point, and boil two minutes. These directions should be strictly complied with.   After the completion of the two minutes' boiling add 100 cc. of cold re-

---

[1] The reducing sugar of the beet and beet products is probably the result of inversion of sucrose.   The methods described for invert-sugar are applicable.

cently boiled distilled water. Determine the copper, in the precipitate, by one of the following methods : (1) Filter immediately under pressure, using the filter-tube described below. The filter-tube, Fig. 40, consists of a 6-inch hard glass tube about ⅝ inch in diameter, into one end of which is sealed a tube about 3 inches long and of convenient size for inserting into the stopper of the filtering apparatus such as that shown in Fig. 49. A perforated platinum disk A, A' is sealed into the bottom of the large tube as a support for an asbestos felt filter. To prepare the tube for filtering, place it in position in the stopper of the filtering apparatus, start the filter-pump, then pour water containing finely divided asbestos in suspension upon the disk. The asbestos forms a film or felt; dry and weigh. Moisten the felt before commencing the filtration. A funnel should be used in pouring the liquid and precipitate into the filter-tube, to prevent the cuprous oxide from adhering to the walls of the tube, near the top.

FIG. 40.

Transfer all of the precipitate to the filter and wash thoroughly with hot water. After washing with water pass a few cc. of alcohol through the filter and finally a little ether. Dry the precipitate. Pass a continuous current of pure, dry hydrogen through the tube, at the same time gently heating the cuprous oxide, with a Bunsen burner, until it is completely reduced to the metallic state; cool in a current of hydrogen and weigh.

(2) Filter immediately after the reduction is completed, using a Gooch crucible. Wash the beaker and precipitate thoroughly with hot water, but without any effort to transfer the entire precipitate to the crucible. Wash the asbestos film and the adhering cuprous oxide back into the beaker, using hot dilute nitric acid. After the copper is all in solution, filter through a Gooch crucible, using a very thin asbestos film, and wash thoroughly with hot water. Add 10 cc. of dilute sulphuric acid, containing 200 cc. acid of 1.84 specific gravity, per litre, to the filtrate and evaporate it until the copper salt has largely crystallized. Heat carefully on a hot iron plate or a sand-bath until the evolution

of white fumes.   Add 8 to 10 drops of nitric acid, specific gravity 1.42, and rinse into a platinum dish of 100 to 125 cc. capacity.   Precipitate the copper on the dish by electrolysis. Wash the copper thoroughly with water before breaking the current ; remove the dish from the circuit, wash with alcohol and ether successively, and dry at a temperature that can easily be borne by the hand, cool and weigh.   A beaker may be substituted for the platinum dish, the copper being deposited upon a platinum cylinder.

When a direct current is used in lighting the sugar-house, it is the most convenient source of electricity for the deposi-

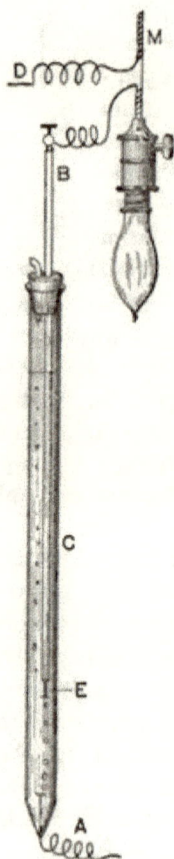

FIG. 41.

tion of the copper.   The current must be passed through a resistance or regulator in addition to the lamp.   A convenient and durable regulator is shown in Fig. 41.   C is a glass tube partly filled with water slightly acidulated with sulphuric acid ; the wire A connects with a platinum wire sealed into the tube ; B is a glass tube through which a copper wire extends and connects with a platinum wire E sealed into this tube.   The tube B may be slipped up or down, thus regulating the distance between the wires E and A and regulating the current.   The twin wire M is separated, severed, and one end, D, connected with the platinum dish in which the copper is to be deposited, and the other with the regulator B, thence through the acidulated water and A with the platinum cylinder suspended in the copper solution.   Sufficient current for a large number of dishes, arranged in sets of four, will pass through a 16 C. P. or 32 C. P. lamp.   The copper should be deposited very slowly.   Usually, if the apparatus be connected when the lights are turned on in the evening, all the copper will be deposited before they are turned off in the morning.

Having determined the weight of copper reduced by one

of the above-described methods, ascertain from Herzfeld's table the per cent of invert-sugar corresponding to the weight of copper.

Herzfeld's Table for the Determination of Invert-sugar in Materials Containing 1 Per Cent or Less of Invert-sugar and a High Percentage of Sucrose.

Copper reduced by 10 Grams of Material.	Invert-Sugar.	Copper reduced by 10 Grams of Material.	Invert-sugar.	Copper reduced by 10 Grams of Material.	Invert-sugar.
Milligrams.	Per Cent.	Milligrams.	Per Cent.	Milligrams.	Per Cent.
50	0.05	120	0.40	190	0.79
55	0.07	125	0.43	195	0.82
60	0.09	130	0.45	200	0.85
65	0.11	135	0.48	205	0.88
70	0.14	140	0.51	210	0.90
75	0.16	145	0.53	215	0.93
80	0.19	150	0.56	220	0.96
85	0.21	155	0.59	225	0.99
90	0.24	160	0.62	230	1.02
95	0.27	165	0.65	235	1.05
100	0.30	170	0.68	240	1.07
105	0.32	175	0.71	245	1.10
110	0.35	180	0.74		
115	0.38	185	0.76		

Gravimetric Method for Materials containing more than 1 *Per Cent of Invert-sugar.* — Prepare a solution of the material to be examined, in such a manner that it contains 20 grams in 100 cc.; clarify and remove the excess of lead with a small excess of sodium carbonate (*see* **74**). Prepare a series of solutions in large test-tubes by adding 1, 2, 3, 4, etc., cc. of this solution successively. Add 5 cc. of the Soxhlet solution (**192**) to each, heat to boiling, boil two minutes and filter. Note the volume of sugar solution that gives the filtrate lightest in tint but still distinctly blue. Place twenty times this volume of the solution in a 100-cc. flask, dilute to the mark and mix well. Use 50 cc. of this solution for the determination, which is conducted as under the preceding method, for materials containing 1 per cent or less of invert-sugar, until the weight of copper is obtained. For the calculation of the result use the following formulæ and table of factors of Meissl and Hiller:

Let Cu = the weight of copper obtained;

$\quad P$ = the polarization of the sample;

$\quad W$ = the weight of the sample in the 50 cc. of the solution used for the determination;

$\quad F$ = the factor obtained from the table for conversion of copper to invert-sugar;

$\quad \dfrac{Cu}{2}$ = approximate absolute weight of invert-sugar

$\qquad = Z$;

$Z \times \dfrac{100}{W}$ = approximate per cent of invert-sugar $= y$;

$\dfrac{100P}{P + y} = R$, relative number for sucrose;

$100 - R = I$, relative number for invert-sugar;

$\dfrac{CuF}{W}$ = per cent of invert-sugar.

Z facilitates reading the vertical columns; and the ratio R to I, the horizontal columns of the table, for the purpose of finding the factor, F, for the calculation of the copper to invert-sugar.

Example.—The polarization of the sugar is 86.4, and 3.256 grams of it, W, are equivalent to 0.290 gram of copper. Then

$$\frac{Cu}{2} = \frac{.290}{2} = .145 = Z;$$

$$Z \times \frac{100}{W} = .145 \times \frac{100}{3.256} = 4.45 = y;$$

$$\frac{100P}{P + y} = \frac{8640}{86.4 + 4.45} = 95.1 = R;$$

$$100 - R = 100 - 95.1 = I = 4.9;$$

$$R : I = 95.1 : 4.9.$$

By consulting the table it will be seen that 150 mg. in the vertical column are nearest the value of Z, 145 mg., and the horizontal column headed 95 : 5 is nearest the ratio R to I, 95.1 : 4.9. Where these columns meet we find the factor 51.2 which enters into the final calculation:

$$\frac{CuF}{W} = \frac{.290 \times 51.2}{3.256} = 4.56 \text{ per cent of invert-sugar.}$$

MEISSL AND HILLER'S FACTORS FOR THE DETERMINATION OF MORE THAN 1 PER CENT OF INVERT-SUGAR.

Ratio of Sucrose to Invert-sugar = $R : I.$	Approximate Absolute Weight of Invert-sugar = $Z.$						
	200 Milligr.	175 Milligr.	150 Milligr.	125 Milligr.	100 Milligr.	75 Milligr.	50 Milligr.
	Per Ct.	Per Ct.	Per Ct.	Per Ct.	Per Ct.	Per Ct.	Per Ct.
0 : 100	56.4	55.4	54.5	53.8	53.2	53.0	53.0
10 : 90	56.3	55.3	54.4	53.8	53.2	52.9	52.9
20 : 80	56.2	55.2	54.3	53.7	53.2	52.7	52.7
30 : 70	56.1	55.1	54.2	53.7	53.2	52.6	52.6
40 : 60	55.9	55.0	54.1	53.6	53.1	52.5	52.4
50 : 50	55.7	54.9	54.0	53.5	53.1	52.3	52.2
60 : 40	55.6	54.7	53.8	53.2	52.8	52.1	51.9
70 : 30	55.5	54.5	53.5	52.9	52.5	51.9	51.6
80 : 20	55.4	54.3	53.3	52.7	52.2	51.7	51.3
90 : 10	54.6	53.6	53.1	52.6	52.1	51.6	51.2
91 : 9	54.1	53.6	52.6	52.1	51.6	51.2	50.7
92 : 8	53.6	53.1	52.1	51.6	51.2	50.7	50.3
93 : 7	53.6	53.1	52.1	51.2	50.7	50.3	49.8
94 : 6	53.1	52.6	51.6	50.7	50.3	49.8	48.9
95 : 5	52.6	52.1	51.2	50.3	49.4	48.9	48.5
96 : 4	52.1	51.2	50.7	49.8	48.9	47.7	46.9
97 : 3	50.7	50.3	49.8	48.9	47.7	46.2	45.1
98 : 2	49.9	48.9	48.5	47.3	45.8	43.3	40.0
99 : 1	47.7	47.3	46.5	45.1	43.3	41.2	38.1

The above methods have been taken, with a few changes in the wording and with additions, from Bulletin No. 46, U. S. Department of Agriculture.

Gravimetric Method using Soldaïni's Solution.[1]—Place 100 to 150 cc. of Soldaïni's solution (**193**) in an Erlenmeyer

---

[1] *Traité d'Analyse des Matières Sucrées*, D. Sidersky, p. 148.

flask; boil five minutes; add a solution containing 10 grams of the material previously clarified with subacetate of lead, if necessary, the excess of lead being removed with small excess of carbonate of sodium (*see* **74**); boil five minutes. In boiling always use the naked flame. Having completed the reduction, remove the flask from the flame and add 100 cc. cold distilled water. Filter immediately through a Gooch crucible and determine the copper in the precipitate by the electrolytic method, or collect the precipitate in a filter-tube, Fig. 40, and reduce in hydrogen. These methods are described on page 79.

The weight of metallic copper $\times$ 0.3546 $\div$ weight of the material used in the determination $\times$ 100 $=$ per cent invert-sugar. It is claimed that this method is very exact and that invert-sugar can be determined to within .01 per cent with certainty.

73. Determination of Reducing Sugars (Glucose, etc.). Volumetric Methods.—*A Modification of Violette's Method.*—This is the rapid method used very generally in cane-sugar-houses. If always conducted under the same conditions as to dilution, method, and time of heating, the results are approximately correct and are comparable with one another.

Take a definite weight of the juice, a multiple of 5 grams is most convenient, varying this quantity with the amount of reducing sugar present, clarify with subacetate of lead, precipitate the excess of lead with small excess of carbonate of sodium (*see* **74**), and dilute to 100 cc.; mix and filter.

A sufficient quantity of the juice should be taken, if practicable, to give a reading on the burette of approximately 20 cc. in the titration to be described. In seed selection, as will be explained, it is unnecessary to adhere strictly to these specifications, but in using this method with other products they should, as far as practicable, be complied with.

It is convenient in this work to use an automatic, zero burette, in measuring the copper solution. Such a burette as designed by Squibb is shown in Fig. 42. This burette is filled by suction, as with a pipette, applying the suction at the mouthpiece shown at the end of the rubber tube. The reagent is drawn into the burette to a point a little above the zero mark, the mouthpiece is then released and the liquid siphons back into the reservoir, leaving the burette filled to exactly zero. A wash-bottle containing caustic soda solution should be connected with the air-inlet near the reservoir to prevent the entrance of carbonic acid. This is one of the most convenient of the many forms of automatic burettes. These burettes may be used with advantage in nearly all the measurements required in volumetric analysis, in the sugar-house laboratory.

Measure 10 cc. of Violette's modification of Fehling solution (195) into a large thin glass test-tube, 1.5 × 9 inches and dilute it with an equal volume of water. If the alkaline copper reagent be prepared with the copper in one solution and the alkali in a second, use 10 cc. of each solution and omit the addition of the 10 cc. of water. Heat the reagent in the tube, over

FIG. 42.

the naked flame of a lamp, to the boiling-point, then add a few cubic centimetres of the sugar solution, and boil two minutes. A sand-glass is convenient for use in timing the boiling. Repeat these operations until the blue color almost disappears, taking care to add the juice very gradually as this point is approached. After the first boiling, it is only necessary to boil the liquid a few seconds each time. Now add the juice, a drop or two at a time, until the blue color disappears. Filter off a small portion of the liquid, using a Wiley or Wiley-Knorr filter-tube, and proceed as described farther on.

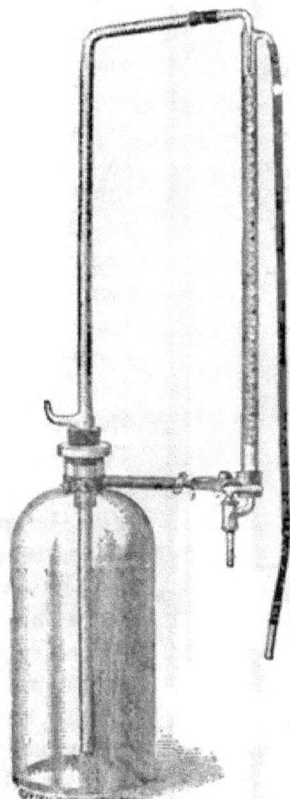

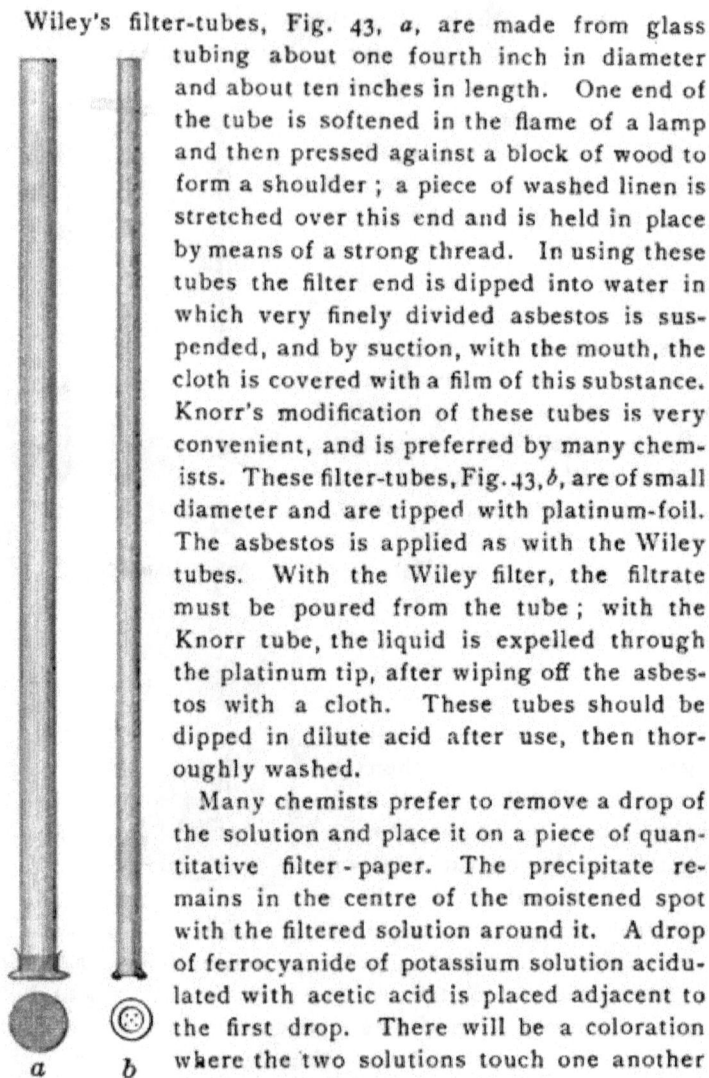

Wiley's filter-tubes, Fig. 43, *a*, are made from glass tubing about one fourth inch in diameter and about ten inches in length. One end of the tube is softened in the flame of a lamp and then pressed against a block of wood to form a shoulder ; a piece of washed linen is stretched over this end and is held in place by means of a strong thread. In using these tubes the filter end is dipped into water in which very finely divided asbestos is suspended, and by suction, with the mouth, the cloth is covered with a film of this substance. Knorr's modification of these tubes is very convenient, and is preferred by many chemists. These filter-tubes, Fig. 43, *b*, are of small diameter and are tipped with platinum-foil. The asbestos is applied as with the Wiley tubes. With the Wiley filter, the filtrate must be poured from the tube ; with the Knorr tube, the liquid is expelled through the platinum tip, after wiping off the asbestos with a cloth. These tubes should be dipped in dilute acid after use, then thoroughly washed.

Many chemists prefer to remove a drop of the solution and place it on a piece of quantitative filter-paper. The precipitate remains in the centre of the moistened spot with the filtered solution around it. A drop of ferrocyanide of potassium solution acidulated with acetic acid is placed adjacent to the first drop. There will be a coloration where the two solutions touch one another if there be still copper in solution.

FIG. 43.

If a portion of the solution be filtered off in one of the tubes above described, pour it into a few drops of acetic acid, to acidity, in a depression in a white porcelain test-plate ; the acid discharges the color from the solution and neutralizes the alkali of the Violette's solution. Add a drop of a dilute solution of ferrocyanide of potassium, yellow prussiate of potash ; a brown coloration shows the copper has not all been reduced, and that more juice must be added. The juice must be added very carefully as the test reaction diminishes in intensity, until finally all the copper is reduced, there being no further brown coloration. The burette reading is now made.

It is advisable to make a preliminary test to guide in the dilution of the juice and to show within a few tenths of a cubic centimetre the volume of juice required for the reduction of the copper, and then add nearly all the sugar solution at one time in a final test.

A porcelain dish may be substituted for the large test-tube, but on account of the small surface exposed for evaporation, the latter is preferred.

Calculations.

$W =$ the weight of juice in 1 cc. of the solution ;

$B =$ the burette reading ;

Per cent reducing sugar $= x = \dfrac{0.05 \times 100}{W \times B}$.

When W is .05 gram the formula reduces to $x = \dfrac{1 \times 100}{B}$

or $x =$ reciprocal of the burette reading multiplied by 100.

A table of reciprocals is given on page 294 to simplify these calculations.

If a multiple of 5 grams of juice be diluted to 100 cc. for this determination, the reciprocal of the burette reading multiplied by 100 is the same multiple of the per cent of reducing sugar.

If 5 grams in 100 cc. should prove a too-concentrated solution, dilute to 200, 300, etc., and multiply 100 times the reciprocal of the burette reading by 2, 3, etc.

If 5 cc. or a multiple of 5 cc. of juice be used for the analysis, the above-mentioned method of calculation may be employed, but the value of x must be divided by the specific gravity of the juice to reduce it to terms of the weight of the juice.

On account of the very small percentage of reducing sugar in beet-juices a much higher burette reading than 20 cc. may be necessary, even using the undiluted juice; further, for the same reason, it may be necessary to use only

5 cc. of Violette's solution.   It is preferable in such cases to use a gravimetric method.

The accurate determination of reducing sugar by this method requires rapid work and considerable practice.

Sidersky's Volumetric Method, using Soldaïni's Solution.[1] Standardize the Soldaïni solution by means of a solution of invert-sugar containing 5 grams of the reducing sugars per litre.   Proceed as in **73**, except that the end reaction is judged by the disappearance of the blue color instead of by the ferrocyanide test.   The method described in **73** is probably applicable, though Sidersky was guided solely by the disappearance of the blue color.

This method has the advantage of freedom from the source of error, due to the presence of sucrose, in the older method of Violette.   For highly colored products, such as molasses, etc., Sidersky has modified his method as follows: Dissolve 25 grams of the material in water, add sufficient subacetate of lead for clarification (*see* **74**), dilute to 200 cc., mix and filter.   To 100 cc. of the filtrate add 25 cc. of a concentrated solution of sodium carbonate, mix and filter; of this filtrate use 100 cc., corresponding to 10 grams of the material, for the reduction.   Boil 100 cc. of Soldaïni's solution five minutes in a flask over a naked flame, then add the sugar solution, little by little, continuing the heating an additional five minutes.   Remove the flask, add 100 cc. cold distilled water, and collect the precipitate upon an asbestos felt in a Gooch crucible, with the assistance of a filter-pump.   Wash the precipitate with hot water until the wash-waters are no longer alkaline.   Three or four washings are usually sufficient.   Wash the cuprous oxide into an Erlenmeyer flask and add 25 cc. normal sulphuric acid (**199**) and two or three crystals of chlorate of potassium, then heat gently until the cuprous oxide is completely dissolved.   Titrate the solution with a standard alkali solution (**201**), determine by difference the volume of the acid saturated, and from this the amount of copper reduced.   It is preferable to use a half-normal solution of ammonia (**201**) for this titration, letting the sulphate of copper act as an indicator.   Check the ammonia solution

[1] *Traité d'Analyse des Matières Sucrées*, D. Sidersky, p. 150.

against the normal sulphuric acid, using 2 cc. of a concentrated solution of sulphate of copper as an indicator to 25 cc. of the ammonia. Continue the addition of the acid until the blue color disappears.

In making the titration proceed as follows: Cool the sulphate of copper solution, resulting from the treatment of the cuprous oxide with the normal sulphuric acid and chlorate of potassium, add 50 cc. half-normal ammonia solution and titrate back with the normal sulphuric acid. The blue color disappears with each addition of the acid, but reappears on stirring the solution so long as any unsaturated ammonia remains. When all the ammonia is saturated the color of the solution is no longer blue, but a faint green. Note the burette reading. Each cc. of the sulphuric acid is equivalent to .0317 gram of copper. Multiply the weight of copper by .3546, Bodenbender and Scheller's factor, to obtain the weight of reducing sugar (invert-sugar), or multiply the burette reading by .1124 to obtain the per cent reducing sugar.

Volumetric Permanganate Method.[1] The saccharine strength of the solution should be approximately one per cent. The solution should be clarified as usual, and the excess of lead removed (**74**). Ten cubic centimetres of this solution are placed in a porcelain dish with a considerable excess of copper solution (**192**). If the saccharine solution contain no sucrose, heat to the boiling-point and maintain this temperature until the reducing sugar is oxidized. When sucrose is present the temperature should not exceed 80° C., and the heating should be continued longer than at the higher temperature. There should be enough of the copper solution used to maintain a strong blue coloration at the end of the reaction. Ervin E. Ewell[2] advises using the following modification of the method of determining the weight of copper reduced: Collect the precipitate on asbestos in a Gooch crucible, with the assistance of a filter-pump, and wash thoroughly with hot recently boiled distilled water. Transfer the asbestos, with

---

[1] *Principles and Practice of Agricultural Analysis*, H. W. Wiley, 3, 134.

[2] *Op. cit.*, 136.

as much of the precipitate as possible, to the beaker in which the precipitation was made, beat it up with 25 to 30 cc. of hot recently boiled distilled water, and add from 50 to 75 cc. of a saturated solution of ferric sulphate in 20 per cent sulphuric acid; pour this solution through the crucible to dissolve adhering portions of the cuprous oxide. The precipitate must be well beaten up with the water to break all large lumps or there may be difficulty in effecting solution with the ferric salt. After the solution is complete, titrate with permanganate of potassium of such strength that 1 cc. is equivalent to .01 gram of copper (**203**), or decinormal permanganate solution (**202**) may be used. In addition to standardizing the permanganate solution with metallic iron or oxalic acid, as is usual for general purposes, it should be standardized, for this method, by titrations with copper, reduced by solutions of invert-sugar which have been standardized by the gravimetric method (**72**). The invert-sugar value of 1 cc. of the permanganate solution is thus ascertained for use in calculating the percentage of reducing sugar in the material.

Ewell's modification of the permanganate method of determining the amount of reduced copper, is also recommended for use in the methods in **72**.

74. Notes on the Determination of Reducing Sugars.—Edson, Pellet and other chemists have shown that a part of the reducing substances in certain sugar-house products is precipitated by subacetate of lead, but not at all or to a very small extent with the normal acetate. Edson advises that the solutions be acidulated with acetic acid before filtering off the lead precipitate, and finds that acidulation practically obviates this source of error. The author's experience confirms Edson's observations. Bornträger [1] states that sodium sulphate is preferable to sodium carbonate for the precipitation of the excess of lead. According to his experiments, an excess of the sulphate is less objectionable than of the carbonate. The carbonate is almost exclusively used by sugar-house chemists for the removal of the excess of lead.

---

[1] *Zeit. Angew. Chem.*, 1892, 333.

75. Determination of the Ash.—*Sulphated Ash.*—
Dry 10 grams of the juice in a tared platinum dish. Add a
few drops of concentrated sulphuric acid to moisten the
residue, and heat over the flame of a lamp or in a muffle at
low redness until the organic matter is charred, then increase the temperature to bright redness and heat until all
the carbon is consumed. In the event of too high a temperature, the ash will melt and thus may vitiate the results.

The ash so obtained is termed the " sulphated ash," since
certain of the mineral constituents are converted into
sulphates by the acids. It is estimated that the average
increase in the weight of the ash, due to the formation of
sulphates instead of carbonates, is 10 per cent, hence a
correction of one tenth is customary to reduce the sulphated
ash to terms of the normal or carbonated ash.

Calculation.—Weight of ash × 9 = per cent normal ash.
It is usually more convenient to measure 10 cc. of the juice
than to weigh 10 grams. In such cases calculate as follows :

$$\frac{\text{Weight of sulphated ash} \times 9}{\text{Specific gravity of the juice}} = \text{per cent normal ash.}$$

The above method of incineration
is usually employed, since there is
usually difficulty in the direct incineration of saccharine materials.

Normal Ash.—The normal or carbonated ash may be obtained by Boy-
er's method, as follows : Dry 10 grams,
or 10 cc., of the juice in a platinum
dish, then heat carefully to caramelize
the sugar, but not enough to char it;
add 2 cc. benzoic acid solution, 25
grams benzoic acid in 100 cc. of 90 %
alcohol, and warm gently to expel the
alcohol. Char the sugar at a low
heat, at the same time volatilizing the
acid ; incinerate at a low red heat.
The ash consists largely of alkaline
carbonates, which, on exposure to the air, quickly absorb
moisture. Cool the ash in a desiccator and weigh quickly.

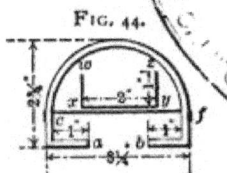

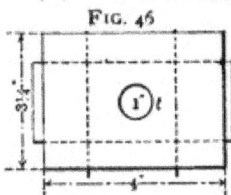

FIG. 44.

FIG. 45.

FIG. 46

The weight of the ash ÷ the weight of the juice × 100
= per cent ash.

The following described muffle, devised by Schweitzer
and Lungwitz,[1] is effective, and may be cheaply constructed
for sugar purposes.

In a French clay muffle a narrow slot is cut the length of
the bottom, Fig. 44, *a*, *b* ; holes are drilled in the walls at *c*,
d, Fig. 45, and heavy platinum wires are inserted. These
wires are supports for a trough of platinum-foil, Fig. 45, *w*,
x, *y*, *z*, upon which the dishes rest during the incineration.
A hole is cut in the dome of the muffle at *t*, Fig. 46. The
muffle is placed on a support and is heated by wing-top
burners.

76. Determination of the Total Nitrogen
Albuminoids.—The beet contains, in addition to albu-
minoid matter, several nitrogenous substances classified as
amido-compounds. Some of these substances may be readily separated, others require complicated analytical processes.   For an extended study of the nitrogenous bodies
in agricultural analysis, Wiley's *Principles and Practice of
Agricultural Analysis* is recommended. Allen gives methods
for several of the amido-compounds in Vol. III, Part III,
Commercial Organic Analysis.   E. O. von Lippmann has
published a very exhaustive study of the nitrogenous constituents of the beet-juice in *Berichte der deutschen chemischen
Gesellschaft*, **29**, 2645. A translation of this paper is published in *Bulletin de l' Association des Chimistes de France*, **14**,
601 and 819. See also this book, page 201. It has long been
customary in plant analysis to multiply the per cent of total
nitrogen by 6.25 and term the product the per cent of
"albuminoids." The figures obtained in this way are often
of value in sugar-house work.

A modification of Kjeldahl's moist combustion process[2]
may be conveniently employed for nitrogen determinations :

(1) *The Digestion.*—Ten cc. of the juice, dried in a small
capsule, are brought into a 550-cc. digestion-flask with
approximately .7 gram of mercuric oxide and 20 cc. of

---

[1] *Journ. Am. Chem. Soc.*, **16**, 151

[2] Adapted from Bulletin 46, Div. Chem., U. S. Dept. Agric.

sulphuric acid. The flask is placed on a frame in an inclined position, and heated below the boiling-point of the acid for from 5 to 15 minutes, or until frothing has ceased. If the mixture froth badly, a small piece of paraffine may be added to prevent it. The heat is then raised until the acid boils briskly. No further attention is required till the contents of the flask have become a clear liquid, which is colorless, or at most has only a very pale straw color. The flask is then removed from the frame, held upright, and, while still hot, potassium permanganate is dropped in carefully and in small quantity at a time, till, after shaking, the liquid remains of a green or purple color.

(2) *The Distillation.*—After cooling the contents of the flask, add about 200 cc. of water, then a few pieces of granulated zinc and 25 cc. of potassium-sulphide solution, 40 grams commercial potassium-sulphide in 1000 cc. water, shaking the flask to mix its contents. Next add 50 cc. of a saturated caustic-soda solution, free from nitrates, or sufficient to make the reaction strongly alkaline, pouring it down the side of the flask so that it does not mix at once with the acid solution. Connect the flask with the condenser, which should be of block-tin, mix the contents by shaking, and distil until all the ammonia has passed over into the standard acid. The first 150 cc. of the distillate will generally contain all of the ammonia. This operation usually requires from 40 minutes to one hour and a half. The distillate is then titrated with standard ammonia, using cochineal as an indicator, and the calculations are made as usual. Previous to use, the reagents should be tested by a blank experiment with sugar, which will partially reduce any nitrates present, which might otherwise escape notice.

77. Determination of the Total Solids.—The degree Brix is usually considered as representing the total solid matter in solution. An accurate determination of the total solids can only be made by actually drying the juice in an oven.

The problem of drying saccharine materials, to a constant weight, is not as simple as may appear at first glance. A number of methods have been devised for this purpose, two of which are given.

Carr and Sanborn's Method for Drying Sugar-house Prod ucts.—This is a modification of the ordinary pumice-stone method. Prepare pumice-stone in two sizes. One size should pass a 1-mm. sieve and the other should pass a 6-mm. sieve, circular perforations. Place a layer 3 mm. thick of the finer pumice-stone on the bottom of a small metal dish, and a layer of the coarse, 6 mm. to 10 mm. thick, upon the first layer, and dry and weigh. Tin caps for bottles are inexpensive and well adapted for use in this determination.

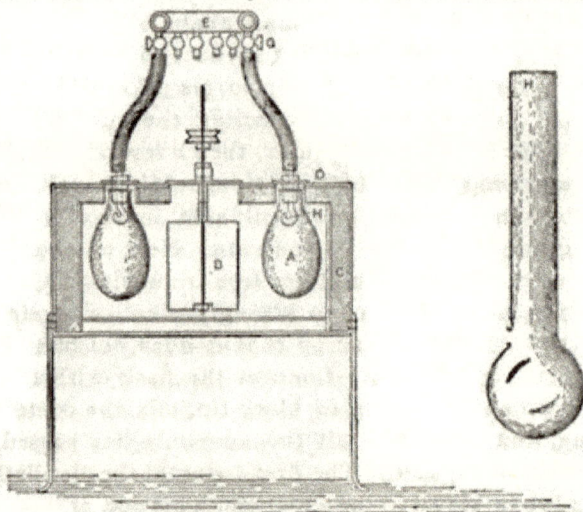

FIG. 47.

Each dish is used but once, then thrown aside. Distribute about 5 grams of juice over the pumice-stone, weighing it accurately from a weighing-bottle. Dry this juice to a constant weight in a water-oven or in a vacuum-oven at 70° C., making trial weighings at intervals of two hours. Calculation: Weight of solid matter ÷ weight of juice employed × 100 = per cent total solids.

Method of Drying Employing a Vacuum Apparatus.—This method was suggested to the author by that of Courtonne,[1] from which it differs in several important particulars, notably in the construction of the oven and drying-bottles. Courtonne heats the bottles by immersion in hot water.

---

[1] *Manuel-Agenda des Fabricants de Sucre*, MM. Gallois and Dupont, 1891, p. 215.

The oven and bottles are shown in section in Fig. 47. The walls of the oven are double and are filled with plaster of Paris, C; the bottom is also double, the space being filled with air. A fan, D, driven by a toy engine, or other suitable means, agitates the air inside the oven and insures a strictly uniform temperature in all parts.

The drying-bottles, A, are connected by means of short tubes with a central vacuum-pipe, E, which is in turn connected with an ordinary filter-pump or the third pan of the triple-effect. Each bottle may be removed by closing the cock G without disturbing the others. A small trap, H, of glass, shown also in detail at the right of the oven, prevents any moisture which may condense in the tubes from falling back into the bottle.

The following procedure is advised : Place a quantity of small fragments of pumice-stone sufficient to absorb 5 cc. of juice, in a weighing-bottle, dry in the oven, cool, insert the glass stopper and weigh ; distribute a definite weight of the juice, approximately 5 grams, upon the pumice-stone. Insert the stopper, provided with the trap, in the bottle, and connect with the vacuum-pipe. A vacuum of 20 inches is usually all that is required, and in fact is preferable to a higher vacuum. The drying is usually complete in one hour; it is advisable to dry to a practically constant weight, weighing at intervals of one hour or more as may be convenient. The calculations are made as in the preceding method.

This apparatus may also be used for drying in an inert gas.

The per cent total solids by the spindle, the degree Brix, and the per cent total solids by drying, are employed in calculating the purity coefficients or quotients (**106**).

78. Acidity of the Juice.—The normal juice of the beet and the diffusion-juice are always acid. This acidity is due to a number of organic acids. It is not often necessary to determine the acidity of the juice. This determination is made by a titration with a decinormal alkali solution (**201**). It is somewhat difficult to determine the end reaction, since the color of the juice obscures the color of the indicator to some extent. Phenolphthalein is usually employed as the indicator. Collier recommended the use of logwood solution as an indicator in determining the acidity of sugar-cane juices, and in the author's experience it has been satisfactory.

The acidity may be expressed in terms of the number of cubic centimetres of normal alkali solution required to neutralize the juice or, for comparative purposes, more conveniently as cubic centimetres of normal alkali per 100 grams of sucrose or 100 degrees Brix.

79. Analysis of Carbonated Juices. — The methods of analysis of the purified juices are the same as for the raw juice, except that the carbonated juice must receive an additional treatment with carbonic acid to precipitate all of the calcium. This is evidently necessary, since these analyses are made in part for the purpose of comparing the purity of these juices with that of the diffusion-juice before treatment.

80. Alkalinity of the Juice.—It is occasionally necessary to determine the total alkalinity of the juice after liming and before carbonatation; it is also necessary at very frequent intervals to determine the total alkalinity of the carbonated juices, in the control of the carbonatation process. In many factories an alkalimetric method is employed in ascertaining when to shut off the carbonic acid gas in the carbonatation of each tankful of juice.

The total alkalinity is usually expressed in terms of the grams of lime (CaO) per litre of juice, although the alkalinity is in part due to the presence of caustic alkalis.

Methods are usually employed, in the control of the carbonatation of the juice, which are very rapid and well adapted to the use of unskilled employés, but which yield only moderately accurate results (**81**).

It is advisable that the rapid methods indicated be occasionally checked in the laboratory. This is necessary in order to know to what extent the results vary from the truth, that the carbonatation may be the more satisfactorily controlled.

81. Rapid Methods of Moderate Accuracy for the Alkalinity of Juices.—(1) *Standard Acid Solution.* —Prepare a standardized solution of sulphuric acid contain-

ing 35 grams of the monohydrated acid (H_2SO_4) in 1000 cc.
(*See* **200**.) The strength of this solution is such that 1 cc.
will neutralize 0.02 gram of lime (CaO).

This solution is used for limed juices and juice from the
first carbonatation. A more dilute acid is employed for the
titration of juice from the second carbonatation. This
acid is prepared by diluting 100 cc. of the above standard
acid to 1000 cc., and contains 3.5 grams of sulphuric acid in
1000 cc.

Indicators.—As great accuracy is not necessary in this
determination, indicators which are more or less affected
by carbonic acid may be employed. Among those most
commonly used are neutralized corallin, phenolphthalein,
cochineal, etc. A few drops of the solution of the indicator
are added to the juice, or in this class of analyses, with certain indicators, more conveniently to the acid solution, when
standardizing it, and before completing the volume to 1000
cc. (*See* **213**.)

Titration.—Measure 20 cc. of the juice into a porcelain
dish or into a small Erlenmeyer flask. If the flask be used,
it should be placed over a sheet of white paper or a porcelain slab during the titration.

Except in the case of the limed juice, before carbonatation, the liquor should be filtered.

Add a few drops of the indicator to the juice, if it be not
already contained in the standard acid, and deliver the acid
cautiously from a burette. Note the point when the
alkalinity is saturated by the change in the color of the
indicator, and read the burette.

Calculation.—1 cc. of stronger acid solution neutralizes
0.02 gram of lime (CaO); hence for each cc. of acid used
there is an alkalinity corresponding to 0.02 gram of lime
per 20 cc. of juice, or to 0.1 gram per 100 cc. of juice, or
1 gram per litre of juice.

Example.—20 cc. of juice required 2.2 cc. of the acid.
∴ 0.02 × 2.2 × 50 = 2.2 grams lime per litre of juice, or the
number of cc. of acid used = grams of lime per litre.

The calculations are the same when using the weaker
acid with second carbonatation juices, except that 1 cc. of
the acid corresponds to 0.002 gram of lime.

(2) *Vivien's Method.*—This exceedingly convenient and simple method is employed very generally in France. Like the preceding method, it only gives approximately correct results. Vivien employs a solution of sulphuric acid containing a small quantity of phenolphthalein, of such strength that one volume of this acid will neutralize one volume of juice containing .05 gram of lime per litre, *i.e.*, total alkalinity expressed as lime (CaO).

A specially graduated tube shown in Fig. 48 is used with this method. This tube is divided into six parts of equal volume. Each part except the bottom one is subdivided into five parts.

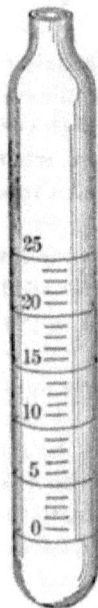

FIG. 48.

Acid Solution.—Prepare a standardized solution of sulphuric acid containing 0.875 gram of the monohydrated acid (H_2SO_4) in 1000 cc.; add a small quantity of phenolphthalein to the solution before completing the volume to 1000 cc. Standardize by titration against decinormal alkali solution; 10 cc. of the alkali should neutralize 56 cc. of this solution.

Manipulations.—Fill the tube, Fig. 48, to the zero mark with juice; add the standardized acid cautiously, placing the thumb over the mouth of the tube and agitating from time to time. The solution turns red at the first addition of the acid, provided it be not added in excess; finally, when the acid is in very slight excess, the color disappears. The reading on the scale is next made. Every ten divisions correspond to an alkalinity due to 1 gram of lime per litre of juice, and each division to 0.1 gram of lime (CaO.) per litre.

For second carbonatation juice, use a much more dilute acid; for example, one half or one fifth the strength of the above. In this case every ten divisions of the scale correspond to 0.5 gram or 0.2 gram of lime per litre.

It is evident that these methods are susceptible of many modifications, but for the purposes of this book those described are sufficient.

These methods must be used with caution in analyzing

the juice from the second carbonatation, for the reasons given below.

It is the practice in the second carbonatation to saturate all the lime; hence this process is often termed the "saturation." If this point be passed, the caustic sodium and potassium, which remain as such in the presence of the caustic lime, are converted into carbonates. This is wrong, from manufacturing considerations, and farther it would be objectionable to leave lime unprecipitated. It is thus apparent that a process should be employed which will show the exact moment at which all the lime has been combined with the carbonic acid. In practice it is usual to ascertain, in the laboratory, approximately the alkalinity the juice should have when the lime has all been precipitated, and be guided by this in the control of the carbonatation.

The use of phenacetoline is said to be an advantage in this test. It is used in the cold. Degener recommends the use of a few drops of a 1 per cent solution of phenacetoline in alcohol.

82. Methods for the Determination of the Total Calcium in the Juice.—Gravimetric Method.— To 100 cc. of the juice add an excess of ammonium hydrate, heat to the boiling-point and filter, should there be a precipitate. Wash the filter with hot water, add an excess of oxalate of ammonium to the filtrate, boil two hours, and let stand several hours ; collect the precipitate in a small quantitative filter and wash with dilute ammonia. The filter and contents are next transferred to a tared platinum crucible, partly dried and the filter charred at a low temperature, then ignited until the carbon is removed. Add a small quantity of sulphate of ammonia solution containing chloride of ammonia (see **136**), dry at a moderate heat, and ignite at a high temperature. The residue consists of sulphate of calcium $(CaSO_4)$. Cool in a desiccator and weigh. The weight of the calcium sulphate multiplied by .41158 is the weight of calcium oxide (lime) per 100 cc. of juice. This number is practically the percentage of calcium oxide (CaO) by weight in the juice, or the correct percentage is this number divided by the specific gravity of the juice.

Fradiss' Volumetric Method.[1]—Treat 100 cc. of juice as described under the preceding method. Decompose the oxalate of calcium with warm dilute sulphuric acid. The acid combines with the calcium and sets the oxalic acid free. The oxalic acid is determined by means of a 1/10 normal solution of permanganate of potassium (**202**).

Titrate the solution without filtering, maintaining a temperature of 60° to 80° C. The addition of the permanganate solution should be continued until a permanent pink color is produced.

Calculation.—Multiply the burette reading, the cc. permanganate solution, by 0.0028 to obtain the weight of calcium oxide (CaO), or by 0.002 to obtain the weight of calcium (Ca). The numbers so obtained are the per cents by volume of the juice. Divide by the specific gravity of the juice to obtain the corresponding per cents by weight.

Soap Method.—This is an application of Clarke's soap test, used in estimating the hardness of water. The total percentage of calcium as calcium oxide (CaO) may be rapidly and closely estimated by this method. As used by the French it is more convenient for sugar-house purposes than the English method.

Chloride of Calcium or Barium Solution.—Dissolve 0.25 gram of pure chloride of calcium or 0.55 gram of pure crystallized barium chloride ($BaCl_2 + 2H_2O$) in water and dilute to 1 litre.

Special Burette.—The burette is so graduated that 2.4 cc. correspond to 23 divisions. The zero of the graduation is placed at the second division to allow for the quantity of soap solution required to produce a permanent lather with 40 cc. of distilled water; the 22 divisions correspond to 0.01 gram of chloride of calcium dissolved in distilled water; hence a division or 1° corresponds to 0.00045 gram of the chloride in 40 cc., or 0.0114 gram per litre.

Special Bottle.—This bottle is graduated at 10, 20, 30, and 40 cc. Only two of these graduations, viz., at 10 and 40 cc., are used in sugar work.

Method of Making the Test.—Introduce 40 cc. of the calcium chloride or barium chloride solution into the special

---

bottle, and add the soap solution (*see* **186**) little by little, with agitation, until a foam 5 mm. deep forms and persists during 5 minutes. The solution must be vigorously agitated by shaking the stoppered bottle after each addition of the soap. If the soap solution be of the correct strength, a volume corresponding to 22 divisions of the burette is required. The burette should always be filled to the division above the zero mark, and the reading should be from zero. If the reading be not 22°, add sufficient cold, recently boiled distilled water to dilute it to this strength.

To 10 cc. of the juice in the special bottle, add sufficient cold, recently boiled distilled water to dilute it to 40 cc. Proceed as above, using the standarized soap solution. Multiply the number of "degrees" read on the burette by 0.0228 to calculate the lime (CaO) per litre of juice. This method may be applied to the sirup, massecuites, and molasses, using 1 gram of the material diluted to 40 cc.

See page 171 relative to the influence of magnesia in this test. The presence of magnesia, resulting from dolomite in the limestone, may vitiate the results obtained. Parallel determinations by the soap and the gravimetric methods, or an examination of the lime, will show whether sufficient magnesia is present to render this process unavailable. This method is not applicable to the juice from the first carbonatation.

83. Free and Combined Lime and Alkalinity Due to Caustic Alkalis. Pellet's Method.[1]—

A. Determine the total alkalinity by titration with sulphuric acid, using litmus as an indicator. The titration must be made at the boiling-point of the juice. Calculate the alkalinity as lime per 100 cc. of juice.

B. Add an equal volume of strong alcohol to a measured portion of the juice; the "free" lime is precipitated as an insoluble saccharate of lime; filter and determine the alkalinity of the filtrate operating upon an aliquot part; calculate as lime per 100 cc. of juice. This alkalinity is, however, due to sodium and potassium hydrates, but is expressed as lime for comparative purposes.

---

[1] *Fabrication du Sucre*, Beaudet, Pellet, etc., **2**, 305.

C. The total lime is determined by one of the methods in **82**, and is also expressed as lime per 100 cc. of juice.

The following example illustrates the calculations:

Example.

As Lime per 100 cc.

(*A*) Total alkalinity.............................. 0.027 gram.
(*B*) Alkalinity due to soda and potassa...... 0.021 "
(*C*) Total lime, including organic salts...... 0.023 "
Free lime (*A* − *B*)............................... 0.006 "
Combined lime, *i.e.*, lime salts (*C*−[*A*−*B*]).. 0.017 "

ANALYSIS OF THE SIRUP.

84. Analysis of the Sirup.—The analysis of the sirup is conducted as that of the juice (**67** to **83**); the same determinations are made, the only variations being in the quantities of the material used for the analysis.

All the portions used for analysis should be weighed, not measured. This is necessary on account of the viscosity of the sirup.

ANALYSIS OF THE MASSECUITES AND MOLASSES.

85. Determination of the Density.—The determination of the density of massecuites presents certain difficulties which cannot well be avoided, and which compel the acceptance of results which are not strictly accurate.

As has been explained, the degree Brix of a solution is the percentage, by weight, of pure sugar which it contains, but it is usually taken as the percentage of solid matter in the solution. The use of a spindle or pyknometer for the determination of the degree Brix, assumes the impurities in the solution, or the non-sucrose, to have the same specific gravity as sucrose. This assumption, unfortunately for the convenience of the chemist, is far from true, especially in the denser products and in those from which a part of the sugar has been removed, viz., the second, third, etc., massecuites and the molasses. The mineral impurities influence the specific gravity very materially, since they differ so widely in specific gravity from the sugars. Since the proportion of inorganic non-sugar increases as one passes from the products of high purity to those of low purity, the difference between the apparent percentage of total solids, as indicated by the density, and the true percentage of total solids, becomes greater.

From this, it is apparent that calculations of the total solids in massecuites, etc., from the density of the product, must be accepted with caution, and then only for comparative purposes, when uniform conditions of analysis are maintained.

The methods by dilution and spindling are given in this book for calculating approximate coefficients, etc., and must not be assumed to give strictly accurate results.

It is customary to term the degree Brix, as deduced from the specific gravity of the material, the "apparent degree Brix," or simply the "degree Brix"; the term "true or real degree Brix" is sometimes applied to the percentage of total solids, when this number is determined by actually drying the material in an oven.

86. Determination of the Density by Dilution and Spindling. Apparent Degree Brix.— Dissolve 250 grams of the massecuite or molasses in water and dilute to 500 cc. Transfer a portion of the solution to a cylinder and determine its degree Brix. Calculate the degree Brix of the product used by the following formula:

$$\text{Apparent degree Brix} = \frac{Sp.\ Gr. \times B \times V}{W},$$

in which B is the degree Brix (corrected) of the solution, $Sp.\ Gr.$ the specific gravity corresponding to the degree Brix of the solution before correction, V the volume of the solution, and W the weight of massecuite used.

The above formula reduces to the following if the weight and volume specified have been used:

$$\text{Apparent degree Brix} = 2 \times Sp.\ Gr. \times B.$$

The following is a very convenient modification of the above method:

Dissolve a definite weight of massecuite in an equal weight of water, mix the solution thoroughly, and spindle.

The degree Brix of the massecuite is two times the degree Brix of the solution. (*See also* **88**, Weisberg's method.)

87. Determination of the Total Solids or Moisture by Drying.—The method of Carr and Sanborn, and the vacuum method given in **77**, are recommended. In the latter case use 1 gram of the massecuite, and in both methods, after weighing the material, dissolve it in a small quantity of distilled water, in order to distribute it evenly. In the Carr-Sanborn method, dilute the sample to content of about 20 to 30 per cent dry matter, using a weighed portion of water. Add such quantity of the diluted material to the pumice-stone, in the tared dish, as will yield approximately 1 gram dry matter.

88. Approximate Determination of the Total Solids and Coefficient of Purity of Massecuite, etc., by Dilution and Spindling. Weisberg's Method.[1]—This is the ordinary method by dilution and spindling, but conducted under certain definite conditions, under which a table of coefficients, deduced by Weisberg from a very large number of experiments, is used.

. Weigh three times the normal weight, or any convenient multiple of the normal weight, of the massecuite and dissolve it in water; transfer the solution to a 300-cc. flask, or to a flask corresponding to the multiple of the normal weight of massecuite used, and dilute to the graduation. Mix the solution thoroughly and determine its degree Brix, using a spindle graduated to twentieths of a degree. Transfer 50 cc. of the solution, corresponding to the half-normal weight of the massecuite, to a flask, clarify with subacetate of lead, dilute to 100 cc., mix and filter. Polarize the filtrate, and multiply the polariscopic reading by 2 to compensate for the dilution. This gives the percentage of sucrose in the massecuite. In materials containing notable quantities of raffinose, etc., use the method of Creydt (**89**) to ascertain the per cent of sucrose in the massecuite. The methods of calculation are most conveniently explained by an example.

---

[1] *Bul. Assoc. Chimistes de France*, **14**, 978.

Example and Formula for Calculations.

Weight of massecuite (2½ times the normal) = 65.12 gram
Volume of the solution..................... = 250 cc.
Degree Brix of the solution = B............ = 22
Specific gravity corresponding to the degree
  Brix (*see* table page 275) = D............ = 1.09231
Polariscopic reading × 2 = R.............. = 55.
Constant (normal weight ÷ 100)............ = .26048

(1) $\dfrac{R \times 0.26048}{D}$ = per cent sucrose in the diluted solution, S;

(2) $\dfrac{S}{B} \times 100$ = apparent coefficient of purity (**106**) of the solution and of the massecuite.

WEISBERG'S TABLE OF COEFFICIENTS.

Apparent Coefficient of Purity.	Coefficients.	Apparent Coefficient of Purity.	Coefficients.
57	1.054	78	1.021
57.5	1.052	79	1.020
58	1.050	80	1.019
58.5	1.048	81	1.018
59	1.046	82	1.017
60	1.044	83	1.016
61	1.042	84	1.015
62	1.040	85	1.014
63	1.038	86	1.013
64	1.036	87	1.012
65	1.034	88	1.011
66	1.033	89	1.010
67	1.032	90	1.009
68	1.031	91	1.008
69	1.030	92	1.007
70	1.029	93	1.006
71	1.028	94	1.005
72	1.027	95	1.004
73	1.026	96	1.003
74	1.025	97	1.002
75	1.024	98	1.002
76	1.023	99	1.001
77	1.022	100	1.000

The letters have the values indicated in the statement of the example and in equation (1).

(3) Multiply the apparent coefficient of purity by the coefficient corresponding to it in Weisberg's table to obtain the true coefficient of purity of the massecuite.

(4) The true per cent total solids of the massecuite is deduced by dividing its percentage sucrose by the true coefficient of purity and multiplying by 100.

Substituting the values of R and D in formula (1) we have

$$\frac{R \times 0.26048}{D} = \frac{55 \times 0.26048}{1.09231} = 13.12 = S;$$

and substituting the values of B and S in formula (2) we have

$$\frac{S}{B} \times 100 = \frac{13.12}{22} \times 100 = 59.64,$$

apparent purity of the massecuite; and from (3),

$$59.64 \times 1.045 = 62.32,$$

the approximately true purity of the massecuite.

From (4), $\frac{R}{62.32} \times 100 = 88.25$, the approximately true per cent of total solids in the massecuite.

In checking this method by actual drying of the above massecuite, Weisberg obtained a true purity of 62.02. This sample was a very severe test of the method owing to the low purity of the massecuite.

Weisberg constructed his table from experimental data, obtained in the examination of massecuites produced without "boiling in" molasses, as is now practised to a considerable extent. With massecuite obtained by "boiling in" molasses on first-sugar, it is possible that the method may not give as satisfactory results as indicated in the example.

89. Determination of Sucrose and Raffinose. Creydt's Formulæ.—This is the official German method;[1] it is that of Clerget, as published by the German Government, except that the acidulation of the solution for direct polarization is recommended. This method is not applicable in the presence of optically active bodies other than sucrose and raffinose. Percentages of raffinose less than 0.33 cannot be determined with certainty by the inversion methods.

[1] *Zeit. Rübenzucker-Industrie*, **38**, 867.

Dissolve the normal weight of the material in water, clarify as usual, and dilute to 100 cc. Filter, and polarize the filtrate at 20° C. Record the polarization as the "direct reading." It is recommended that this solution be slightly acidulated with acetic acid before diluting to 100 cc.

Dissolve 13.024 grams of the substance in 75 cc. of water, in a 100-cc. flask, and add 5 cc. hydrochloric acid containing 38.8 per cent of the acid, mix the contents of the flask by a circular motion, and place it on a water-bath heated to 70° C. The temperature of the solution in the flask should reach 67° to 70° C., in two and one half to three minutes. Maintain a temperature of as nearly 69° C. as possible for seven to seven and one half minutes, making the total time of heating ten minutes. Remove the flask and cool the contents rapidly to 20° C., and dilute the solution to 100 cc. If necessary treat the solution with 1 gram of dry bone-black (**189**) to decolorize it. Polarize in a tube provided with a lateral branch for the insertion of a thermometer. A tube, provided with a jacket, through which a current of water of 20° C. circulates, should be used. The invert reading should be made at 20° C., and be multiplied by 2. If a preliminary calculation, using the formula, per cent sucrose $= 0.7538 \times$ algebraic sum of the direct and invert readings, give a percentage which is more than 1 per cent higher than the direct reading, raffinose is probably present, and the following formulæ by Creydt should be used in making the calculations :

$P =$ the direct reading, *i.e.*, the polarization before inversion ;

$I =$ the invert reading , multiplied by 2.

$S =$ the percentage of sucrose ;

$R =$ the percentage of anhydrous raffinose.

$$ S = \frac{0.5188 P - I}{0.845}; \quad R = \frac{P - S}{1.85}. $$

It is very important in this process that the time and temperature conditions be strictly complied with. The amount of material used should be varied, according to the nature of the substance, that the invert solution may have a concentration of approximately 13.7 grams in 100 cc., *i.e.*, the invert-sugar produced in the inversion of 13.024

grams of sucrose. The value of the constants varies considerably with the concentration (*see* **267**).

90. Determination of Sucrose and Raffinose. Lindet's Inversion Method as Modified by Courtonne. — Courtonne[1] has slightly modified the method of Lindet[2] in order to facilitate the manipulations.

Dissolve the normal weight of the material in water and dilute to 100 cc. Transfer 50 cc. of this solution to a 50-cc. flask and add sufficient dilute subacetate of lead solution (**207**); acidulate with acetic acid; mix, filter, and polarize the filtrate. Increase the polariscopic reading one tenth and record as the direct reading (*A*).

Transfer 20 cc. of the original solution of the material to a 50-cc. flask, and add to it 5 grams of zinc-dust. The dust must be weighed. Heat the flask and contents by immersion in boiling water or in the steam from a water-bath. Add 10 cc. of dilute hydrochloric acid, in portions of about 2 cc. at a time, being careful that none of the liquid is lost through a too rapid addition of the acid. The portions of acid may be added as frequently as convenient. The dilute acid is prepared by adding an equal volume of distilled water to pure hydrochloric acid of 1.2 specific gravity.

In the original method of Lindet, it is specified to heat the contents of the flask on the boiling-water bath about 20 minutes. In the modified method, it is only necessary to heat a few minutes after the last addition of acid.

The quantity of acid is so gauged that a portion of the zinc is left undecomposed and occupies a volume of .5 cc., for which a correction must be made in the calculations.

After the inversion is completed, cool the solution, either by immersing the flask in cold water or by setting aside to cool slowly. When the temperature reaches 20° C. complete the volume to 50 cc., mix and filter. Polarize the filtrate in an observation-tube provided with a lateral branch for the insertion of a thermometer. Multiply the reading by 2.475 if a 20-centimetre observation-tube were used. This factor includes a correction for the volume of the excess of zinc used. The polarization should be made

[1] *Bul. Assoc. Chimistes de France*, 7, 232.
[2] *Op. cit., supra*, 7, 432.

ιt 20° C. Caculate the percentages of sucrose and raffinose by the following formulæ:

A = the direct reading, i.e., before inversion;

B = the invert reading, corrected to terms of the normal weight;

C = the algebraic sum of the direct and the indirect reading;

S = the per cent sucrose;

R = the per cent raffinose.

The first set of formulæ is for the Laurent polariscope, instruments whose normal weight is 16.29 grams, and the second set for the Schmidt and Haensch polariscope, instruments whose normal weight is 26.048 grams:

Set No. 1:

$$(1) \ S = \frac{C - 0.489A}{0.81} \ ; \ (2) \ R = \frac{A - S}{1.54}.$$

Set No. 2:

$$(1) \ S = \frac{C - 0.493A}{0.827} \ ; \ (2) \ R = \frac{A - S}{1.57} = 1.017A - \frac{C}{1.298}.$$

In the formulæ, R is the percentage of hydrated raffinose. To obtain the percentage of anhydrous raffinose substitute 1.84 for 1.54 in the denominator in the first set of formulæ and 1.85 for 1.57 in the second set.

The invert solutions by this method are perfectly colorless and require no bone-black or other treatment preparatory to polarization.

This process is only applicable to materials containing no optically active bodies other than sucrose and raffinose. As beet products, under normal conditions, rarely contain reducing sugars, this process is generally applicable in all beet work.

The formulæ given in the set No. 1 are those of Creydt, modified by Lindet.

The correction for the space occupied by the undecomposed zinc-dust, is based upon the fact that only enough hydrochloric acid is used to decompose a certain quantity of zinc. If the quantities of acid and zinc indicated be used, there will be sufficient excess of zinc to occupy a volume of nearly .5 cc.

The author prefers to use a solution of hydrochloric acid, standardized by means of a normal alkali solution. The acid should be measured from a burette. It is conducive to accuracy to use a flask graduated at 50.5 cc., and an observation tube 50 centimetres long. To insure an observation at 20° C. a tube, provided with a water-jacket, through which water of that temperature flows, is necessary.

The great advantage claimed for Lindet's method is that it permits the inversion at the boiling-point of water without decomposition of the resultant products. Further, the matter of the time element is very much simplified, since while the inversion is complete in less than twenty minutes, there is no perceptible decomposition of the invert-sugar on heating a much longer time.

This method, in common with other inversion methods, has been the subject of much investigation and discussion. The evidence appears to be largely in favor of the methods of inversion given in **89** and **92.**

91. Determination of Sucrose and Raffinose in the Presence of Reducing Sugars.—J. Wortman [1] recommends the following method for this determination :

The reducing sugar is determined by the method with alkaline copper solution on page 81, using the following formulæ for the calculation :

$$N = \text{per cent reducing sugar;}$$
$$Cu = \text{the weight of copper reduced;}$$
$$q = \text{the weight of material employed;}$$

$$N = \frac{47 Cu}{q}.$$

The value of N is substituted in the following equations:

I. Per cent sucrose $= S = \dfrac{0.9598 P - 1.85 P' - 0.277 N}{1.5648}$;

II. Per cent raffinose $= R = \dfrac{P - S + 0.3103 N}{1.85}$;

in which P is the direct polarization and P' is the invert reading. These formulæ are based upon the work of Herzfeld and are for the normal weight of 26.048 grams.

---

[1] *Zeit. Rübenzucker-Industrie*, 39, 766.

The inversion, as far as concerns time, temperature, and acid, is made as in section **89.**

92. Determination of Sucrose in the Presence of Reducing Sugars. Clerget's Method.— In this modified method of Clerget, as adopted by the Association of Official Agricultural Chemists, the direct and invert readings are obtained as in **89.** The readings, especially if much reducing sugar be present, should both be made at very nearly the same temperature. This temperature should not vary more than two or three degrees from 20° C. The readings should be made at 20° C. if practicable, in which case the following formula is used :

$$\text{Per cent sucrose} = \frac{100\,S}{142.66 - \frac{20}{2}} = 0.7538 S,$$

in which S is the algebraic sum of the direct and invert readings.

Should the temperature (t) vary from 20° C., use the following formula :

$$\text{Per cent sucrose} = \frac{100\,S}{142.4 - \frac{1}{2}t}.$$

Certain important precautions are given in **89** in connection with this method.

Since beet products obtained under normal conditions rarely contain appreciable quantities of reducing sugars, and the very low products probably always contain raffinose, the methods of Creydt, preferably, or of Lindet (**89, 90**) are usually employed.

All inversion methods are usually spoken of as "Clerget's method," from the chemist who devised the original process of which all are slight modifications.

93. Determination to be made in the Analysis of Massecuites and Molasses.— All the determinations required in the analysis of the sirups are also to be made in the massecuite and molasses. The methods for sucrose and raffinose are those given in **89** to **92.**

The scheme given in the next paragraph is convenient for use in this class of work.

94. Scheme for the Analysis of Massecuites and Molasses, Adapted from Sidersky's Method.

—In order that Weisberg's table of coefficients may be available (see page 104), the quantity of material he advises should be used.

Dissolve 78.144 grams of massecuite in distilled water, dilute the solution to 300 cc., and use portions of it for the various determinations.

Determine the per cent sucrose, the apparent and true coefficients of purity (**106**), and the apparent and true degrees Brix by Weisberg's method (**88**).

For the determination of the ash, evaporate 19.2 cc. of the solution (5.0012 grams of the material) nearly to dryness. Multiply the corrected weight of the ash by 20 to obtain the per cent of ash. The slight excess of material used over 5 grams does not introduce an appreciable error even in low-grade molasses.

For the determination of the alkalinity, use a measured volume of the solution, remembering that each cubic centimetre corresponds to 0.26048 gram of the material. Use the methods in **95**.

95. Alkalinity of Massecuites and Molasses.— It is often necessary to determine the alkalinity of massecuites, and occasionally of the molasses.

These products are often very dark, rendering it difficult to employ a volumetric method. The following method devised by Buisson[1] gives satisfactory results in very highly colored products: Transfer 25 cc. of a solution of the material to be titrated, to a glass-stoppered flask, add one drop of a neutral solution of corallin and 10 cc. of washed ether. The ether must be neutral. After each addition of the standard acid (see **81** *et seq.*), agitate thoroughly and wait a few seconds for the ether to separate and rise to the surface. The slightest excess of acid reacts upon the corallin and colors the ethereal solution yellow. This reaction is very sharp.

The alkalinity is calculated as lime (CaO), percentage by weight. The methods given in **80** to **83** are also applicable to these products.

96. Estimation of the Proportion of Crystallized Sugar.—Many of the methods of estimating the

---

proportion of crystallized sugar, in sugars and massecuites, were suggested by Scheibler's modification of Payen's method for estimating the refining values of raw sugars.

In this method the crystals, in the weighed sample, are washed with successive portions of the following solutions: (I) 85 per cent alcohol containing 50 cc. acetic acid per litre and saturated with sugar; (II) and (III) 92 and 96 per cent alcohol, respectively, saturated with sugar; (IV) absolute alcohol; and (V) one third ether and two thirds absolute alcohol. The residual sugar is washed into a sugar-flask and its content of sucrose determined by the polariscope.

This method is no longer used, but is given in outline for historic reasons and because it suggested other methods which are in use.

Pellet devised a somewhat similar method, using first a saturated solution of pure sugar and afterwards saturated alcoholic sugar solutions, of increasing alcohol content, to wash the crystals. The sugar crystals are finally dried and weighed.

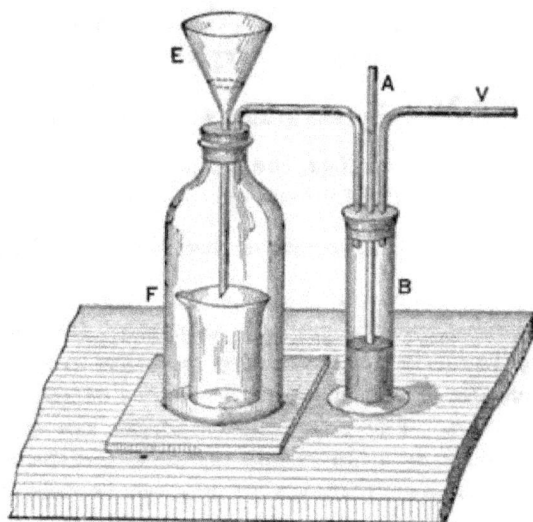

FIG. 49.

The following described methods, of which the author prefers Dupont's, are the most practical:

Vivien's Method.—Place a weighed quantity of massecuite in the funnel E, Fig. 49, of the pressure filtering apparatus; for example, 200 grams. The funnel is fitted with a perforated filtering-cone, as indicated by the dotted line. Connect the apparatus with a filtering-pump, Chapman's or other simple model, by the tube V. Wash the crystals with a cold solution of sugar containing 2 parts of pure sucrose to 1 part of distilled water. The pressure is regulated by raising or lowering the tube A, which dips into mercury in the cylinder B. The material in the funnel should always be covered with the wash-liquor. Continue the washing until all the crystals are free from molasses, then transfer them to a tared dish, mix thoroughly and weigh. Determine the moisture in 10 grams of the crystals by drying as usual in an oven. Since the wash-liquor contained 1 part water and 2 parts of sucrose, the loss in weight on drying multiplied by 3 gives the weight of the liquor adhering to the crystals.

Example and Calculations.

Weight of massecuite............... 200 grams
Weight of moist crystals........... 176.5 "
Moisture in 10 grams of the crystals.   0.56 "

Then $\dfrac{0.56 \times 176.5}{10} = 9.884$ grams water in the moist crystals, and $9.884 \times 3 = 29.652$, the weight of the wash-liquor adhering to the crystals.

$\dfrac{176.5 - 29.652}{2} = 73.43$ grams of dry, crystallized sugar in 100 grams of massecuite.

Kracz' Method.[1]—This method, as applied to raw sugar, consists in dissolving the adhering molasses in pure anhydrous glycerine and filtering off a portion of the solution for polarization. The polarizations of the raw sugar and of the glycerine solution supply the data for the calculations. The apparatus shown in Fig. 50 is used for the filtration. The application of the method to massecuite is given farther on.

Since anhydrous glycerine is very hygroscopic, it must

---

[1] *Zeit. Rübenzucker-Industrie*, **31**, 500.

be protected from the moisture in the air at each stage of the analysis.

Weigh 30 to 50 grams of the sugar and transfer to a glass dish containing an equal weight of glycerine. Mix intimately with a glass rod, and place in a desiccator containing fused calcium chloride or concentrated sulphuric acid. Repeat the mixing from time to time, until the crystals are well separated and the molasses uniformly distributed in the glycerine solution. This requires fifteen minutes and upwards. Place a plug of dry filtering-cotton in the funnel of the apparatus (Fig. 50), transfer the mixture to the funnel, and replace the cover. Filter under pressure, using a filter-pump. The mixture is protected from moisture, during filtration, by chloride of calcium tubes, as shown in the figure.

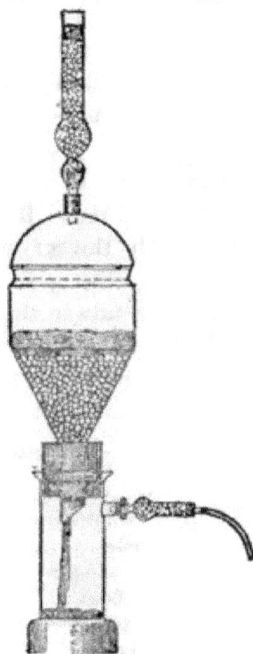

Fig. 50.

Polarize the normal weight of the filtrate. Kracz'[1] formula has been shown to be inexact; hence the corrected formula is given.

Formula for the Calculations.

x = sucrose in the molasses attached to the crystals;

P = per cent sucrose in the raw sugar;

p = per cent sucrose in the glycerine filtrate;

$$x = \frac{200 - P}{100 - p}\, p, \quad \text{and} \quad P - x = \text{the percentage of crystal-}$$

lized sugar.

Example.—Polarization of the raw sugar = 95.6; polarization of the filtrate = 6.75.

$$x = \frac{200 - 95.6}{100 - 6.75} \times 6.75 = 7.55, \quad \text{and} \quad 95.6 - 7.55 = 88.05, \text{ the}$$

percentage of crystallized sugar.

---

[1] *Zeitschrift f. Zuckerindustrie* Bohem,. Jan. 1895.

Perepletchikow [1] recommends the following procedure with massecuites :

Transfer the normal weight of the massecuite treated with an indefinite quantity of pure anhydrous glycerine, as described above, to the funnel of Kracz' apparatus, and filter off the glycerine solution. Wash the crystals with repeated portions of glycerine, until the filtrate is no longer colored. Remove the funnel from the apparatus and wash the crystals into a sugar-flask, dissolve, and polarize. The polariscopic reading is the percentage of sugar crystals in the massecuite.

Perepletchikow made comparative tests of the various methods, with the results given in the following table :

Method.	Crystals per cent Massecuite. Massecuite No. 1.	Massecuite No. 2.	Time required for Making the Analysis. Minutes.
1. Washing with sugar solution ..	71.6	62	45
2. Pellet................	69.8	60.7	60
3 Washing with glycerine .	70	60.8	60
4. Dupont	71.1	61.6	45
5. Sidersky [2]	71	61.5	150
6. Kracz	70	60.6	60
7. Perepletchikow........ ..	70.1	61.3	30
Actual percentage of crystals present ..	70.4	61.1	

Dupont's Method.[3]—Heat a quantity of massecuite of known polarization, 500 grams, for example, to 85° C. and centrifugal in a small machine, such as is constructed for laboratory purposes. The wire sieve of the centrifugal machine should be covered with thin flannel. Dry the sugar as thoroughly as possible. Determine the percentage of sugar in the molasses with the polariscope. Calculate the percentage of crystallized sucrose by the following formula, in which a = the polarization of the massecuite ; p = polarization of the crystals ; p' = polarization of the

---

[1] Zapiski, 1894, **18**, 346; Abstract in *Bulletin de l'Association des Chimistes,* **12**, 407.

[2] $x : 100 = a : b$, in which x = sugar adhering to the crystals; a = per cent ash (sulphated) in the massecuite; b = per cent ash in the molasses, obtained by filtration ; $100 - x$ = per cent crystallized sugar in the massecuite.

[3] *Manuel-Agenda des Fabricants de Sucre,* 1891, p. 293.

molasses; and $x =$ the weight of crystallized sucrose in a unit of the massecuite :

$$x = \frac{a - p'}{p - p'}, \text{ and } 100x = \text{the per cent of crystallized sucrose}$$

in the massecuite.

Example.

Polarization of the massecuite $= 84.5 = a$
Polarization of the molasses $= 60.6 = p'$

The crystals may be considered to be pure sugar; hence $p = 100$.

Substituting in the formula, we have

$$x = \frac{84.5 - 60.6}{100 - 60.6} = 0.6066, \text{ and } 100x = 100 \times 0.6066 = 60.66,$$

the percentage of crystals in the massecuite.

Dupont's formula is applicable to the calculation of the crystallized sugar in the massecuite, on the basis of the data obtained by the analysis of the massecuite, and of the molasses flowing from the sugar-house centrifugals, provided the sugar is not washed in the machines. Further, it is necessary to filter the molasses through flannel, to remove fine crystals which may have passed the centrifugal sieves.

The above is one of the most practical methods yet proposed for the estimation of the proportion of crystallized sugar in massecuites.

97. Notes on the Estimation of the Crystallized Sugar.—This estimation is of great practical value in the control of the vacuum-pan work and the centrifugals. The reduction in the yield of first-sugar in many sugar-houses, through careless centrifugal work, or by sugar-crystals passing into the molasses through holes in the sieves, "too small to amount to anything," is undoubtedly often quite large. Dupont's method affords an easy control of this part of the manufacture, and should be systematically applied. Loss in the centrifugals may also be due to a very fine grain.

ANALYSIS OF SUGARS.

98. Analysis of Sugars.—The usual determinations to be made in sugars are the percentages of sucrose and ash. The latter is determined as in **75**. The moisture is occasionally required. It is determined as usual by drying a weighed portion of the sample, in an oven, to constant weight. For high-grade sugars the temperature of the oven may be 105° C., and for very low grades 100° C., or, preferably, these sugars should be dried in the vacuum-oven, page 94, at a temperature below 95° C.

White sugars can be polarized without clarification of the solutions; filtration is necessary, however, to remove dust and mechanical impurities. Raw-sugar solutions must be clarified with a few drops of dilute subacetate of lead solution.

Aluminic hydrate will sometimes facilitate the clarification of low-grade sugars.

With compensating instruments the polarization of sugars should be effected at moderate temperatures. The Schmidt and Haensch instruments give correct percentages, when the normal weight of the sugar is contained in 100 Mohr's cubic centimetres of the solution, and is observed in a 200-mm. tube at $17\frac{1}{2}$° C. With the Laurent apparatus, the normal weight of the sugar should be contained in 100 true cubic centimetres. Creydt's method, **89**, should be used with low-grade sugars.

The chemist is occasionally called upon to estimate the refining value or "titrage" of a raw beet-sugar. The method adopted in Germany for this calculation is as follows: Deduct 5 times the per cent of ash from the polarization of the sugar to obtain the titrage. If a saccharate process have been used, an additional allowance of 1 per cent of the titrage, as calculated above, is made. This method is not entirely satisfactory to the refiners, who claim that with modern methods of raw-sugar manufacture the allowance is too small. They suggest that 2 times the per cent of non-sugar be deducted.

The French deduct 4 times the per cent of ash and 2 times the per cent of reducing sugar from the polarization. For white first sugars, the deduction is 5 times the per cent of ash. Fractions in the polarization are not counted. The French Government, in its calculations, uses only the per cent soluble ash and not the total ash.

These methods are purely arbitrary and are based solely upon refining experience.

99. Notes on the Analysis of Massecuites, Sugars, and Molasses.—In the event of obtaining very dark-colored solutions which are difficult to polarize, shake the solution with about one gram of finely powdered dry bone-black, and filter. To avoid an error, due to the absorption of sugar by the bone-black, it is advisable to use the latter in small quantity, or to filter the solution through a very small quantity of bone-black, rejecting the first half of the filtrate.

In the clarification of the solution with subacetate of lead, the reagent should be added as long as a precipitate forms. In solutions which contain invert-sugar or raffinose acetic acid should be added to restore the normal rotatory power. Since the rotatory power of raffinose is modified by subacetate of lead, it is advisable that the direct polarization, in the inversion methods, be made in a solution acidulated with acetic acid as advised by Pellet.

There is much room for improvement in the existing methods for the analysis of the low products, especially of molasses.

ANALYSIS OF FILTER PRESS-CAKE.

100. Determination of Moisture.—Dry 5 grams of the press-cake at 100° C. to constant weight. The loss in weight × 20 = percentage of moisture.

101. Determination of the Total Sucrose.— The sucrose in the press-cake is partly in combination with the lime, as a saccharate, and partly in water solution. The saccharate must be decomposed and the sucrose set free. Several processes for the decomposition of the saccharate, in this analysis, have been suggested, a few of which are given in the following methods :

Stammer's Method.—Place 100 grams of the press-cake in a mortar and beat to a smooth cream with water ; transfer to a large tared Erlenmeyer flask, and add sufficient water to make about 200 cc., including that used in beating up the press-cake. Treat with an excess of carbonic acid ; raise the temperature to the boiling-point, and expel the excess of carbonic acid. Cool the flask and contents, place it upon a scale, and add sufficient water to complete the quantity to 200 grams. Mix the water and press-cake thoroughly, filter off 50 cc. of the solution, add 5 cc. subacetate of lead for clarification, and filter. Polarize the filtrate, using as long an observation-tube as the instrument will admit, and increase the reading by 1/10, to correct for the dilution.

Example indicating the Calculations.

Weight of press-cake used...................100 grams.
Water in the sample as determined by drying.. 40 per cent.
Polariscopic reading, 400-mm. observation-tube, corrected for the 1/10 dilution (Schmidt and Haensch polariscope)... 4
Water in the press-cake = 100 × 40........... 40 grams.
Water added................................... 200 "
                                               ─────
     Total water............................. 240 "

The volume of the total water therefore is 240 cc.

Formula.

Let R = polariscopic reading in a 200-mm. tube;

V = total volume of water added and the water in the press-cake;

F = the normal weight divided by 100.

Then $\dfrac{F \times R \times V}{100} = \begin{cases} \text{the per cent of sucrose in the press-} \\ \text{cake.} \end{cases}$

Substituting the values of R, V, and F in the formula,

$\dfrac{.26048 \times 2 \times 240}{100} = \begin{cases} 1.25, \text{ the per cent of sucrose in the} \\ \text{press-cake.} \end{cases}$

The saccharates of lime may be decomposed by one of the methods given below.

There is an inappreciable error in this method in considering the volume of the water added and that of the water in the press-cake as the volume of the sugar solution in Mohr's units.

Sidersky's Method.—This is one of the most convenient methods for the analysis of well-formed press-cake. It is based upon the fact that the volume of the insoluble matter in 26.048 grams of press-cake is approximately 5 cc.

Beat 25 grams of press-cake, 15.7 grams for the Laurent polariscope, and a small quantity of cold water to the consistence of a cream, using a glass mortar and pestle, and transfer the mixture to a 100-cc. flask. Add a few drops of a solution of phenolphthalein as an indicator, then sufficient dilute acetic acid, drop by drop, to discharge the color. Clarify with subacetate of lead, filter and polarize. It is advisable to use a 400-mm. or 500-mm. observation-tube for the polarization, and divide the polariscopic reading by 2 or 2.5 to obtain the per cent of sucrose.

Various Methods.—Other methods usually differ from Sidersky's in the reagent used for the decomposition of the saccharates of lime. Among these reagents may be mentioned boracic acid, carbonate of sodium, bicarbonate of magnesium and sulphate of magnesium. The object of this treatment is to decompose the saccharates without decomposing salts of optically active bodies which may be present in the press-cake. Herzfeld does not consider magnesium sulphate suitable for this purpose.

102. Determination of the Free and Combined Sucrose.—Proceed by one of the above methods for the sucrose in press-cakes, except to use neither acetic acid nor other reagent which will decompose the saccharates. Add a few drops of acetic acid to the solution before polarizing. This gives the free sucrose.

The combined sucrose, *i.e.*, the sucrose in the saccharates, is obtained by deducting the free from the total sucrose.

ANALYSIS OF RESIDUES FROM THE MECHANICAL FILTERS.

103. Composition and Analysis.—The composition of the residues from the sirup filters is quite variable, and, aside from the sucrose, is of interest on account of incrustation in the multiple-effect and the difficulty sometimes experienced in the filtration. The composition depends somewhat on the quality of the stone and coke used in the lime-kiln, and upon the method of conducting the carbonatation and the saturation. The difficulties in the filtration may often be traced to the presence of gelatinous silica.

The important constituents of the residues are sucrose, oxide, carbonate, sulphite and sulphate of calcium, iron, alumina and silica.

The moisture and sucrose are determined by the methods in sections **100** and **101**. After the removal of the organic matter, by ignition, the inorganic constituents may be determined by the methods given for the analysis of limestone, page 148.

ANALYSIS OF WASH AND WASTE WATERS.

104. Determination of the Sucrose.—The sucrose is usually the only determination required in wash and waste waters.

The water used in washing the filter press-cake is analyzed in the same manner as carbonated juices.

The waste waters from the diffusion-battery contain exceedingly small quantities of sucrose. The determination may be made either by the optical or the chemical method. In the former add one or two drops of concentrated subacetate of lead solution (**207**) to 100 cc. of the water, mix and filter. Polarize in 400-mm. or 500-mm. observation-tube. Obtain the percentage of sucrose by inspection, from the following table (Schmitz):

Tenths of the Polariscopic Reading.	Per Cent Sucrose.	Tenths of the Polariscopic Reading.	Per Cent Sucrose.
0.1	0.03	0.6	0.15
0.2	0.05	0.7	0.17
0.3	0.07	0.8	0.20
0.4	0.11	0.9	0.22
0.5	0.12		

The chemical method (**37**) is applicable to all waste waters, especially to those containing little more than traces of sucrose. Proceed as follows:

Concentrate a measured volume of the water to small volume, invert with hydrochloric acid, neutralize with caustic soda, and determine the reducing sugar by one of the methods in **72** or **73**. Multiply the percentage of reducing sugar obtained by .95 to obtain the percentage of sucrose. Tartaric acid may be added to the water before the concentration, thus inverting the sucrose and dispensing with the hydrochloric acid.

In the examination of the ammoniacal waters from the multiple effect, by the chemical method, the ammonia should be driven off by boiling.

ANALYSIS OF THE EXHAUSTED COSSETTES.

105. Indirect Method.—Cut the sample of well-drained cossettes into very small fragments by means of a meat-chopper, or, preferably, reduce to a cream with a mill. This machine should be one which will not press the cossettes and whose construction permits easy access for cleaning.

Express the thin juice from the cossettes with a powerful press. It is essential that as great pressure as practicable be exerted, in order that a fairly representative sample of the juice may be obtained. Several models of powerful presses are made for this purpose, one of which is shown in Fig. 38.

To 100 cc. of the juice, in a sugar-flask, add sufficient subacetate of lead for the clarification, dilute to 110 cc. and filter. Polarize the filtrate, using as long an observation-tube as the instrument employed will accommodate. Calculate the percentage of sucrose by Schmitz' table, page 285. In order to calculate the percentage of sucrose upon the weight of the exhausted cossettes and then to terms of the weight of the beets, it is necessary to know the percentage of water in the cossettes and the weight of the latter per 100 pounds of beets. The water is determined by the usual method, of drying a sample to constant weight in an oven. The weight of cossettes per cent beets is ascertained by actual experiment.

The well-drained exhausted cossettes, when working by water-pressure, contain approximately 95 per cent of thin juice; hence the percentage of sucrose in the thin juice × 95 ÷ 100 = the percentage of sucrose in the cossettes.

Direct Method (Stammer's slightly modified).—Grind a sample of the well-drained exhausted cossettes to a cream in a cylindro-divider, Fig. 36, or other suitable milling device. To 300 grams of the cream add 10 cc. dilute solution of subacetate of lead for clarification, mix thoroughly, and filter. Polarize the filtrate, using as long an observation-tube as the polariscope will accommodate.

Example and Calculation.—Three hundred grams of the creamed cossettes, containing 90 per cent of water, were treated as above described. The polariscopic reading, Schmidt and Haensch instrument, was 1.6, corrected for tube length.

Whence $300 \times 90 = 270$ grams of water in the cream $= 270$ cc., and 270 cc. $+$ 10 cc. subacetate of lead solution $= 280$ cc., the total volume of the solution, exclusive of the marc (**113**), and $1.6 \times .26048 = 0.417$ gram of sucrose per 100 cc. of the solution; $0.417 \times 280 \div 100 = 1.168$ grams, the sucrose in 300 grams of the exhausted cossettes ; 1.168 $\div 300 \times 100 = 0.39$, the per cent sucrose in the exhausted cossettes.

The error due to calculating the percentage of water as the percentage of thin juice in the cossettes is inappreciable.

DEFINITIONS OF THE COEFFICIENTS AND TERMS USED IN SUGAR ANALYSIS.

106. Coefficient of Purity, True and Apparent.—The *true coefficient of purity* is the percentage of sucrose contained in the total solid matter in the product, and is calculated by dividing the percentage of sucrose by the percentage of total solids, as determined by drying, and multiplying the quotient by 100. The *apparent coefficient of purity* is calculated as above, except that the degree Brix, as determined by spindling or from the specific gravity, is substituted for the percentage of solids, as ascertained by drying.

This coefficient is also often termed the "quotient of purity," the "degree of purity," or the "exponent."

The calculations may be much simplified by the use of Kottmann's table, page 295. It will be noticed that this table advances by .2 per cent sucrose. Intermediate values may be obtained by interpolation. This is sufficiently accurate for all calculations based upon the degree Brix as ascertained by spindling, since this degree itself only approximates the true percentage of solids.

107. Glucose Coefficient, or Glucose per 100 Sucrose.—This coefficient is frequently termed the "glucose ratio."

Calculation.

$$\frac{\text{Per cent reducing sugars}}{\text{Per cent sucrose}} \times 100 = \left\{ \begin{array}{l} \text{the glucose (reducing} \\ \text{sugars) coefficient.} \end{array} \right.$$

This coefficient is useful in detecting inversion. An increase in the glucose coefficient at different stages of the manufacture, provided there has been no removal of sucrose or decomposition of reducing sugars, shows that a portion of the sucrose has been inverted.

108. Saline Coefficient.—The saline coefficient is the quantity sucrose per unit of ash.

Calculation.

$$\frac{\text{Per cent sucrose}}{\text{Per cent ash}} = \text{saline coefficient.}$$

109. Proportional Value.—This coefficient is employed in comparing the manufacturing value of different samples of beets.

Calculation.

$$\frac{\text{Per cent sucrose} \times \text{coefficient of purity}}{100} = \text{proportional value.}$$

110. Apparent Dilution.—The apparent dilution is the amount of water added to the normal juice to increase its volume to that of the diffusion-juice. This is expressed in percentage terms of the normal juice.

111. Actual Dilution.—The actual dilution is the proportion of water added to the normal juice to reduce its percentage of sugar to that of the diffusion-juice; hence the actual dilution represents the evaporation necessary, per cent normal juice, to remove the added water. In calculating the dilution we use either the percentage of sucrose or the degree Brix. In figuring coal consumption all statements should be based on the actual dilution. The nearer we approach a perfect extraction, the nearer the apparent dilution approaches the actual.

112. Coefficient of Organic Matter.—This coefficient is the quantity of sucrose per unit of organic matter other than sucrose. The true coefficient and the apparent coefficient are calculated as follows, using the solids by drying for the former and the degree Brix as the per cent solids in the latter:

$$\frac{\text{Per cent sucrose}}{\text{Per cent total solids} - (\text{per cent sucrose} + \text{per cent ash})} = \text{coefficient of organic matter.}$$

The apparent coefficient of organic matter is of doubtful value.

DETERMINATION OF THE MARC.

113. Determination of the Marc.—The marc is that portion of the sugar-beet which is insoluble in water.

Direct and indirect methods are used for its determination. In the direct methods, the soluble matter is removed with water, under certain temperature conditions. The indirect methods assume that a juice can be obtained, by heavily pressing the pulp, which has the mean composition of all the juice in the beet. The investigations of distinguished chemists indicate the presence of water that holds little if any sugar in solution, and which is termed, by the Germans, "*Coloïd-wasser.*" In view of this fact, indirect methods cannot be depended upon for other than approximate results, hence are not given in this work.

Method of von Lippmann.—Place 20 grams of the finely ground sample in a basket of wire netting. The mesh must be very fine, and any portions of the pulp which pass it, in the subsequent operations, must be returned. Insert the basket in a current of water heated to 65° to 70° C. for 30 to 35 minutes, or until the pulp yields no more soluble matter. Drain the exhausted pulp, then complete the washing with a mixture of alcohol and ether. This last washing is for the displacement of part of the remaining water, and thus facilitates the drying. Dry the exhausted pulp, at first slowly at a temperature of 80° to 90° C., and then complete the drying at 100° C. to constant weight. Cool in a desiccator and weigh quickly. Weight of residue ÷ 20 × 100 = per cent marc, and 100 — per cent marc = per cent juice contained in the beet.

Method of Pellet.—For convenience in the manipulations,

Pellet uses the apparatus shown in section in Fig. 51;
c, c is a small cylinder of finely
perforated metal that fits snugly
into an outer vessel or envelope
of copper, V, V, V, perforated
at the lower part ; a perforated
disk, a, a, provided with a stem
or rod, S, fits snugly into the
cylinder. The cylinder is large
enough to hold 25 to 50 grams
of beet-pulp.

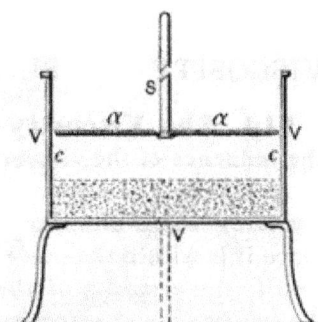

FIG. 51.

Tare the cylinder and disk
and place 25 to 50 grams of
very finely divided pulp in it. The pulp should be such as
is suitable for Pellet's diffusion method, 62. Place the
cylinder in the envelope, the disk on top of the pulp, and
the entire apparatus in a funnel. Wash with cold water,
i.e., at the laboratory temperature. With 25 grams of
pulp, allowing 10 to 12 minutes for the filtration, the extraction is complete with 500 cc. of water. It is simpler to
direct a stream of water upon the disk, maintaining a
uniform level, and using about 2000 cc. for the extraction.
Return the first portions of the extract, since it may contain fine particles of pulp. After the extraction is completed, press the exhausted pulp, by means of the disk,
then loosen the residue, with the stem, to facilitate the
drying. It is convenient to pass a few cubic centimetres
of strong alcohol through the pulp after pressing, to
economize time in the desiccation. Dry the marc, and calculate its percentage as in the preceding method.

VISCOSITY OF SUGAR-HOUSE PRODUCTS.

114. The Viscosity of Sirups, etc.—The study of the influence of the viscosity of the liquors upon the rate of evaporation, and upon the crystallization of the sugar, is receiving some attention in the European sugar-houses. Since it is within the power of the manufacturer to slightly modify the viscosity of the sirups, etc., in the purification, the importance of viscosity-tests is evident.

Several models of viscosimeters are made, all of which are designed primarily for testing oils, but which may be readily applied to the examination of sugar-house products.

These instruments may be divided into two classes—viz., the torsion-viscosimeter and flow-viscosimeters. There is but one torsion-instrument, that devised by Doolittle; there are many models of flow-viscosimeters, ranging from a simple pipette to complicated instruments with devices for controlling the temperature and flow.

It is evident in viscosity comparisons that the same products should be compared with one another at the same densities, *i.e.*, juices should be reduced to a common degree Brix, and sirups to a degree common to all.

So little of this work has been done with sugar-house products that there have been few opinions published relative to a method of procedure. The following densities are recommended as standards: Juices, 10° Brix; sirups, 40° Brix; molasses, 65° Brix.

Doolittle Viscosimeter.—This instrument is well adapted for sugar-work. The following description is from that published by Doolittle in the *American Engineer and Railroad Journal:*

"Having experimented with a number of these viscosimeters (flow-instruments) in the laboratory of the Philadelphia & Reading Railroad Company, we found them so very unsatisfactory where rapid and accurate work was required that we abandoned them all and designed an instrument on the above-mentioned principle (torsion-balance). In the torsion-viscosimeter we have an instrument which, during the year and a half we have had it in daily use, has proved itself reliable, accurate, and satisfactory in every way. It is very easy to clean and manipulate, is adapted to oils of all ranges of viscosity, and reduces personal error to a minimum.

"A glance at the cut will show how the principle has been applied. A steel wire is suspended from a firm sup-

port and fastened to a stem which passes through a graduated horizontal disk, thus allowing us to measure accurately the torsion of the wire. The disk is adjusted so that the index-point reads exactly o, thus showing that that there is no torsion in the wire. A cylinder 2 in. long by 1½ in. in diameter, having a slender stem by which to suspend it, is then immersed in the oil and fastened by a thumb-screw to the lower part of the stem of the disk. The oil-cup is surrounded by a bath of water or high fire-test oil, according to the temperature at which it is desired to take the viscosity. This temperature being obtained, while the disk is resting on its supports, the wire is twisted 360° by rotating the milled head at the top. The disk being released, the cylinder rotates in the oil by virtue of the torsion of the wire.

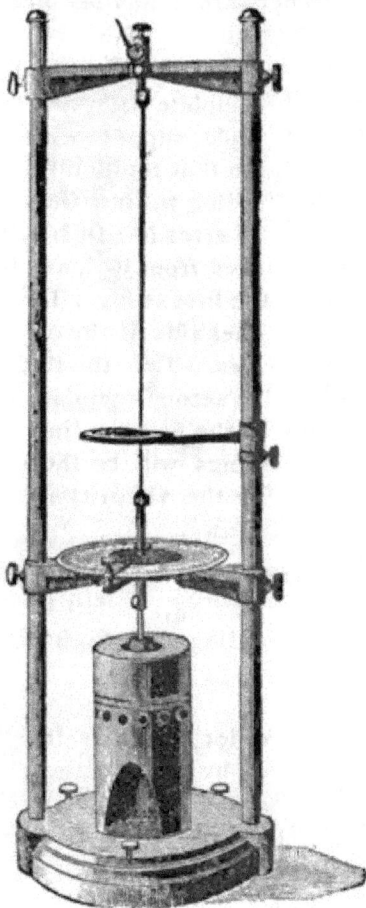

Fig. 52.

"The action now observed is identical with that of the simple pendulum.

"If there were no resistance to be overcome, the disk would revolve back to o, and the momentum thus acquired would carry it 360° in the opposite direction. What we find is, that the resistance of the oil to the rotation of the cylinder causes the revolution to fall short of 360°, and that the greater the viscosity of the oil the greater will be the resistance, and hence the retardation. We find this retardation to be a very delicate measure of the viscosity of the oil.

"There are a number of ways in which this retardation may be read, but the simplest we have found to be directly in the number of degrees retardation between the first and second complete arcs covered by our rotating pendulum. For example, suppose we twist the wire 360° and release the disk so that rotation begins. In order to obtain an absolute reading to start from, which shall be independent of any slight error in adjustment, we ignore the fact that we have started from 360°, and take as our first reading the end of the first swing. Ignore the next reading, which is on the other side of the o point, as it belongs in common to both arcs. Take the third reading, which will be at the end of the second complete arc, and on the same side of the o point as the first reading. The difference between these two readings will be the number of degrees retardation caused by the viscosity of the oil. Suppose the readings are as follows:

First reading,	right hand	355.6°
Second reading,	left hand—ignore	
Third reading,·	right hand	338.2°

17.4° retardation.

"In order to secure freedom from error, we make two tests: one by rotating the milled head to the right and the other to the left. If the instrument is in exact adjustment these two results will be the same; but if it is slightly out the mean of the two readings will be the correct reading."

Flow-viscosimeter.—Of the many efficient forms of flow-viscosimeters a brief description of Engler's, which is preferred by Dupont,[1] will suffice.

---

[1] *Bulletin de l'Association des Chimistes*, **14**, 948.

Engler's apparatus is shown in Fig. 53. The inner or oil chamber has an accurate arrangement for measuring the

FIG. 53.

liquid. This chamber is surrounded by a water-bath. A plug at the centre closes the exit-tube.

The apparatus is so arranged that the flow will be under the same conditions in comparative tests.

In making a test, the inner chamber is filled to the mark with water at 20° C., this temperature being maintained by means of the water-bath. The plug is lifted and the time noted, in seconds, required for 200 cc. of the water to flow into the graduated flask. A stop-watch should be used in timing the flow.

The inner chamber and the tube are thoroughly dried, and the chamber is filled with the liquid to be tested. The temperature of the water in the bath is again maintained at 20° C. for some time, to insure a corresponding temperature of the liquid in the inner chamber. The plug is lifted as before, and the time in seconds required for the flow of 200 cc. of the liquid is again noted. This time, divided by that required for the flow of the water is the specific viscosity of the liquid. It is usual, in testing oils, to state the viscosity as the number of seconds required for a given volume of the oil to flow through an orifice which will pass the same volume of a standard oil at the same temperature in a given time.

It would be more convenient to state the viscosity of a sugar solution in this way and operate at a higher temperature than 20° C.

CONTROL OF THE OSMOSIS PROCESS FOR THE TREATMENT OF MOLASSES.

115. Analytical Work.—The object of the osmosis process (dialysis) is the reduction of the proportion of the saline and organic impurities, so that an additional quantity of sugar can be removed from the molasses by crystallization. The proportion of saline matter in the molasses and in the by-products from the osmosis is so high that the apparent coefficients of purity are of but comparatively little value, and the time required for the determination of the true coefficients is so long that they cannot be made available for the immediate control. Notwithstanding the objections mentioned, manufacturers are compelled to be guided largely by the apparent purities in conducting the osmosis. The saline coefficient is a more reliable guide, but, unfortunately, its determination also requires much time. In actual practice, the following determinations are usually made in the molasses, before and after osmosis, and in the osmosis water: Degree Brix, percentage of total solids by drying, percentage of sucrose, percentage of ash, the percentage of organic matter not sucrose by difference, the percentage of reducing-sugars, and the alkalinity due to lime.

The following coefficients, true and apparent, should be calculated : Coefficient of purity, saline coefficient, glucose coefficient, and coefficient of organic matter.

Practice and the expense of the application of the process, as compared with the value of the sugar recovered, must be the guides in determining the improvement to be made in the above coefficients.

Gallois and Dupont give the following advice, in their manual,[1] relative to the character of molasses that may

---

[1] *Manuel-Agenda*, 1891, p. 383.

be treated by this process with profit: "It is useless to dialyze molasses whose saline coefficient is higher than 6°, since it will yield a satisfactory quantity of sugar on further concentration. Molasses containing more than 1 per cent of reducing-sugars cannot be treated with profit. Molasses containing much lime, especially organic salts of lime, are difficult to dialyze. Such molasses should receive a preliminary addition of carbonate of soda or acid phosphate of barium to precipitate the lime, which must be removed. Molasses containing as much as 0.2 per cent of lime (CaO) should be treated as indicated. If there are indications of fermentation, or if the molasses is but slightly alkaline, neutral, or acid, caustic soda should be added."

ANALYSIS OF SACCHARATES.

116. Saccharates.—The various chemical processes for the extraction of the sugar from the molasses, usually depend upon its precipitation as a saccharate of lime or strontium. The precipitation of lead and barium saccharates has been proposed, and the latter used to a limited extent. These saccharates possess the requisite properties, but for commercial reasons lime and strontium are the precipitants almost exclusively employed.

The following are the chemical formulæ of the lime saccharates, of which the tribasic is the most important:

Monobasic saccharate, $(C_{12}H_{22}O_{11})$. CaO
Dibasic saccharate, $(C_{12}H_{22}O_{11})$. 2CaO
Tribasic saccharate, $(C_{12}H_{22}O_{11})$. 3CaO

Strontium forms two saccharates, the monobasic and the dibasic; barium forms a monobasic saccharate. The formulæ of these saccharates are similar to those of the corresponding lime compounds.

117. Determination of the Sucrose, Lime, Strontium, and Barium.—Mix a large quantity of the saccharate thoroughly to obtain a uniform sample, transfer a portion to a mortar and rub to a smooth paste. Titrate ten grams of this paste with normal hydrochloric acid solution (**176**), using phenolphthalein as an indicator. It is advisable to reduce the paste, before titration, with a few cc. of water. Calculation: 1 cc. normal hydrochloric acid will saturate .028 gram calcium oxide (CaO), 0.07671 gram barium oxide (BaO), or .0518 gram strontium oxide (SrO). Multiply the burette reading by the factor for calcium oxide, barium oxide, or strontium oxide, as given above, and this product by 10, to obtain the percentage of calcium oxide, etc., in the saccharate.

To determine the sucrose: To the normal weight of the

saccharate add acetic acid to slight acidity, using phenolphthalein as an indicator. Transfer the solution to a
sugar-flask, add a few drops of subacetate of lead solution,
dilute to 100 cc., mix the contents of the flask and filter.
Polarize the filtrate in a 200-mm. tube. The polariscopic
reading is the percentage of sucrose in the saccharate.

**118. Apparent and True Coefficients of Pu-
rity.**—Beat an indefinite quantity of the saccharate, to a
cream, in a mortar with distilled water, add sufficient oxalic
acid to combine with the greater part of the lime, being
cautious not to add sufficient to decompose all of the saccharate. Transfer the mixture to a strong flask, add a few
drops of phenolphthalein solution, and saturate with carbonic-acid gas from a suitable generator. The discharge
of the color indicates the termination of the process. It is
advisable to saturate under moderate pressure. Close the
flask with a 2-hole stopper; pass the gas-delivery tube
through one hole, nearly to the bottom of the flask, and in
the other hole insert a short piece of tubing closed with a
rubber tube and a pinch-cock. The cock should be opened
from time to time at the beginning of the operation for the
escape of air. If the gas from the lime-kiln be used for
this saturation, the apparatus should be placed in a fume-
chamber, and a regulator, on the principle of that shown in
Fig. 49, should be connected with the short tube.

On the completion of the saturation, boil the mixture to
expel the excess of carbonic acid, and filter off the solution.
Cool the filtrate and determine its density (Brix) and per
cent sucrose as in juices, and calculate the apparent and
true coefficients of purity.

**119. Analysis of "Mother-liquors" and
Wash-waters.**—The analysis of these products is made
as described for saccharates (**117**), except that in the sucrose determination, only sufficient acetic acid should be
added to neutralize the alkalinity, phenolphthalein being
used as an indicator.

EXAMINATION OF BONE-BLACK.

120. Limited Use of Bone-black in Sugar Factories.—The use of bone-black, or char, as it is often termed, in sugar factories, is now very limited, it having been almost entirely replaced by sulphurous acid. In view of this fact, only a few essential tests are given.

121. Revivification.—The practical test to determine whether the revivification has been properly conducted is qualitative, and employs a caustic-soda solution, as follows :

Boil about 50 grams of bone-black two or three minutes in 50 cc. of a solution of caustic soda ($9°$ Brix or $5°$ Baumé). Decant or filter the solution into a test-tube, using an asbestos filter, and note its color. A faint tinge of color indicates a good revivification; a yellow or brown color indicates insufficient revivification; a colorless or greenish solution indicates over-revivification. This test is of great importance, and should be made frequently. A reddish-tinged char indicates imperfect revivification; gray, leakage of air into the retorts; and white, an overburned bone-black.

122. Weight of a Cubic Foot of Bone-black.—Bone-black increases in weight each time it is used, by the absorption of impurities which are not removed in the revivification. This gradual increase in weight is a measure of the deterioration from usage.

The weight per cubic foot depends in a large measure upon the size of the grains, and in new char will range from about 43 to 48 lbs. On commencing work with new char its weight per cubic foot should be recorded, and this weight employed in future comparisons. According to Gallois and Dupont,[1] the weight of bone-black of good quality, while in use, should not exceed 1.23 times its weight when new; at 1.47 it is in a very bad condition; and at 1.50 times the original weight it should be rejected.

---

[1] *Manuel-Agenda des Fabricants de Sucre*, 1891, p. 307, Ch. Gallois and F. Dupont.

123. Sulphide of Calcium.—A greenish color on treatment with caustic soda, as in **121**, is an indication of the presence of sulphide of calcium. Occasional tests should be made for this substance, since its presence very materially affects the quality of the bone-black, and when it is present in more than very small quantities, the char should be rejected. This salt may be tested for qualitatively by treating the char with strong acid and testing the gas liberated for sulphuretted hydrogen.

124. Moisture.—The moisture should be determined in new bone-black, since this substance is very hygroscopic. An increase over 6 per cent moisture is an excess chargeable to the dealer, and for which the sugar manufacturer should not pay. Bone-black may absorb 20 per cent of moisture without showing external indications of this increase.

125. Decolorizing Power of the Bone-black. —The decolorizing power of bone-black is determined by means of a colorimeter. Stammer's instrument for this purpose is a very convenient form, and the results obtained by different operators are comparable. This instrument consists essentially of an arrangement for comparing the depth of color of a column of sugar solution with standard-colored glass plates. An ocular is so arranged that the color of the solution under examination appears upon one half of a disk, and that of the standard glass on the other. The eyepiece and a tube containing the glasses are raised and lowered by means of a rack and pinion, the length of the column of solution being varied at the same time; this length is shown on a scale by means of a pointer, carried by a slide. The theory of this instrument depends upon the variations in the intensity of the color of the solution, which is proportionate to the length of the column. In using the colorimeter, the object is to equalize the intensities of the colors as seen on the disk through the ocular, by lengthening or shortening the column of the solution under examination. The strength of solution being known, a comparative statement of depth of color in terms of the sucrose present may be made, or the reading on the scale may easily be reduced to an expression showing the depth of color as compared with the standard.

This instrument may be used in determining the decolorizing power of a char in the following manner:

A standard-color solution should be prepared, using caramel, a definite quantity being taken. Duboscq recommends 2 grams per litre for his instrument. Prepare the caramel by heating pure cane-sugar to about 215° C., until all the sugar is decomposed. In examining bone-black, determine the depth of color in the standard solution, then heat a measured volume of this solution with a weighed portion of the char a certain length of time, for example, half an hour, filter, and again determine the intensity of color. The difference in the depth of color referred to the standard represents the efficiency of the bone-black in decolorizing. In sugar-house work, a standard bone-black of a known decolorizing capacity is convenient for comparison. Comparable results can only be obtained by adopting certain conditions and adhering to them in all experiments.

The decolorizing power may be roughly determined, in the absence of a colorimeter, as follows: Treat a measured volume of a standard-color solution as described above. Fill a cylinder, similar to those used in Nesslerizing, to a certain depth with the decolorized and filtered solution; place the same volume of the standard solution in a similar cylinder, and add water to the latter from a burette until a portion of the same depth as that of the decolorized solution shows the same intensity of color when examined over a white background. The volume of water added is inversely proportional to the decolorizing power of the char.

126. Determination of the Principal Constituents. — The constituents which it is sometimes necessary to determine in the examination of bone-black are the following: moisture, carbon, carbonate, sulphate, sulphide, and phosphates of calcium, sand and other foreign substances. •

The moisture is determined by drying 2 grams of the powdered bone-black at 140° to 150° C.

The other constituents are determined by the usual methods of quantitative analysis as given in the various text-books.

ANALYSIS OF THE LIME-KILN AND CHIMNEY GASES.

127. Analysis of the Gas from the Lime-kiln.
—Carbonic acid (CO_2) is the valuable constituent of the gas. The other gases which are present and should be quantitatively determined in the control of the lime-kiln are carbonic oxide (CO) and oxygen (O). Hydrosulphuric acid (H_2S) and sulphurous acid (SO_2) are sometimes present in traces. The following analyses[1] illustrate the usual composition of the gas, as reported by six sugar-houses, and will be a guide in the analysis :

	A	B	C	D	E	F
Carbonic acid, pr. ct.,	31.2	30	27	25 to 30	29.5	33
Carbonic oxide, "	1.6		0.5	3		0.5
Oxygen, "	1.3		1	2.5		1.5
Nitrogen, by difference, pr. ct.,	65.9		71.5		69.75	
Sulphurous acid, pr. ct.,					0.75	

A modification of Orsat's apparatus, Fig. 54, is convenient for use in this analysis. It consists essentially of a burette for measuring the gases and a series of pipettes or U-tubes for their absorption. The burette A has a capacity of 100 cc.; the lower part is divided into cubic centimetres and tenths; it is enclosed in a cylinder, which is filled with water, in order to avoid variations in the temperature of the gas during measurement. The U-tubes B, C, and D are filled with the reagents for absorbing the gases. The surfaces exposed to the gases are increased by filling the U-tubes with small glass tubes.

The connecting tubes are of very small internal diameter and are made of heavy glass. Prepare the solutions for

---

[1] *Bul. Assoc. des Chimistes de France.*

absorbing the gases as follows and fill each U-tube about
half full:

For tube B: Use a concentrated solution of caustic potassium (KHO) of about 60° Brix.

For tube C: Dissolve 25 parts of pyrogallic acid in 50 parts
of hot water and add 100 parts of caustic potassium solution

FIG. 54.

of approximately 50° Brix. The volume of the U-tube
should be ascertained, and only sufficient of this solution
should be prepared to half fill the tube.

For tube D : Cuprous chloride for filling this tube may
be quickly prepared by the following method : Dissolve
35 grams of cupric chloride in a small quantity of water and
add stannous chloride in excess as indicated by the change
in the color of the solution. Cuprous chloride, insoluble in
water, separates as a white crystalline precipitate. Wash
the cuprous chloride several times, by decantation, with
distilled water. Avoid exposing the precipitate to the action
of the air more 'han is strictly necessary. In the last
washing, pour off the water close to the precipitate, then dis-

solve the latter in concentrated hydrochloric acid, and wash it into a bottle with this acid, using in all 200 cc. ; dilute the solution with approximately 120 cc. of water.   Place a few pieces of copper wire or turnings in the bottle, stopper it, and set it aside until required for filling the tube.

The following method may be used instead of the above : Place 35 grams of cupric chloride in a glass-stoppered bottle, add 200 cc. concentrated hydrochloric acid and a quantity of copper turnings or fragments of copper-foil. Stopper the flask and set aside for two days, shaking occasionally ; add 120 cc. water (Wagner).   The excess of this solution over the quantity required for filling the tube should be preserved as described above.

Fill the bottle F with distilled water and set it on the table.   Close the cocks on the tubes B, C, and D, and turn the 3-way cock G so that it connects the burette with the outer air.   Lift the bottle F until the water fills the burette, then close the pinch-cock.

A U-tube, partly filled with water, and having each branch loosely plugged with cotton, is connected with the tube E, and has a tube for connection with the pipe leading from the carbonic-acid pump.

Having filled the burette, close D and place the bottle F on the table ; open the cock on the U-tube B, then cautiously open the pinch-cock, and as the water flows out of the burette the caustic potash solution will rise in B ; close the pinch-cock when the potash solution reaches the mark on the tube just below the stop-cock.   Fill the burette again and repeat these manipulations with C and D successively. Care must be exercised not to let the liquids rise to the stop-cocks.   Pour a little kerosene oil on the surface of the solutions in those limbs of the U-tubes at the back of the apparatus, to protect them from the action of the air.

A small pipe should be led from the gas-main to a convenient place in the laboratory, where the apparatus can be permanently arranged for these analyses.   The pipe should terminate in a pet-cock for drawing the samples of gas.

Connect the open branch of the U-tube at the inlet E with the pet-cock.   The apparatus is now ready for testing. If there be no leaks, the cock G being open to the gas-inlet, the pet-cock closed, and the pinch-cock open, the water-column in the burette should sink a little, then remain stationary. Having satisfactorily tested the connections, proceed to the analysis as follows : Fill the burette to the 100-cc. mark with water, close the pinch-cock, and place the bottle F on the table ; turn the 3-way cock G so that it connects with the open air and with the capillary tube leading to the burette, then open the pet-cock carefully and let the gas bubble through the U-tube connected with E and expel the air from the tubes, including that leading from the gas-main. Disconnect the apparatus from the open air, and let the gas flow slowly into the burette, by opening the pinch-cock. Hold the bottle F so that the level of the water in it is the same as that in the burette; when the latter reaches the lowest graduation, which should be zero, close the 3-way cock. Relieve any pressure there may be in the apparatus by manipulating the 3-way cock G, opening it to the air. To determine the carbonic acid (CO_2) in the gases, open the cock on the U-tube B, containing caustic potash solution, and by raising the bottle F force the gases into the tube, and close the pinch-cock ; lower the bottle to the table, and, by gradually opening the pinch-cock, let the potash solution rise to the mark. Repeat this manipulation two or three times, holding the bottle F close to the burette, each time, with the water in the two at the same level. As soon as there is no further decrease in the volume of the gas, close the cock on B and read the burette, being careful to hold the bottle so that the water-level is the same in both it and the burette. The rise of the water in the burette corresponds to the percentage by volume of carbonic acid (CO_2) in the gas. Repeat these manipulations, using the U-tube C containing the pyrogallate of potassium. The burette reading is the sum of the percentages of carbonic acid and oxygen. Again repeat, using the U-tube D containing the cuprous chloride. This burette reading is the sum of the percentages of carbonic acid, oxygen, and carbonic oxide (CO). The separate percentages are obtained by subtraction. The residual gas, consisting almost entirely of nitrogen, is expelled by lifting the water-bottle after opening

the cock *G*. The burette should be left filled with water
to the 100-cc. mark, and it is then always ready for a new
analysis. The cocks should be well greased with a mixture of vaseline and mutton tallow.

The pipe leading from the main should always be well
rinsed with the gases before each test.

The solutions in the U-tubes should be changed as soon
as the absorption becomes sluggish, which will be after
150 determinations or more. The order indicated of absorbing the gases should be followed.

The lime-kiln gases may also contain small quantities of
hydrosulphuric acid (H_2S) and sulphurous acid (SO_3),
from sulphur in the coke. The former gas is tested for
with filter-paper dipped in subacetate of lead solution, or
by passing the gases into the lead solution. A black
precipitate of sulphide of lead is formed in the presence of
this gas.

Sulphurous acid is tested for by shaking a little of the
gas, in a test-tube, with iodized starch solution. In the
presence of sulphurous acid the blue color is discharged.

**128. Simple Apparatus for the Determina-
tion of Carbonic Acid in Lime-kiln Gases
(Stammer).**—This apparatus consists of a 50 cc. gas-
burette, in 1/10 cc., with glass stop-cock, the measurement
from the stop-cock; also a large glass cylinder of sufficient
depth to immerse the greater part of the burette. Fill the
cylinder and the burette with water; connect the delivery-
tube of the burette with the gas-main, as with Orsat's apparatus, holding the burette vertically with the mouth
under the water in the cylinder. Fill the burette with the gas
and close the stop-cock. Immerse the burette in the water to
the lowest graduation, then open the cock for the escape of
gas until the water stands at the same level inside and outside
the tube. Pass a small piece of caustic sodium or potassium
under the surface of the water and into the burette, placing
the thumb over the mouth of the latter. Remove the burette
from the water and shake it, to bring the caustic soda
in contact with the gas. Place the mouth of the tube under
the water, move the thumb a little to one side to permit
the rise of the water to take the place of the carbonic acid

absorbed. Repeat these manipulations until there is no further rise of water in the burette. Lower the burette in the cylinder until the level of the water is the same in both, then note the rise in the water. Multiply this number by 2 to obtain the percentage of carbonic acid. The results by this method are sufficiently accurate for practical purposes.

129. Analysis of the Chimney Gases.—The analysis of the chimney gases is conducted in the same manner as that of the gas from the lime-kiln (**127**), except that it is necessary to use a pump or a double-acting rubber bulb in drawing the samples.

It is usually only necessary to determine the carbonic acid (CO_2), carbonic oxide (CO), and the oxygen (O). With a good boiler-setting and satisfactory firing, the proportion of carbonic acid should be large, and that of the carbonic oxide very small.

ANALYSIS OF LIMESTONE.

130. Preparation of the Sample.—Fragments should be chipped from a large number of pieces of the stone, and reduced to a uniform size, then mixed and sub-sampled by "quartering." The small sample should be reduced to a very fine powder in an iron mortar or on a grinding-plate. Particles of metallic iron, from the mortar or plate, should be removed by stirring the powder with a magnet. Sift the powder through an 80-mesh sieve, and mix thoroughly by sifting or otherwise.

131. Determination of Moisture.—Dry 2 grams of the powdered stone to constant weight in a tared flat dish or a watch-glass. The oven should be heated to 110° C. The loss of weight divided by 2 and multiplied by 100 is the percentage of moisture.

132. Determination of Sand, Clay, and Organic Matter.—Treat 1 gram of the powdered limestone, in a beaker, with a few cubic centimetres of hydrochloric acid, being cautious, in adding the acid, to prevent the projection of particles of the material from the glass. Cover the beaker with a watch-glass and heat the liquid a few minutes. Collect the residue on a tared quantitative filter, wash it thoroughly with hot water, and reserve the filtrate (A) for further treatment. Dry the filter and residue to constant weight at 110° C. The weight of the residue multiplied by 100 is the percentage of sand, clay, and organic matter. Place the filter and residue in a tared platinum crucible and incinerate.. The weight of the residue (A) multiplied by 100 is the percentage of sand and clay (silica and combined silica and alumina). The difference between this percentage and that obtained before incineration is the percentage of organic matter.

133. Determination of Soluble Silica.—Evaporate the filtrate (A) from the preceding determination to strict dryness, on the water-bath, using a platinum or porcelain dish. Moisten the residue with hydrochloric acid and again evaporate to dryness. It is advisable to continue the heat for an hour or longer after apparent dryness, to insure the insolubility of the silica. Treat the residue with dilute hydrochloric acid ; collect the insoluble portion on a small quantitative filter and wash it thoroughly with hot water until free of chlorides. Reserve the filtrate (B) for further use. Partially dry the filter and contents, then insert it in a tared platinum crucible, and char it by the application of a very gentle heat. If charred too rapidly, there may be difficulty in subsequently burning off the carbon. Increase the heat until the filter is completely incinerated, and then raise to bright redness. Cool in a desiccator and weigh. The weight of the ash of good quantitative filters, or of the so-called "ashless filters," is so small that it need not be taken into account.

The weight of the residue, multiplied by 100, is the percentage of silica, SiO_2, in the soluble silicates of the stone.

134. Determination of Total Silica.—Mix the residue A (**139**), in the platinum crucible with four or five times its weight of [1] mixed carbonates of sodium and potassium, and fuse at a red heat. Continue the heat about 30 minutes after the contents of the crucible are in a quiet state of fusion.

Remove the bulk of the mass from the crucible, while still warm, with a platinum wire, to facilitate the subsequent solution. Place the crucible and the material removed from it in a beaker and treat with dilute hydrochloric acid, being careful to avoid loss by the projection of the liquid from the glass. Use heat, if required. Wash and remove the crucible. Filter the solution and evaporate to strict dryness, as under soluble silica in the preceding paragraph (**133**). Treat with dilute hydrochloric acid, collect the residue as before, and reserve the filtrate (C). Incinerate and heat to bright redness, weigh, and calculate the per-

---

[1] Use strictly chemically pure, dry carbonate of sodium and potassium, mixed in molecular proportions and finely powdered. The proportions are 106 parts sodium carbonate to 138 parts potassium carbonate.

centage of silica as described in the preceding paragraph. Subtract the percentage of soluble from that of the total silica, to obtain the percentage of silica present as sand in insoluble silicates.

135. Determination of Iron and Alumina.— Combine filtrates A, B, and C from the preceding operations and concentrate to a convenient volume. Add a slight excess of pure ammonia while the solution is still hot, boil until only a slight odor of ammonia can be detected, collect the precipitate on a small filter, filtering rapidly while the solution is hot. If there be considerable iron and alumina present, it is advisable to dissolve the precipitate with dilute hydrochloric acid and reprecipitate with ammonia as directed above, uniting the filtrates (D). Partly dry both filters, and incinerate as advised in **133.** If the so-called ashless filters be used, no correction need be made for the weight of the ash from the two filters.

The residue consists of the mixed oxides of iron and alumina (Fe_2O_3, Al_2O_3). Multiply the weight of the residue by 100 to obtain the percentage.

It is not usually necessary to determine the iron and alumina separately. If required, however, proceed as follows: Treat 1 gram of the powdered limestone with concentrated hydrochloric acid, most conveniently in a platinum dish. Evaporate to strict dryness, moisten with hydrochloric acid, and again dry on the water-bath, as described in **133,** in the silica determination. Treat the residue with dilute hydrochloric acid, with heat, and filter; wash the filter with hot water, and treat the filtrate with ammonia, as described above, to precipitate the iron and alumina. Wash the precipitate into a small dish, dissolve it in sulphuric acid, and evaporate the solution nearly to dryness. Wash the residue into an Erlenmeyer flask, being cautious in the first addition of water.

The iron is now most conveniently determined by titration with a standardized solution of permanganate of potassium (**202**).

Add a small quantity of pure zinc-dust to the solution in the flask, to reduce the iron from the ferric to the ferrous state, and titrate with the decinormal permanganate solution. This solution is added until a faint permanent pink color is produced. Multiply the burette reading by .008 to obtain the weight of ferric oxide in 1 gram of the stone. Multiply this weight by 100 to obtain the percentage of ferric oxide (Fe_2O_3); subtract this number from the combined percentages of iron and alumina, as obtained above, to obtain the percentage of alumina.

136. Determination of Calcium.—To the filtrate from the iron and alumina determination (D), corresponding to 1 gram of the stone, add sufficient hydrochloric acid to render it slightly acid. Concentrate this solution to a convenient volume, neutralize with ammonia, heat to boiling, and add an excess of boiling-hot oxalate of ammonium solution. Set aside for 12 hours, then collect the precipitate of oxalate of calcium on a quantitative filter, wash with cold water (filtrate E), dry and incinerate the filter in a tared platinum crucible, then ignite the residue strongly. The residue consists of almost pure calcium oxide (CaO), and may be weighed as such, or, more accurately, it may be converted into the sulphate ($CaSO_4$) or carbonate ($CaCO_3$), and weighed as such. It requires less time and labor to convert into the sulphate, using the following solution:

Dilute one volume of sulphuric acid with an equal volume of water, and neutralize three parts of stronger water of ammonia with this acid, then add two parts of ammonia. Dissolve 2 grams of ammonium chloride in each 100 cc. of this solution. Filter, if necessary, and preserve for use in calcium determinations. Strictly chemically pure reagents must be used in preparing this solution.

Add an excess of the ammonium sulphate solution, prepared as above, to the residue in the crucible, evaporate to dryness, ignite strongly, cool and weigh. The weight of the residue multiplied by .41158 gives the weight of calcium oxide (CaO), and by .73416 the weight of calcium carbonate ($CaCO_3$), in 1 gram of the stone, and these numbers multiplied by 100 give the percentages of calcium oxide (quicklime) and calcium carbonate, respectively.

The residue may be converted directly into calcium carbonate, if preferred, as follows: Mix it with finely powdered ammonium carbonate, moisten with water, heat

some time to expel the ammonia at a temperature between
50° and 80° C., then below a red heat. Repeat this operation until a constant weight of carbonate of calcium is
obtained. The weight of the carbonate of calcium multiplied by .56 gives the weight of calcium oxide in 1 gram
of the stone. The weight of calcium carbonate multiplied
by 100 is the percentage of calcium carbonate, or that of
the calcium oxide multiplied by 100 is the percentage of
this substance.

137. Determination of Magnesium.—To the
filtrate E, from the calcium determination (**136**), after
concentration to approximately 100 cc., corresponding to
1 gram of the stone, add a slight excess of ammonium
hydrate, then add, drop by drop, with vigorous stirring,
sodium phosphate solution in excess to precipitate the
magnesium as a phosphate. After 15 minutes add a
decided excess of ammonia. Set aside several hours,
preferably overnight, to insure the complete precipitation.
Collect the precipitate in a Gooch crucible, wash with
dilute ammonia, 1 part stronger ammonia, 0.96 specific
gravity, to 3 parts water. The washing should be continued until a drop of silver nitrate solution added to a
drop of the filtrate, acidulated with nitric acid, produces at
most only a faint opalescence. The precipitate is ammonium-magnesium phosphate; dry it, first at a gentle heat,
then increase the temperature to expel the ammonia, and
finally ignite a few minutes in the flame of a blast-lamp to
convert the residue into pyrophosphate of magnesium.
Cool the residue in a desiccator and weigh it. The weight
of the magnesium pyrophosphate ($Mg_2P_2O_7$) multiplied by
.36208 gives the corresponding weight of magnesium oxide.
The magnesium is present in limestone as carbonate. To
obtain the weight of the carbonate, multiply the weight of
the pyrophosphate by .7574. Multiply by 100 to obtain the
percentages of the weight of the stone.

In limestones which contain very little magnesium, the
method proposed by Prinsen Geerligs and modified by
Herzfeld[1] and Forster may be used.   Dissolve 2 grams of

---

[1] *Zeit. Rübenzucker-Industrie,* 1896.

the powdered stone in concentrated hydrochloric acid in a porcelain dish. Evaporate to dryness on a hot-plate or sand-bath, then heat over a naked flame, to render the silica insoluble. Treat the residue with hydrochloric acid, boil, add a few drops of nitric acid, and evaporate to small bulk, to expel the greater part of the acid. Dilute the solution with water, and add an excess of calcium carbonate, to precipitate the iron and alumina and filter into a flask, washing the precipitate with hot water. Add lime-water in excess to the filtrate, mix, then fill the flask to almost the top of the neck with water. Stopper the flask and set it aside for the precipitate to settle, then decant the supernatant liquid through a filter, and wash the precipitate by decantation as before. Dissolve the precipitate, including any particles which may adhere to the filter, using hydrochloric acid. Precipitate the calcium from the solution, as described in **136**, with oxalate of ammonium, and remove it by filtration; precipitate the magnesium as ammonium-magnesium phosphate, and convert it into the pyrophosphate as already described.

138. Determination of Carbonic Acid.—It is not usually necessary to determine the carbonic acid, as it may be calculated from the quantity required to combine with the lime and magnesia, except when sulphates are present.

The gravimetric determination is made with one of the various forms of alkalimeters. Knorr's apparatus, Fig. 55, is one of the best of these. The method of using this apparatus is as follows: A weighed quantity, 5 grams or more, of the finely powdered limestone, is introduced into the flask (Fig. 55), with 50 cc. or more of distilled water. The tube G is connected with a filter-pump to draw a current of air through the apparatus during the entire process. The bulb B contains the acid for decomposing the stone, preferably concentrated hydrochloric. C is a guard-tube, filled with fragments of caustic soda or potash, or with soda-lime, to prevent the entrance of carbonic acid, with the air. Open the stop-cock on the bulb-tube B and admit the acid slowly; the liberated gas passes into the tube D, where most of the moisture is condensed, thence through the bulbs *E*, containing concentrated sulphuric acid, which removes every trace of moisture ; the dry gas bubbles through the tared bulbs *F*, containing a caustic potash solution of 1.27 specific gravity, which absorbs the carbonic acid, and the residual air, containing moisture from the potash solution, passes on through the guard-tube *F*, which absorbs the moisture, and escapes through *G* and the filter-pump.   The gas should flow at the rate of

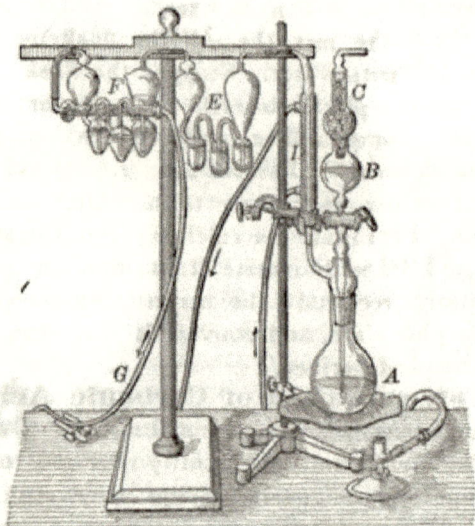

FIG. 55.

4 to 5 bubbles per second.   When the bulb *B* is empty, heat the contents of the flask carefully, finally boiling the liquid slowly, to expel the carbonic acid.   Air should be passed through the apparatus for a few minutes after boiling, to insure the removal of all the carbonic acid. Caps should be placed over the inlet and outlet tubes of *F* while making the weighings, to prevent the absorption of carbonic acid or moisture.   When the operation is com-pleted, place the bulbs and guard-tube *F* in the balance-case, and after a few minutes, weigh.   The increase in weight divided by the weight of material used and multi-plied by 100 is the percentage of carbonic acid.

A similar apparatus may be fitted up, using an ordinary

flask, with cork connections and an empty U-tube, as recommended by Gladding, instead of the condenser D.

In the determination of carbonic acid with Schroetter's or similar apparatus, proceed as follows: The description refers to Fig. 56. Fill the tube on the left, to above the upper bulb, with concentrated sulphuric acid, and that on the right with dilute hydrochloric acid. Weigh the flask and contents, then introduce approximately 1.5 to 2 grams of the powdered limestone by the opening at the left and weigh again. The difference in the two weights is the weight of the powder used. Lift the stopper on the hydrochloric-acid tube, and open the stop-cock and admit a little acid. In the

FIG. 56.

decomposition of the stone, the carbonic acid is given off and bubbles through the sulphuric acid, which retains any watery vapor that would otherwise pass off with the gas. Repeat this operation from time to time until no more carbonic acid is disengaged. Add small excess of the hydrochloric acid. Heat gently, to expel the carbonic acid from the solution, cool, and weigh. After cooling and wiping the apparatus, it should be placed inside the balance case a few minutes before weighing. The loss in weight is the weight of carbonic acid set free. Divide this weight by the weight of limestone used and multiply by 100 to obtain the percentage of carbonic acid.

The carbonic acid in the limestones, used in sugar-manufacture, is almost entirely combined with the calcium; a small portion is usually combined with magnesium, and occasionally the stone contains a vein of dolomite, a carbonate of calcium and magnesium. In the absence of gypsum, sulphate of calcium, if either the percentages of calcium or magnesium and carbonic acid are given, the percentages of the two carbonates may be calculated: The percentage of calcium oxide (CaO) $\times$ 1.7857 = percentage of calcium carbonate ($CaCO_3$); the percentage of carbonic acid in the magnesium carbonate ($MgCO_3$) multiplied by 1.916 = the percentage of magnesium carbonate.

Example.—A sample of limestone contains 54.8 per cent

calcium oxide and 43.4 per cent carbonic acid; required, the percentages of calcium carbonate and magnesium carbonate.

Calculation.

$54.8 \times 1.7875 = 97.96$, per cent calcium carbonate.
$97.96 - 54.8 = 43.16$, carbonic acid in the calcium carbonate.
$43.4 - 43.16 = 0.24$, carbonic acid in the magnesium carbonate.
$0.24 \times 1.916 = 0.46$, the per cent magnesium carbonate.

Many sugar-house chemists calculate the carbonates in this way in order to economize time. In many cases this method will supply all the information necessary relative to the purity of the stone, but it is not usually advisable to depend entirely upon it. A serious objection to this process is the fact that there may be slight errors in the determinations of the calcium and carbonic acid which would lead to false deductions. It is advisable, as a rule, to determine both the bases and the acid.

139. Determination of Sulphuric Acid.—The limestone may contain small quantities of sulphate of calcium, which is calculated from the percentage of sulphuric acid. Digest 5 grams or more of the powdered limestone with hydrochloric acid, using heat. Dilute the solution, filter, and wash the residue thoroughly with hot water. Evaporate the filtrate to a very small bulk, to remove the greater part of the acid. Precipitate the sulphuric acid with barium chloride, as described in the analysis of coke (**149**). The weight of barium sulphate $(BaSO_4) \times .34271 \div$ weight of limstone used $\times$ 100 = percentage of sulphuric anhydride (SO_3); the weight of barium sulphate $\times .5828 \div$ weight of limestone used $\times$ 100 = percentage of calcium sulphate.

140. Notes on the Analysis of Limestone.— It may be necessary in some of the determinations to use a larger portion of the stone than 1 gram. If so, it is more convenient to use a multiple of 1 gram, and dissolve and dilute to a definite volume 5 grams to 500 cc., for example, and use measured portions of this solution for the determinations.

A Gooch crucible will often be found much more convenient for the filtrations and ignitions than filter-paper and an ordinary crucible.

In the methods of analysis, only those determinations are given which are necessary in judging a limestone for sugar-house purposes. A number of analyses of limestones is given on page 213, with remarks on the values of the stones for use in sugar manufacture.

Sundstrom [1] has suggested a method for the rapid analysis of a limestone, an abstract of which follows:

(A) Weigh two portions of 1 gram each of the finely-powdered sample, transfer to a small dish and add about 100 cc. of distilled water to each. To one portion add 25 cc. of normal hydrochloric acid (**197**), cover the dish with a watch-glass until all action ceases ; heat to boiling, cool, and titrate with normal sodium hydrate (**201**), using methyl orange as an indicator. The number of cc.'s of normal hydrochloric acid — the number of cc.'s of normal soda solution = cc.'s of normal hydrochloric acid required to saturate the carbonates of lime and magnesia.

(B) To the second portion of 1 gram cautiously add 5 cc. of concentrated hydrochloric acid, keeping the dish covered to avoid loss. After all effervescence ceases, evaporate the material to complete dryness over a low flame. When dry, cool, take up with a little hot water and a few drops of hydrochloric acid ; heat to boiling, filter through an ashless filter, washing all insoluble portions into the filter, and wash free of all traces of chlorides with boiling water.

(C) Dry the filter and contents; ignite in a platinum crucible to bright redness, cool under a desiccator and weigh for silica. (SiO_2).

(D) Neutralize the filtrate and washings from (B) with ammonium hydrate, in slight excess ; heat to boiling, collect the precipitate and wash free of chlorides. Dry and ignite the filter and contents; cool and weigh for oxides of iron and aluminum (Fe_2O_3 and Al_2O_3).

(E) Heat the filtrate and washings from (D) to boiling, add a concentrated solution of oxalate of ammonium, also heated to boiling. Allow to stand until clear, which, if the analysis have been rightly conducted, requires two or three minutes; decant the clear solution into a filter, dissolve the precipitate in hydrochloric acid and reprecipitate with ammonium hydrate. Allow to settle and decant as before, and then wash the whole precipitate into the filter and wash with hot water until free of chlorides and oxalates. Dry the filter and contents, ignite in a platinum crucible, at first cautiously, then over a blast-lamp, until the residue is converted into calcium oxide (CaO); cool under a desiccator, weigh and calculate the weight of calcium carbonate (CaCO$_3$). Titrate the residue with normal hydrochloric acid as a check.

Divide the percentage of calcium carbonate by 5 (= cc. of normal hydrochloric acid required for calcium carbonate), subtract the quotient from the number of cc. of normal hydrochloric acid required for (A), and multiply the remainder by 4.2 to obtain the percentage of MgCO$_3$.

Sundstrom states that this method is very rapid and sufficiently accurate for practical purposes.

ANALYSIS OF LIME.

141. Determination of the Calcium Oxide (Lime).—Add sufficient water (30 cc. *ca.*) to 10 grams of lime, in a mortar, to form a thick milk. Add an excess of pure sucrose in the form of a solution of 35–40° Brix and mix intimately with the lime which dissolves, a soluble saccharate being formed. Transfer the solution and residue to a 100-cc. flask, using a sugar solution of the above composition to wash the last portions from the mortar and to complete the volume to 100 cc.; mix and filter. Titrate 10 cc. of the filtrate with a normal solution of hydrochloric acid (**197**), using phenolphthalein or lacmoïd as an indicator. The burette reading × .028 = the weight of calcium oxide (CaO) in 1 gram of the lime, and × 100 = percentage of calcium oxide.

142. Determination of the Proportion of Unburned and Slaked Lime.—Slake 1 gram of lime with water, add an excess of normal sulphuric acid (**178**) and heat to expel carbonic acid if present; add a few drops of cochineal solution (**215**) or other suitable indicator, and ascertain the excess of sulphuric acid used, by titration with normal sodium hydrate (**180**). Calculation: (cc. of normal sulphuric acid − cc. of normal soda solution) × .028 = the total weight of calcium, as calcium oxide, in 1 gram of the lime, and × 100 = the percentage of total calcium as calcium oxide. This number − percentage of calcium oxide (**141**) = percentage of unburned and slaked lime as calcium oxide.

143. Determination of Calcium Oxide, etc. Degener-Lunge Method.—Both the above determinations may be made with one titration, using phenacetoline as suggested by Degener and applied by Lunge.

Slake a weighed portion of the lime with water, add a few

drops of phenacetoline solution and titrate with normal hydrochloric acid. Add the acid until the yellow color changes to a red, and read the burette. This reading multiplied by .028 gives the weight of calcium oxide. Continue the addition of the acid; the solution remains of a red color until all the calcium is saturated, then changes to a golden yellow. It is advisable to make this titration a few times for practice with material of known composition. The burette reading multiplied by .028 gives the total weight of calcium as calcium oxide. The unburned and slaked limes are determined by difference.

144. Complete Analysis.—The methods described for limestones, page 148, may be applied for a further analysis of the lime if required.

ANALYSIS OF SULPHUR.

145. Estimation of the Impurities.—Transfer 0.5 gram of the powdered sulphur to a flask provided with a well-fitted glass stopper. Add at one time an excess of saturated bromine-water and shake thoroughly. Water dissolves 2 to 3.25 per cent of bromine at ordinary temperature, and, as at least 15 parts bromine are required for 1 part of sulphur, it is advisable to use from 275 to 400 cc. of the bromine water to insure sufficient of the reagent for the oxidation of the sulphur to sulphuric acid. Boil the solution to expel the excess of bromine, collect the residue and wash with hot water; dry and weigh. A Gooch crucible is convenient for collecting the residue. The weight of the residue × 200 = percentage of impurities. The percentage of sulphur may be determined directly from the proportion of sulphuric acid in the filtrate (**149**), or, with sufficient accuracy for practical purposes, by subtracting the percentage of impurities from 100.

Commercial roll-sulphur is usually very pure. Its quality can generally be satisfactorily determined from its color and relative freedom from dust and small fragments.

ANALYSIS OF COKE.

146. Preparation of the Sample.—The sample should be obtained as with limestone (**130**), and be very finely powdered.

147. Determination of the Moisture.—Heat 2 to 3 grams of the powdered coke in a tared flat dish or a watch-glass in an oven at a temperature of 110° C. Three hours' heating is usually sufficient for drying the sample. Loss of weight ÷ weight of material used × 100 = per cent moisture. The author is of the opinion, though not based upon experiment, that more satisfactory results would be obtained in drying coke or coal in a vacuum-oven.

148. Determination of the Ash.—Place 2 grams of the finely-powdered coke in a flat platinum dish and heat in a muffle, first at a moderate temperature and finally at a high temperature. Cool and moisten the ash with strong alcohol (Muck's method), then repeat the heating in the muffle until all the carbon is burned off. The weight of the ash ÷ weight of material used × 100 = per cent ash.

149. Determination of the Sulphur.—Mix 1 gram of the powdered coke intimately with 1 gram of calcined magnesia and 0.5 gram anhydrous sodium carbonate. Heat over a lamp in an open platinum crucible, inclined so that only its lower half may be brought to a red heat. The ignition requires forty-five to sixty minutes; the mixture should be stirred with a platinum rod every five minutes. The process is complete when the ash is yellowish or brownish. Let the mixture become quite cold, mix intimately with the ash, by means of a rod, ¼ to 1 gram of ammonium nitrate, and heat to redness for five to ten minutes, the crucible being covered with its lid.[1] The sodium carbonate may be advantageously replaced by carbonate of

---

[1] Crooke's *Select Methods*, 3d ed., 588.

potassium.[1] Transfer the residue to a beaker and cover with distilled water. Detach adhering portions of the residue from the crucible with hot water, aided by a rod, and wash into the beaker. Heat to dissolve the sulphate formed, filter and wash the residue with hot water. Determine the sulphuric acid in the filtrate, as barium sulphate : Concentrate the filtrate in a beaker, to a volume of approximately 50 cc., if necessary acidulate with hydrochloric acid, heat to boiling and add a solution of barium chloride. Add the barium solution gradually, a few drops at a time, keeping the liquid at the boiling-point. Remove the beaker from the lamp after each addition, for the subsidence of the barium sulphate. Add a drop of the barium chloride solution and note whether a precipitate forms in the clear supernatant liquid. Continue the boiling and addition of the reagent until there is no further separation of the sulphate. Collect the precipitate in a tared Gooch crucible, wash with hot water, dry and heat to redness. The weight of the residue, barium sulphate ($BaSO_4$), $\times .13734$ = the weight of sulphur ; this weight $\div$ weight of coke used $\times$ 100 = per cent sulphur in the coke.

---

[1] *Chemiker-Zeitung*, 1892, 60.

LUBRICATING OILS.

150. Tests Applied to Lubricating Oils.—A few oil tests may be made in the sugar-house laboratory without expensive or special apparatus. Some of the methods given here, while not assuring the greatest accuracy, will generally answer for sugar-house purposes. The usual tests are the "cold test," viscosity, the acidity or alkalinity and the purity.

151. Cold Test.—Pour a portion of the oil, to the depth of approximately one and a half inches, into a test-tube one and three-eighths inches in diameter. Plunge the tube into a freezing mixture and stir with a thermometer until the paraffine begins to separate, or until the oil ceases to flow, on inclining the tube. Remove the tube from the mixture and hold it between the eye and the light and note the temperature at which the paraffine disappears. The oil must be stirred during the entire test. Repeat the test two or three times and record the mean of the two readings which agree best with one another as the temperature of the cold test. With very dark oils, and with certain other oils, the beginning of the separation of the paraffine cannot be noted with accuracy, hence the reading is made at the temperature at which the oil ceases to flow.

152. Viscosity Test.—The viscosity test is best made with a viscosimeter, such as described in **114**. The method of making tests with these instruments is sufficiently described in the sections cited. In the absence of a viscosimeter, a moderately accurate test may be made with a large pipette. The pipette should be inclosed in a water-jacket so that the oil may be heated to 15.5° C., or 100° C., as its nature requires. The pipette is standardized with pure rape-oil or other oil that may easily be obtained in a state of great purity. The time, in seconds, required for the flow of 50 cc. of the rape-oil is noted by means of a stop-watch. The pipette is then filled with the sample to be tested and its flow noted under the same conditions as before. According to Redwood, the average time required for the flow of 50 cc. of rape-oil, with his viscosimeter, is 535 seconds at 60° F., and the viscosity of the oil under examination in terms of the viscosity of rape-oil is calculated as follows : Multiply the number of seconds required for the flow of 50 cc. of the oil by 100 and divide the product by 535 (seconds required for the flow of 50 cc. of rape-oil at 60° F.) ; multiply this quotient by the specific gravity of the oil under examination, at the temperature of the experiment, and divide by .915, the specific gravity of rape-oil at 60° F.

It is very difficult to graduate the orifice of a pipette to give the desired flow. For houses of large size using considerable quantities of oil, it is desirable to provide a viscosimeter (Figs. 52 and 53). The viscosity test is the most important in judging the suitability of the oil for the required purpose.

153. Tests for Acidity and Alkalinity.—Shake a portion of the oil with hot distilled water in a test-tube. After the oil and water separate on standing, test the latter for acidity and alkalinity. It should be neutral to test-paper. Oils are usually treated with sulphuric acid followed by washing with water and caustic soda. The acid especially should be completely removed, otherwise the bearings of the machinery may be injured.

154. Purity Tests.—Boil a portion of the oil with distilled water, and, after allowing the two substances to separate, examine the latter, which should remain clear and transparent.

In testing a mineral oil for admixture with animal or vegetable fats and oils, proceed as follows by the saponification method : Transfer a weighed portion of the oil (*e.g.*, 2 grams) to a pressure-bottle, and heat it in a water- or steam-bath with 25 cc. of alcoholic potash solution. This solution is prepared by dissolving 40 grams of good caustic potash in one litre of 95 per cent alcohol. The solution must be filtered if not perfectly clear. The flasks used in the

Kjeldahl nitrogen determination are suitable for pressure-bottles. The stopper of the bottle must be tied down with strong twine. Continue the heating about one hour, revolving the flask from time to time to mix its contents. A parallel experiment should be made in blank, with the reagent only. Cool the bottles to the room temperature and titrate the contents with half-normal hydrochloric acid (197), using phenolphthalein as an indicator. In the absence of animal and vegetable fats and oils, the results of the two titrations should be the same. Should a saponifiable oil be present as indicated by the titration, remove the alcohol by distillation, transfer the residue to a separatory funnel, and extract several times with ether to remove the mineral oil ; evaporate the ether solution and weigh the residue. The saponifiable oil, *i.e.*, animal or vegetable, is determined by difference.

The saponification test may also be conducted, as described above, in a closed flask, but without alcohol. Pour 2 cc. of a solution, containing 100 grams of the pure potassium hydroxide in 58 grams of hot distilled water upon 2 grams of the oil; heat one hour as before; cool, and transfer the contents of the flask to a separatory funnel and extract the mineral oil with ether ; evaporate the ether extract and weigh the residue, consisting of the mineral oil. Should the residue weigh less than 2 grams saponifiable bodies are present.

ANALYSIS AND PURIFICATION OF THE WATER USED IN SUGAR MANUFACTURE.

155. Characteristics of Suitable Water.—The condensation-waters from the multiple effects, vacuum-pans, etc., form an abundant and very satisfactory supply of water for the boilers.

The water for the diffusion-battery should be as pure as possible and should contain a minimum amount of calcium and magnesium salts and of the salts mentioned below as melassigenic. The calcium and magnesium salts, notably the bicarbonates and the sulphate of calcium, foul the heating surfaces of the battery and evaporating apparatus. The bicarbonates decompose to some extent in the diffusers and deposit the normal carbonates upon the cossettes and probably influence the diffusion unfavorably. The water should not contain more than 10 parts per 100,000 of calcium sulphate, otherwise incrustations may form at some stage of the concentration of the liquors.

Pure water should also be used in slaking the lime, though for economy of sugar and in the evaporation certain wash-waters containing sugar, etc., are used for this purpose.

The most important melassigenic salts are sulphates, alkaline carbonates, and nitrates. The chlorides are rather indifferent as regards the formation of molasses.

156. Analysis.—*Collection of Samples.*—When practicable, samples should be collected in large glass-stoppered bottles. The bottles should be thoroughly washed and finally rinsed with the water to be examined. It is advisable to use new bottles for this purpose. When ordinary corks must be used they should be new and thoroughly washed with the water. From two quarts to one gallon of the water will usually be a sufficient quantity of the analyses.

Total Solids.—If the water contain a small quantity of suspended matter, it should be filtered.   Evaporate 100 cc. of the water to dryness in a tared platinum dish over a steam- or water-bath which should have porcelain rings. The residue should be dried to constant weight in an oven at 100° C.   The weight of the residue in milligrams corresponds to the parts of total solids per 100,000 parts of water.   Test the residue for nitrates as follows: Place a drop of a solution of brucia in concentrated sulphuric acid on a white porcelain surface and add a fragment of the residue.   In the presence of nitrates a deep-red color appears, which soon changes to reddish yellow.   Two drops of aniline sulphate solution, with one drop of concentrated sulphuric acid, give a rose-red to a brown-red color on a porcelain plate, with nitrates.   If nitrates be present, the proportion may be estimated with moderate accuracy by the following method:

Nitrogen of Nitrates.—Mix 25 cc. of the water in a small Erlenmeyer flask with 50 cc. pure concentrated sulphuric acid.   Titrate immediately with a solution of indigo prepared as described farther on.   The indigo solution should be added until the color changes to a bluish green.   The flask must be shaken during the entire titration.   Repeat the operation with a fresh portion of the water and acid, adding at one time the nearly full volume of the indigo solution that was required to produce the green color in the preliminary titration; continue the addition of the indigo in small portions until the bluish-green color is produced. The flask must be shaken as in the preliminary titration. The indigo solution is prepared by dissolving 1 part of powdered indigo in 6 parts of pure concentrated sulphuric acid, heating on the water-bath, if necessary, to promote solution.   Add 240 cc. of distilled water to this solution, cool and titrate, as above, against distilled water having a known content of nitric acid.   Dilute the indigo solution so that 6 to 8 cc. correspond to 0.001 gram nitric anhydride (N_2O_3).   Should the water contain more than 0.003 to 0.004 gram of N_2O_5 in 25 cc. as indicated by the preliminary titration, it should be diluted to approximately this content before the final titration.

It requires a great deal of practice for accurate work by this method, and in the presence of much organic matter the results are too low.

Chlorine.—If much chlorine be present, as indicated by a considerable precipitate, on the addition of nitrate of silver solution to the water, in the presence of nitric acid, proceed as follows: Concentrate a convenient volume of the water, *e.g.*, 100 cc. to a small volume, and add 2 cc. of pure concentrated nitric acid and a solution of nitrate of silver in slight excess. Heat to the boiling-point and maintain this temperature a short time, avoiding violent ebullition. Stir during the heating to collect the precipitate, chloride of silver, in a granular form. Wash the chloride, by decantation with 200 cc. hot water containing 8 cc. of concentrated nitric acid and 2 cc. of a 1 per cent nitrate of silver solution. Pass the decanted solutions through a tared Gooch filter.[1] Use small portions of the washing solution at a time and break up the lumps of silver chloride with a glass rod. A filter-pump is used in making the filtration. The arrangement shown in Fig. 49 is a convenient one for this purpose. On the completion of this washing remove the filtrate and filter it a second time through the Gooch filter, rinsing the vessel with cold water. Wash the precipitate by decantation as before, except using about 100 cc. cold water, and finally transfer it to the filter and wash with 100 cc. cold water. After washing, pass a few cubic centimetres of strong alcohol through the precipitate and dry it at a temperature between 140° and 150° C. for 30 minutes; cool and weigh. The weight of the silver chloride × .24726 = the weight of chlorine in the quantity of water used.

Hardness.—The hardness is determined by Clark's soap method and is expressed in terms of the volume of a standard soap solution required to form a permanent lather with a given volume of the water. The soap solution is prepared as indicated in **186**.

Measure 50 cc. of the water into a glass-stoppered flask

---

[1] A platinum crucible, with perforated bottom, which supports a filtering film of asbestos.

of 250 cc. capacity. Shake thoroughly, then remove any carbonic acid that may be given off by suction with a glass tube. Add a small quantity of the soap solution, not exceeding 1 cc., and shake vigorously. Repeat the additions of soap solution and the shaking until the foam remains unbroken over the entire surface of the liquid during five minutes. As in ordinary titrations, the quantity of the standard solution added must be gradually decreased until, at the last, but a drop or two are added at a time. Should the quantity of soap solution used exceed 16 cc., less water should be taken and diluted to 50 cc., with cold, recently boiled distilled water. The calculation is made with the aid of the following table:

TABLE FOR THE CIRCULATION OF HARDNESS OF WATER.
(SUTTON.)

(Parts per 100,000, using 50 cc. of water.)

Cc. Soap Solution.	$CaCO_3$ per 100,000.	Cc. Soap Solution.	$CaCO_3$ per 100,000.	Cc. Soap Solution.	$CaCO_3$ per 100,000.	Cc. Soap Solution.	$CaCO_3$ per 100,000.	Cc. Soap Solution.	$CaCO_3$ per 100,000.	Cc. Soap Solution	$CaCO_3$ per 100,000.
.7	.00	.3	3.64	.9	7.29	.5	11.05	.1	15.00	.7	19.13
.8	.16	.4	3.77	6.0	7.43	.6	11.20	.2	15.16	.8	19.29
.9	.32	.5	3.90	.1	7.57	.7	11.35	.3	15.32	.9	19.44
1.0	.48	.6	4.03	.2	7.71	.8	11.50	.4	15.48	14.0	19.60
.1	.63	.7	4.16	.3	7.86	.9	11.65	.5	15.63	.1	19.76
.2	.79	.8	4.29	.4	8.00	9.0	11.80	.6	15.79	.2	19.92
.3	.95	.9	4.43	.5	8.14	.1	11.95	.7	15.95	.3	20.08
.4	1.11	4.0	4.57	.6	8.29	.2	12.11	.8	16.11	.4	20.24
.5	1.27	.1	4.71	.7	8.43	.3	12.26	.9	16.27	.5	20.40
.6	1.43	.2	4.86	.8	8.57	.4	12.41	12.0	16.43	.6	20.56
.7	1.56	.3	5.00	.9	8.71	.5	12.56	.1	16.59	.7	20.71
.8	1.69	.4	5.14	7.0	8.86	.6	12.71	.2	16.75	.8	20.87
.9	1.82	.5	5.29	.1	9.00	.7	12.86	.3	16.90	.9	21.03
2.0	1.95	.6	5.43	.2	9.14	.8	13.01	.4	17.06	15.0	21.19
.1	2.08	.7	5.57	.3	9.29	.9	13.16	.5	17.22	.1	21.35
.2	2.21	.8	5.71	.4	9.43	10.0	13.31	.6	17.38	.2	21.51
.3	2.34	.9	5.86	.5	9.57	.1	13.46	.7	17.54	.3	21.68
.4	2.47	5.0	6.00	.6	9.71	.2	13.61	.8	17.70	4	21.85
.5	2.60	.1	6.14	.7	9.86	.3	13.76	.9	17.86	.5	22.02
.6	2.73	.2	6.29	.8	10.00	.4	13.91	13.0	18.02	.6	22.18
.7	2.86	.3	6.43	.9	10.15	.5	14.06	.1	18.17	.7	22.35
.8	2.99	.4	6.57	8.0	10.30	.6	14.21	.2	18.33	.8	22.52
.9	3.12	.5	6.71	.1	10.45	.7	14.37	.3	18.49	.9	22.69
3.0	3.25	.6	6.86	.2	10.60	.8	14.52	.4	18.65	16.0	22.86
.1	3.38	.7	7.00	.3	10.75	.9	14.68	.5	18.81		
.2	3.51	.8	7.14	.4	10.90	11.0	14.84	.6	18.97		

The permanent and temporary hardness of waters may be determined by the French modification of the Clarke soap method, as described on page 100. To calculate the hardness in parts per 100,000, as calcium carbonate ($CaCO_3$), multiply the "degrees" of the special burette by 1.03 since 1° corresponds to .0103 part calcium carbonate per 1000 cc. of water.

The presence of magnesia is indicated by the formation of a peculiar curd, and also a lather which disappears on further addition of soap solution.

Permanent Hardness.—Boil gently for thirty minutes a weighed quantity of water in an Erlenmeyer flask. Cool, and add sufficient recently boiled distilled water to compensate for the evaporation. Filter off a portion of this water, and determine the hardness as before.

Calcium, Magnesia, Iron, Silica, Sulphuric Acid, etc.— Evaporate a large measured volume of the water to dryness and determine these constituents in the residue as indicated in the methods for the analysis of limestone.

Notes on Water-analysis.—The results of water-analyses are usually stated in terms of grains per U. S. gallon, parts per 100,000, or parts per 1,000,000.

157. Purification of Water.—To water containing the bicarbonates of calcium and magnesium, add milk of lime in slight excess (Clark). The normal carbonates are formed and precipitated, and may be removed by sedimentation or filter-pressing. If the water be exposed to the air, the excess of lime is quickly precipitated by the carbonic acid. Lime-water in slight excess may be used instead of the milk of lime. The following is the equation:

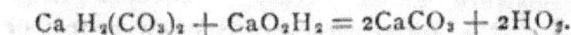

$$Ca\ H_2(CO_3)_2 + CaO_2H_2 = 2CaCO_3 + 2HO_2.$$

The magnesium bicarbonate is decomposed with the production of the hydroxide.

Water containing sulphate of calcium (gypsum) may be improved by treatment with sodium carbonate in slight excess according to the following equation:

$$CaSO_4 + Na_2CO_3 = CaCO_3 + Na_2SO_4.$$

The carbonate of lime may be removed by filtration through a press or by subsidence. In this case a substance which is melassigenic is substituted for one which fouls the heating-surfaces.

Waters containing bicarbonates of calcium and magnesium, and the chlorides and sulphates of these bases, may be improved by the addition of milk of lime or lime-water and caustic soda. Equations:

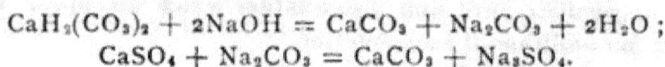

$$CaH_2(CO_3)_2 + 2NaOH = CaCO_3 + Na_2CO_3 + 2H_2O ;$$
$$CaSO_4 + Na_2CO_3 = CaCO_3 + Na_2SO_4.$$

Other calcium and magnesium compounds are decomposed by this treatment with the formation of similar compounds. Silica is precipitated by this process.

Water may usually be improved, especially if it contain organic impurities, by the addition of traces of alum or of chloride of iron and filtration through sand, or coke and sand, as in the Hyatt process.

The economical treatment of the waste waters from the sugar-houses, especially if these waters must be returned to a very small stream, presents many difficulties. There is a tendency on the part of public officials to require the purification of these waters. In some locations where the water-supply is deficient it is an object to purify the waste water for use in the factory.

The water from the condensers of the multiple-effect and vacuum-pans may be sufficiently cooled and purified for use again by means of a "cooling-tower." A tower such as usually is constructed in Cuba and in the beet countries, consists of a framework several stories in height. The framework, at each story, is covered with willow branches. The entire structure is often 30 feet or more in height. The water is pumped to the top of the tower, and then drips from floor to floor, and is finally collected in a pond. This treatment lowers the temperature of the water and improves its quality by oxidation of many of the impurities.

The simplest disposition of the waste waters from the beet-washers, diffusion-battery, and pulp-presses is their use for irrigation. They should be distributed over very large areas.   The organic matters are oxidized, and such water as finally reaches the streams, through drainage, is sufficiently pure.

Where very large areas are available for settling and decantation, the waste water may be improved by treatment with lime or with lime and an iron salt.   The sediment is removed from time to time and distributed over the fields.

SEED-SELECTION.

CHEMICAL METHODS AND APPARATUS.

158. General Remarks.—It is not in the province of this book to deal with the methods of seed-selection, except with a view to the chemical manipulations involved. It is nevertheless necessary that some of the principles upon which such selection is based should be mentioned, that the chemist may have a clear understanding of the purposes of his work and go about it intelligently.

As is true with many plants, a beet may be produced having certain features which persist through many generations and which may be said to have been inherited from its ancestors. The continued selection of individuals, grown under normal conditions and having certain peculiarities, generation after generation, tends to fix these distinctive features, and a "race," as it may be termed, is developed in which the majority of the individuals inherit the race-characteristics. Since the only reason for the improvement of the beet is a commercial one, it is essential that the plants for the production of the seed be grown under commercial conditions, that their progeny may have the same qualities, in field-culture.

Since the good qualities of a race may be developed by selection, one naturally assumes that the opposite qualities may be developed; hence in the selection of "beet-mothers" those roots are chosen in which the valuable features are predominant.

Some varieties have a tendency to run to seed, to produce fibrous roots, to deteriorate early in the silos, or have other bad qualities ; others have a tendency to the production of roots containing a rich, pure juice and a satisfactory tonnage per acre. Experience has demonstrated that these tendencies may be fixed, to a great extent, by a rigid system of selection, and that seed may be grown that will produce roots true to the characteristics of the parent.

It is desirable for the beet to have a certain shape, that it may be easily " lifted " in harvesting ; it should be as free as possible from side roots, since such beets are usually deficient in sugar and are difficult to free from adhering earth in the washers ; the root should be firm, and dense, and contain a high percentage of sugar in a juice of great purity. Relative productiveness and keeping qualities are also considered in the selection of beets. •

It should not be assumed that all seed-growers adopt the same methods of selection. Some select a rather large beet, others, a small one. On some farms the beets are planted very closely together, on others they are well separated to give the plant plenty of light and air.

The following is a brief outline of the methods adopted by the majority of seed-growers : The beets, when ripe, are carefully "lifted" from the soil ; those roots which show imperfections in their development and are too large or too small, are thrown aside and sent to the sugar-factory ; those of satisfactory shape and size are placed in piles and covered with earth to protect them from frost, until the time for siloing them. In the early spring, the beets are taken from the silos preparatory to removal to the laboratory for analysis.

The roots are again sorted, and those which have kept imperfectly, or for other reasons are not suitable, are rejected. The sound beets which fill the necessary physical requirements are now taken to the laboratory and a small cylinder or a portion of pulp is removed from each for analysis. Those beets which contain the desired percentage of sugar are stored for planting at the proper season, and the others are discarded.

In general, the percentage of sugar in the beet is inversely proportional to the size of the root. The small beets of regular shape are usually rich in sugar. Typical sugar-beets are shown in Figs. 57 and 58. Vilmorin's white improved (Fig. 57) and the Kleinwanzlebener (Fig. 58) are favorite varieties abroad and in this country.

FIG. 57.

FIG. 58.

159. Distribution of the Sugar in the Beet.

—The sugar is not uniformly distributed throughout the beet. It varies materially in different parts of the root as is shown in the diagrams, Figs. 59, 60, and 61, after Slassky.[1]

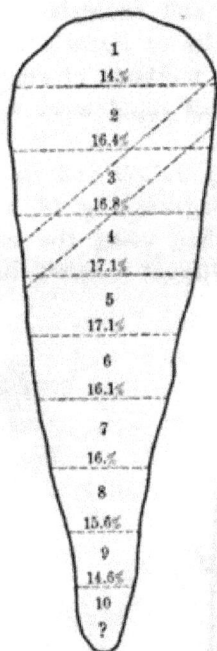 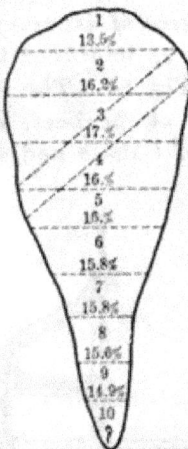 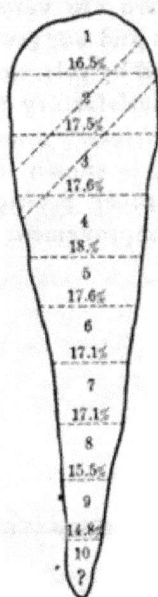

FIG. 59.   FIG. 60.   FIG. 61.

In view of this unequal distribution, care is necessary in removing the sample that the analyses may be comparable with one another.

160. Methods of Removing the Sample for Analysis.

—The dotted lines, inclined from right to left, in Figs. 59, 60, and 61, indicate the usual direction taken by the sound in removing the sample. Slassky[2] punctured a number of beets and analyzed the cylinders, and afterwards the entire beet, with the following results: Sucrose

---

[1] *Zapiski*, 1893, 12; abstract in *Bulletin de l'Association des Chemistes de France*, 12, 277.

[2] *Op. et loc. cit., supra.*

in the juice of the entire beet, 16.4 per cent ; sucrose in the juice from the cylinders, 16.76 per cent, a difference of 0.36 per cent.   It is not so important in seed-selection that the analysis represent the mean sucrose content of the beet as that the analyses shall be comparable with one another.

The present methods of sampling and analysis in seed-selection are very largely the results of Pellet's experiments and suggestions.   The direct methods of analysis, devised by this eminent expert, permit rapid work with a very satisfactory degree of accuracy.

Lindeboom's Sound, as improved by Gallois and Dupont, Paris, is shown in Fig. 62.   The diagram, Fig. 18, shows the proper position of the beet, when using the sound. The improvement of Gallois and Dupont consists in im-

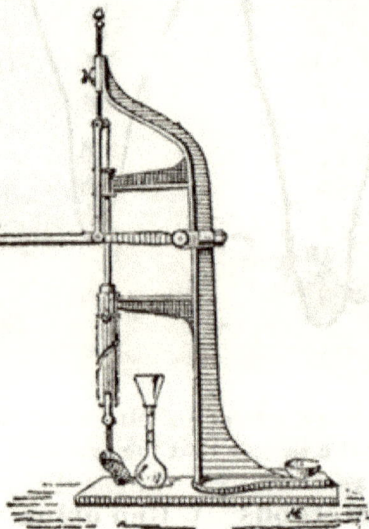

FIG. 62.

parting a rotary motion to the sound, insuring a clean cut, which is necessary to the further preservation of the "beet-mother."   The cylinder which remains slightly projecting from the beet is prepared for analysis by one of the methods described later.

Boring-rasp (*Keil & Dolle*).—This machine differs from

that shown in Fig. 19 only in the method of removing the pulp. The rasps of the two machines are interchangeable, thus making but one machine necessary for both classes of work.

The rasp, as shown in Fig. 63, is provided with a rod, carrying a disk which fits snugly inside the tool. The method of fastening the rasp to the body of the tool is the same in both machines.

FIG. 63.

The pulp passes through the opening shown in Fig. 20, and is held by the disk. The machine is stopped, the rasp unfastened, and the pulp withdrawn by means of the rod and disk. Except in very careful work, it is not necessary to wash the apparatus after each perforation. Each sample pushes any remaining portion of its predecessor against the disk, where it is usually well defined by slight differences in appearance. The rejection of about one-fifth of the cylinder of pulp insures the removal of all portions of the preceding sample.

161. Analysis of the Sample.—The pulp may be analyzed by any of the direct methods, or by the indirect method, *i.e.*, analysis of the expressed juice and calculation to terms of the weight of the beet. The cylinder may be reduced to a pulp in Hanriot's apparatus, or in the cylindro-divider (Fig. 36), and analyzed by the instantaneous-diffusion method. The cylinder may also be rasped and the juice expressed for analysis, if desired. The direct methods require so little time and labor, and so excel in accuracy, that it is advised that one of them be used, preferably the instantaneous-diffusion method.

In using Hanriot's apparatus (Fig. 64), one fourth the normal weight (6.512 grams or 4.075 grams) is cut from the cylinder and placed in the feed-tube of the apparatus ; the lever L is depressed, and the sample is forced against the rasp, which is driven at 2,000 revolutions, and is reduced to a fine pulp. The rubber bulb P contains about 80 cc. of water, with which the pulp is washed into the sugar-flask. The volume of this flask should be sufficient to allow for the volume of the marc, as explained in **64**, *i.e.*, for the one fourth normal weight (6.512 grams), 50.3 cc. The flask is removed, and the analysis is completed by Pellet's in-

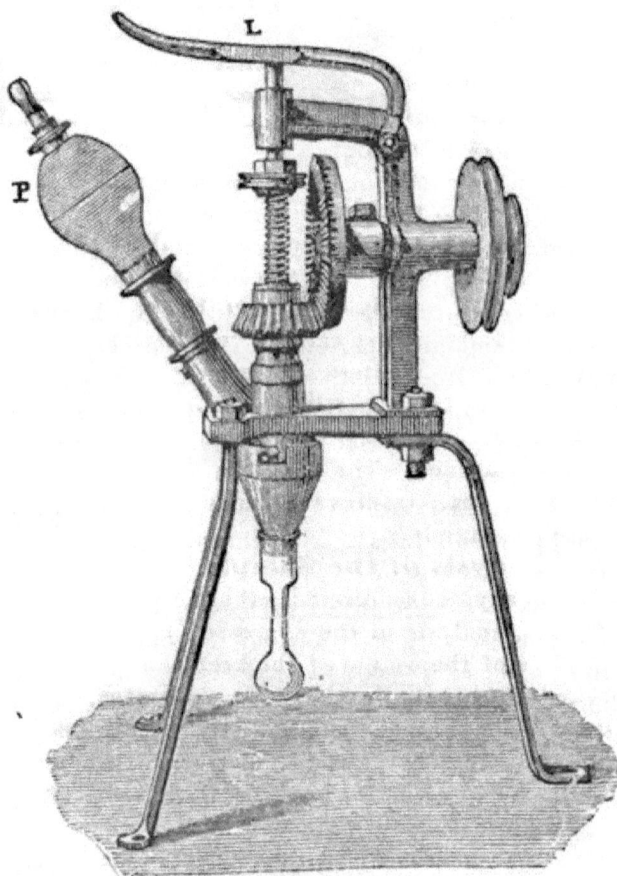

FIG. 64

stantaneous-diffusion method, **62**. A 400-mm. Pellet continuous tube should be used in making the polarization. The polariscope reading must be multiplied by 2 to obtain the per cent of sucrose in the beet. The apparatus is ready for a second polarization without further washing.

Pellet and Hanriot have further improved this method of analysis, as regards speed, by the use of a two-bladed knife for cutting the required weight of material from the cylinder. This knife has parallel adjustable blades, and removes the required weight with sufficient accuracy for seed-selection, thus doing away with the use of a balance.

The knife should be adjusted to cut a cylinder of the required weight from a beet of average density (1.038); a few trials will soon effect this adjustment. In an experiment made by Pellet, sixteen cylinders were cut; the difference between the extreme weights was 0.15 gram, corresponding to a difference of 0.35 per cent sucrose in a beet containing 15 per cent sucrose, and a mean error of 0.15 to 0.2 per cent. These results are sufficiently accurate for the purpose.

The cylinder is placed upon a grooved block when using the knife, and should not be washed or otherwise treated before placing it in the Hanriot apparatus.

Modification of Pellet's Direct Method.—Fr. Sachs[1] and

FIG. 65.

A. Le Docte of Brussels, acting upon the suggestions in the published experiments of a number of chemists, have very materially improved Pellet's diffusion method, as

---

[1] Paper read before *Congrès Internationale de Chimie Appliquée*, Paris, 1896.

follows : The normal weight (26.048 grams) of the finely rasped pulp is weighed directly in the planished copper capsule, shown in Fig. 65, the bottom of which is flat, and the corners rounded.

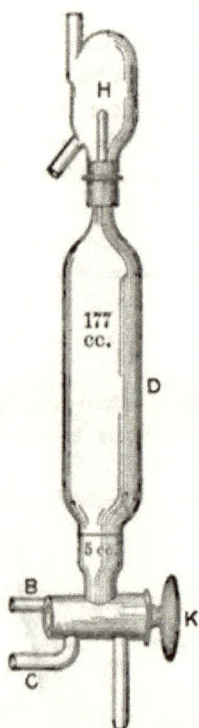

FIG. 66.

The capsule containing the pulp is held under the overflow pipette D (Fig. 66); a quarter-turn of the stop-cock K admits subacetate of lead solution from the reservoir through B to the pipette; when this reaches the 5-cc. mark, a further quarter-turn admits water through the tube C; the instant the water overflows at H a further quarter-turn is given the stop-cock, and the accurately measured contents of the pipette are discharged into the capsule. The capsule is then covered with a glass disk enclosed in a rubber cap, and is held as shown in Fig. 65, and agitated vigorously. The sugar is uniformly distributed throughout the solution within three minutes. The disk is smeared with vaseline previous to placing it upon the capsule. In removing the cover, it should be slipped to one side, not lifted, thus leaving it in readiness for another determination.

For the German polariscopes, the normal sugar-weight of which is 26.048 grams, the pipette should deliver 177 cc.; for the Laurent instrument it should deliver 171.4 cc., twice the normal weight of pulp being used for the determination. It is not advisable to deviate from these proportions of pulp and water, otherwise the diffusion is slow and the results may be uncertain.

Pellet advises acidulating the solutions with acetic acid, and in the opinion of the author this should not be omitted. Sachs' method would be somewhat further simplified by using only one liquid, viz., water containing 3 per cent by volume of subacetate of lead solution of 54.3° Brix.

An ordinary automatic or overflow pipette with a three-way cock can be used if the solution be prepared as indicated above. These pipettes deliver the specified volume of liquid with accuracy and rapidity. The Sachs-LeDocte modification reduces the possible errors in Pellet's method to a minimum, and permits extremely rapid work.

· **162. Pellet's Continuous Tube for Polarizations.**—One of the most important improvements that has been made in several years in polariscopic apparatus is Pellet's continuous observation-tube. This tube permits the rapid polarization of solutions without its removal from the instrument, each solution being displaced by that following it.

The first descriptions of this tube which came into the author's hands were very meagre, and a number of experiments were made before a satisfactory tube was constructed. These experiments were not made with a view to improving the construction of the tube as made for Pellet, but to construct a tube for immediate use. The displacement of the solutions was studied by means of colored liquids in a tube having brass heads and a glass body. The form shown in Fig. 67 was finally accepted as

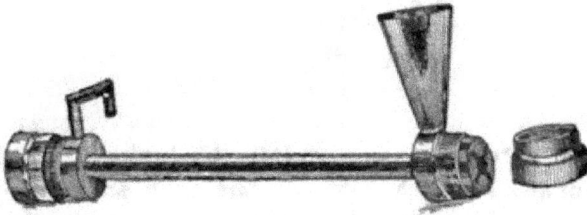

FIG. 67.

in every way satisfactory. The funnel delivers the solution into an annular canal, which connects by separate openings with each of the four grooves shown at the end of the tube. This arrangement insures the equal distribution of the displacing solution. At the opposite end the construction is the same, except that an outlet-tube carries off the waste solution.

In very rapid work, Pellet substitutes a siphon-tube for the funnel. While the chemist is making an observation

the assistant dips the inlet-tube into the fresh solution
ready for the next polarization. The opening of a pinch-
cock permits the old solution to flow out and the fresh one
to take its place. Owing to differences in the density of
the two solutions, striæ form where they come in contact
with one another, rendering it impossible to make a clear
observation so long as any of the old solution remains in
the tube. When the day's work is finished the last solution should be displaced with pure water, and the tube
should be left filled with water ready for the next day's
work.

There is usually some difficulty in filling the tube the first
time on account of air-bubbles. This difficulty may be
overcome by passing a strong current of water from the
hydrant through the tube, until the bubbles are removed.

The author has tested this apparatus with a great variety
of solutions, many varying but slightly in density from one
another, and has always obtained identical results with the
Pellet and ordinary tubes.

The Pellet tube permits extremely rapid work. An expert observer, using a good half-shadow polariscope, can
easily make 500 polarizations per hour. This number may
even be increased under favorable conditions. An assistant makes the entries in the note-book.

Pellet advises that the diameter of all continuous tubes
should be 5 millimetres, thus reducing the quantity of the
solution required for displacement to a small volume.

163. Polariscope with Enlarged Scale. —
Schmidt and Haensch make a polariscope especially for use
in seed-selection (Fig. 68). The instrument is of simple
construction, and is well adapted to the purpose. The
scale is graduated from 0 to 35 per cent.

The enlarged scale, by the same makers, shown in
Fig. 69, is exceedingly convenient for this class of work.
The percentages may be easily read to tenths at a distance
from the instrument, thus enabling an assistant to relieve
the observer of this portion of the work.

Hanriot has arranged a system of electric bells which are
rung by turning the milled screw of the polariscope. If the
sample be of a certain richness, a bell rings automatically.

164. Pellet's Estimate of the Laboratory Apparatus and Personnel Required for a Seedfarm.—The following estimate is based upon the analysis of 2500 to 3000 beet-mothers per day of 10 hours:

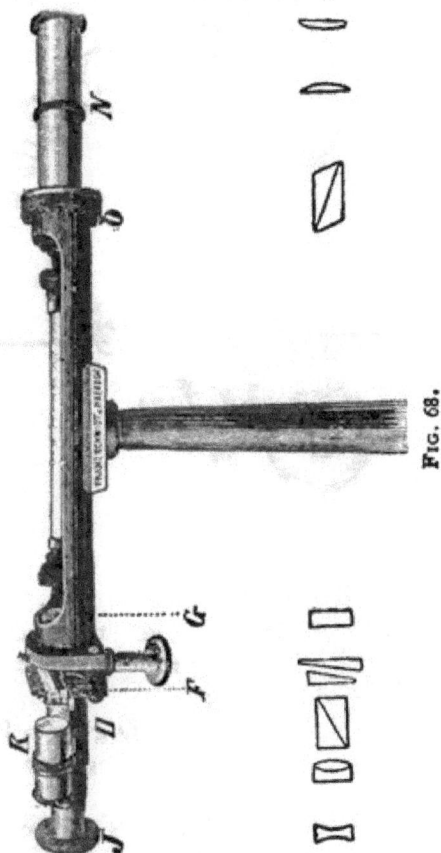

FIG. 68.

1 boring-rasp ;
1 motor (gas, electric, etc.);
1 polariscope ;
200 capsules, numbered, for pulp ;
4 balances ;
8 nickel weighing capsules ;
4 ½-normal weights ;
4 nickel funnels ;
4 wash-bottles or one large bottle with 4 tubes ;
500 50-cc.–55-cc. sugar-flasks ;

500 funnels ;

500 test-glasses for filtered solutions (see Fig. 70) ;

200 numbered clamps, for numbering sugar-flasks and test-
glasses ;

FIG. 69.

1 polariscope (half-shadow) ;

2 continuous polariscope-tubes, 400 mm. each ;

3 baskets divided into compartments for carrying solutions to the polariscope ;

6 dropping-glasses for ether ;

6 dropping-glasses for acetic
acid.

The Sachs-LeDocte apparatus
may be conveniently substituted
for the sugar-flasks (*see* page 181)
and the filtering arrangement

FIG. 70.

FIG. 71.

shown in Fig. 70, for the funnel-racks.   The numbered

clamps (Fig. 71) are made of copper or brass, and are transferred from the sugar-glass to the test-glass after filtration.

The personnel varies with the convenience of the laboratory arrangements. Exclusive of employés who sort and carry the beets to the laboratory, the following are usually necessary with the above equipment :

1 laborer at the rasp ;
1 assistant to arrange the samples in order ;
1 assistant to rasp the beets ;
1 laborer to distribute the capsules of pulp to the balances ;
4 weighers at the balances ;
4 assistants to transfer the pulp to the flasks ;
4 assistants to clarify the solutions and complete the volume to the mark on the flasks ;
2 assistants for filtrations ;
2 observers (polariscopic);
2 assistant observers ;
2 charwomen.

When using the Hanriot apparatus and Pellet's double-bladed knife, the number of laborers is much smaller for a given amount of work, and 3 balances, etc., may be dispensed with. The following is the personnel with this method for making 4000 to 5000 analyses per day:

1 laborer at the sound (Lindeboom) ;
1 "sounder";
1 cutter ;
2 laborers at the rasps;
2 laborers to carry the cylinders, cut from the beets, to the rasps ;
2 assistants to clarify, etc., the solutions ;
2 assistants for filtrations ;
2 observers ;
2 assistant observers ;
2 charwomen.

With this method, seventeen employés can accomplish nearly double the number of analyses that twenty-one can with the borer-rasp and balances.

165. Chemical Method for the Analysis of Beet-mothers.—The chemical method of analysis of beet-mothers is used in a number of laboratories. This application of the alkaline-copper method is due to Violette. The process is very simple, and is well adapted to seed-farms of moderate size which would not justify the outlay for the expensive apparatus described in the preceding pages.

Cut a small cylinder from the beet with a sound (Fig. 62), or on a small scale with a cork-borer ; remove the skin and rapidly cut the cylinder into small fragments. Transfer 5 grams of the fragments to a 100-cc. sugar-flask, add approximately 40 cc. water and 10 cc. normal sulphuric acid (**199**), mix, and heat upon a water-bath, at the boiling-point, cool, add 10 cc. of a normal solution of caustic soda (**201**), to neutralize the solution and complete the volume to 100 cc. The sucrose is inverted by the acid treatment and converted into invert-sugar. Determine the percentage of invert-sugar by Violette's method given in **73**. Multiply the percentage of invert-sugar by .95 to obtain the percentage of sucrose. In seed-selection, it is unnecessary to determine the sucrose with great accuracy, hence the analyst may be guided entirely by the disappearance of the blue color, instead of using a test solution to ascertain the end of the reaction.

In this work the burette readings will vary between 5 and 8 cc. If, for example, the beets are to be divided into three classes, viz., (1) those containing 15 per cent or more of sucrose, (2) those containing between 13 and 15 per cent, and (3) those containing less than 13 per cent, the following is a convenient method of procedure: Heat 10 cc. of Violette's solution (**195**) in a test-tube to boiling, add 6.3 cc. of the invert-sugar solution, and boil ; a complete disappearance of the blue color shows that the beet contains more than 15 per cent of sucrose; if the blue color persist, continue the addition to 7.3 cc. and boil ; a disappearance of the blue color shows that the beet contains 13 per cent sucrose, or more, and less than 15 per cent ; if the blue color persist, the percentage is below 13. The following table may be conveniently used in calculating the percentages to the nearest tenth :

Burette Reading.	Per Cent Sucrose.	Burette Reading.	Per Cent Sucrose.	Burette Reading.	Per Cent Sucrose.
5.0	19.0	6.0	15.8	7.0	13.6
.1	18.6	.1	15.6	.1	13.4
.2	18.3	.2	15.3	.2	13.2
.3	17.9	.3	15.1	.3	13.0
.4	17.6	.4	14.8	.4	12.8
.5	17.3	.5	14.6	.5	12.7
.6	17.0	.6	14.4	.6	12.5
.7	16.7	.7	14.2	.7	12.3
.8	16.4	.8	14.0	.8	12.2
.9	16.1	.9	13.7	.9	12.0

In the application of this method upon a large scale a number of labor-saving devices may be used with advantage: A large sand-bath or a hot plate may be used in making the inversions. The alkaline-copper solution, preferably Violette's modification (**195**), and the sulphuric-acid and soda, are most conveniently measured from an automatic pipette (Fig. 72). The pipette for the Violette reagent should be graduated with that solution, and not with water. This is necessary on account of the viscosity of the reagent. The solutions may be measured with great rapidity and accuracy with this pipette. Several burettes should be arranged in a rack over a corresponding rack holding the large test-tubes containing the copper solution. These latter are heated by an easily adjustable multiple-burner lamp, Lecq[1] uses a revolving rack having four arms, each carrying five burettes and five test-tubes. An arm of the rack is revolved to a position over the lamp, and the contents of the tubes are heated, the sugar solutions are added and heated to boiling, and then a second arm is brought into position. By the time a complete revolution is made, the sub-

FIG. 72.

oxide of copper in the first set of tubes will have settled, and the color of the supernatant liquid may be noted with ease.

Much labor may be economized by the use of a boring-rasp, Fig. 19, for removing the sample from the beet.

[1] Aimé Girard in *Journal des Fabricants de Sucre*, 1883, 10.

SEED-TESTING.

166. Beet-seed.—The "seed" of the beet, as it is commonly termed, is, properly speaking, the fruit of the plant, and is usually called the "seed-ball" by seedsmen. Each ball contains from one to five embryos. For brevity, the expression "ball" or "seed" will be used.

167. Sampling.—The seed should be sampled with a trier or sound, similar to that shown in Fig. 23, designed for use with sugars. The trier for seed-sampling should be provided with a cover, which may be revolved into position before removing the instrument from the sack of seed, and thus retain the entire sample. A quantity of seed should be drawn systematically from the lot, removing a portion from each sack, or from every second sack, etc., according to the amount of seed to be sampled.

The large sample should be thoroughly mixed, distributing the impurities through the seed as uniformly as possible. A convenient method of subsampling is that of Maercker, of the experiment station Halle/a/Saale, Germany, as follows: Cut a disk of cardboard to fit easily inside of a crystallizing-dish. The dish should have vertical walls and a flat bottom. Cut slots *A, A, A, A*, as shown in Fig. 73, in the cardboard, and place it on the bottom of the dish. Two wires should be attached to the disk, to lift it vertically from the dish without jarring. Cover the disk to a uniform depth with the sample of seed, then lift it by means of the wires; the required subsample will pass through the slots and remain in the dish. A few experiments will determine convenient shape and dimen-

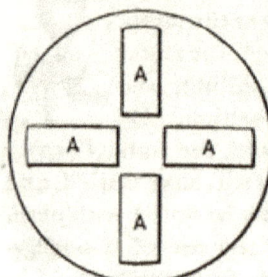

FIG. 73.

sions for the slots. Should an experiment require less seed

than it is convenient to remove in this way, further reduce the quantity by quartering.

168. Moisture.—Dry an entire subsample, removed as above described, and containing approximately 10 grams. The balls should be distributed evenly in a large flat dish and dried in an oven at 105° C. The loss in weight ÷ weight of the sample × 100 = percentage of moisture. Care must be observed in cooling and weighing the dry seed, since it quickly absorbs moisture from the air.

169. Proportion of Clean Seed.—It is difficult to determine the proportion of clean seed, largely through difficulty in distributing the impurities and in removing the foreign matters.

Remove approximately 10 grams of seed from the large sample, as described in **167**; weigh, and transfer to a sheet of paper. Hold each seed in a pair of forceps, brush carefully, and remove foreign matter. Weigh the clean seed; this weight divided by the gross weight and multiplied by 100 is the percentage of clean seed. This determination should be made in duplicate or triplicate, since it is difficult to obtain concordant results.

170. Number of Seeds per Pound or Kilogram.—Beet-seed is sold by the pound or ton in this country. It is, however, more convenient to make the calculations on a metric basis and afterwards reduce them to the customary weights.

In the determining the proportion of clean seed (**169**) time may be economized by counting the balls into the weighing-capsule. The number of seeds per 10 grams is then readily calculated to terms of a kilogram, and thence to pounds (**220**).

The seed should next be placed in a sieve of $\frac{1}{18}$ inch square mesh. The balls which pass this sieve are termed "small," and those which remain, "large." The number of large seeds and small seeds per kilogram and pound is calculated as before.

This is a purely arbitrary classification, and is an outgrowth of the various opinions of authorities relative to the value of large and small seeds.

It is generally conceded that the large, heavy balls are of

greater value than the small ones, so far as germinative power is concerned, and many investigators consider that the heavy seed-balls produce more thrifty plants and beets richer in sugar. It has also been observed that germinative ability varies directly as the size of the seed-balls. The larger the percentage of small or medium-sized balls, however, the larger the number of plants that will be produced, and the more evenly the rows will be filled.

As in other experiments with beet-seed, it is advisable to make the tests in duplicate or triplicate and report the mean of the results.

171. Germination Test.—There are two methods of making this test. viz.: (1) determination of the weight of the seed that germinates in a given weight; (2) determination of the number of seed-balls per 100 which germinate. In view of the wide variations in the size of the seed-balls, and the fact that beet-seed is bought by weight, in is the opinion of many authorities that the test should be by weight and not by count. Both methods have their advocates, and probably a large proportion of the tests is made by count. The simplicity of this method is in its favor.

The plan adopted by Vivien[1] is one of the simplest for testing by weight.

Sift 4 or 5 grams of the seed in a sieve of 5 mm. ($\frac{3}{16}$ inch *ca.*) mesh ; count the number of large and of small balls; weigh and calculate the mean weight. Soak the seed two days in a 5-per-cent solution of sodium nitrate.

Sift fine soil into a flat-bottomed dish of porous earthenware to a depth of approximately 1 cm. ($\frac{3}{8}$ inch *ca.*); place a piece of wire netting of approximately 1 cm. ($\frac{3}{8}$-inch *ca.*) mesh on the earth, and in each mesh place one large ball. Press the earth slightly, and cover with 2 to 4 mm. ($\frac{1}{16}$ to $\frac{1}{8}$ inch *ca.*) of soil. The small seed-balls are similarly planted, using a part of the same dish, and separating one lot from the other by suitable means. The dish is placed in a hothouse, or in a warm place in the laboratory,

---

[1] *Bulletin de l'Assoc. des Chimistes de France*, **12**, 13.

and is occasionally watered with a fine spray of rain or distilled water.

From day to day, as the plantlets appear, the seeds are removed from the soil and counted; it is also usual to count and record the number of embryos which show signs of vitality. A splinter of wood or a piece of a match is substituted for each seed removed.

The average weight of each size of ball is taken into account in calculating the percentage by weight of seed that germinates.

Example.

Per 100 *kilos.*

Large seed......... 22.29 kilos, corres. to 1,311,000 seeds.

Small seed.......... 73.67 kilos, corres. to 5,262,000 seeds.

Foreign matter..... 4.04 kilos.

Average weight of the large seeds........... 0.017 gram.

    Number per gram....................... 13.11

Average weight of the small seeds.......... 0.014 gram.

    Number per gram....................... 52.62

4 grams were used in the germination test, corresponding to 3.84 grams of clean seed:

    13.11 × 4 = 52 large seeds, weighing 0.89 gram.

    52.62 × 4 = 211 small seeds, weighing 2.95 grams.

                                     —————

                               3.84 grams.

At the end of the test 48 large and 182 small seeds had germinated—*i.e.*, 92.3 large seeds per 100 seeds, and 86.24 small seeds per 100 ; to reduce these numbers to percentages by weight (referring to the statement of the example), we have:

Per 100 *kilos.*

    22.29 × 92.3 = 20.57 kilos large seed germinated.

    73.67 × 86.24 = 63.53 kilos small seed germinated.

    (By difference) 11.83 kilos worthless seed.

        4.04 kilos foreign matter.

Substituting the word " pounds " for " kilos," we have the percentages in the customary weights of this country.

(2) In the second method 100 seeds are selected at random, and the number which germinate is counted. In

the above example it is easy to calculate that this method would give 87.65 per cent of good seed instead of 84.1 per cent, as determined by weight.

Authorities differ as to the advisability of soaking the seed prior to the test.

The question of the use of sand, soil, or other material, or of the necessity of sterilization of the culture-bed, is not discussed by Vivien in the paper cited.

In the methods adopted by the Association of American Agricultural Colleges and Experiment Stations[1] blue blotting-paper is used. In supplementary tests, sand which has been heated to destroy organic matter, and sterilized previous to use, is recommended. In sand tests, the sprouts which appear above the ground are counted. The sand and blotters should be kept well moistened with water, but not saturated, during the test. Only potable water of a temperature approximating that of the seed-beet should be used. The temperature should be kept at 20° C. eighteen hours out of each twenty-four, and should in no case fall below 15° C. or rise above 32° C. The seed should be kept in a dark place during the germination test.

Pieters,[2] of the Division of Botany, describes a convenient apparatus for testing seed on a small scale in a report published by the U. S. Department of Agriculture:

"Use a large dripping-pan or an ordinary frying-pan. Paint it to prevent rusting. Put four supports in the pan (inverted porous saucers are good), and place a tin or wire frame upon them, as shown in Fig. 74. The seeds are laid between folds of blotting-paper or cloth, which are then placed on the frame. A flap of paper or cloth hangs down into the water, which half fills the tray and keeps the folds moist.

"If glass can be had to put over the pan, evaporation will not be so rapid; otherwise the water will need replenishing frequently.

"The tin or wire tray need not be expensive, and can be

---

[1] Circular 34, Office of Experiment Stations, U. S. Department of Agriculture.

[2] Yearbook, 1896, p. 183.

replaced by anything the operator may have. It is only necessary that a flap should dip into the water to provide moisture.

"In testing seed some trouble will be experienced from the growth of mold. If the cloths and dishes are used many times this trouble will become worse, unless the spores

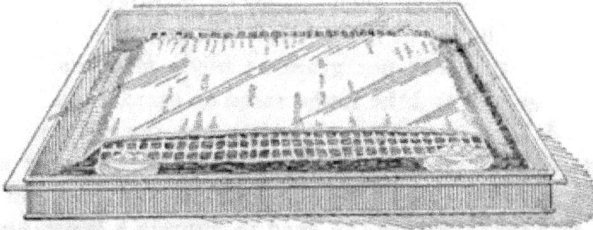

FIG. 74.

of the fungi are killed. This can easily be done by boiling all cloths and washing the dishes in boiling water after each test."

172. Characteristics of Good Seed.—The seed should be clean, containing as much as 95 per cent of clean balls. As much as 75 per cent by weight of the gross seed should germinate in the testing-apparatus within 15 days. As much as 85 per cent of extra good seed will germinate in this time.

The following are the German sugar-manufacturers' specifications relative to beet-seed :

A kilogram of seed should produce 70,000 sprouts; of this number, 46,000 should appear within six days. Seventy-five per cent of the seed (by count), at least, should germinate. Seed containing up to and including 14 per cent moisture may be considered normal; that containing 14 to 17 per cent may be accepted, but a deduction will be made for the excess of moisture above 14 per cent. Foreign matter to the extent of 3 per cent is admissible, and seed containing up to 5 per cent may be accepted, but a deduction will be made for the quantity in excess of 3 per cent. Seed not fulfilling all of these conditions may be rejected. Provision is made for check-analyses in the event of disagreement.

In Austria, seed is considered normal, as regards mois-
ture-content, which, after 24 hours' exposure in an open
flask at a temperature of 18° C. to air of 52 per cent relative humidity, contains 10 per cent of water.

It is usual to state that in good seed 50 to 80 embryos
per gram should show signs of vitality, and 80 to 110 embryos in extra-good seed. This is manifestly not a fair
test of the quality of the seed, since, for example, in a lot
containing 50 seeds per gram 25 seeds may contain more
than 100 germs and the lot be rated as extra good, whereas
the seed is poor. After thinning out the plants, but 25
beets would be produced from seed that this method would
rate high. There can be no question, however, but that
it is an advantage for the seed to contain a large number
of vital embryos, thus insuring greater certainty of having
the rows well filled.

MISCELLANEOUS NOTES.

173. Cobaltous Nitrate Test for Sucrose.[1]— To about 15 cc. of sugar solution add 5 cc. of a 5 per cent solution of cobaltous nitrate. After thoroughly mixing the two solutions, add 2 cc. of a 50-per-cent solution of sodium hydrate. Pure sucrose gives by this treatment an amethyst-violet color, which is permanent. Pure dextrose gives a turquoise-blue color which soon passes into a light green. When the two sugars are mixed, the coloration produced by sucrose is the predominant one, and one part sucrose in nine parts dextrose can be distinguished. If the sucrose be mixed with impurities, such as gum-arabic or dextrine, treat with alcohol or subacetate of lead before applying the test.

174. Test for Sucrose, Using α-Naphthol.—Mix the solution supposed to contain sucrose in a test-tube with 2 to 3 drops of an alcoholic solution of α-naphthol; then, by means of a pipette or other device, let concentrated sulphuric acid flow to the bottom of the tube without mixing with the solution. In the presence of sucrose a violet zone appears at the line of demarkation of the two liquids and gradually spreads. A solution containing 1 part of sucrose in 10,000,000 parts of water shows a pale lilac coloration. When more than 0.2 per cent sucrose is present the sugar is charred by the acid.[2] A similar method of making the test, and probably the original method, was described by H. Molisch.[3] Also the following: Thymol used instead of α-naphthol in the above test yields a deep-red coloration, which, on dilution with water, gives at first a fine carmine, then a carmine flocculent, precipitate.

175. Nitrous Oxide Set Free in Boiling

---

[1] Agricultural Analysis, H. W. Wiley, vol. iii. p. 189.

[2] Rapp and Besemfelder, *Deutsche Zucker.*, 1892, 538.

[3] *Monatsch Chem.*, 6, 198, abstract in *Journ. Chemical Society*, Abs. 50, 923.

Sugar.—Maumené called attention to the non-decomposition of nitrates in general when boiled with sugar, and to the exception that nitrate of ammonium is decomposed under these conditions, with the evolution of nitrous fumes. He observed this phenomenon in boiling sugar, in the vacuum-pan, and also that sugar is decomposed in the presence of nitrate of ammonium, or of other salts of ammonium with nitrates in general. Ewell made similar observations in evaporating sorghum-cane juice in the multiple-effect in a Kansas factory.

176. Relative to the Precipitate Formed on Heating Diffusion-juice.—The precipitate obtained by heating diffusion-juice contains 10 to 20 per cent of the weight of the dry matter of proteïds, also large quantities of pectous substances, fatty acids, oxalic acid, lime, magnesia, and occasionally phosphoric acid. It contains no optically active bodies, though they are probably originally present.[1]

177. Spontaneous Combustion of Molasses.—The feeding of cattle with a mixture of molasses and forage is extending in beet-sugar countries. The storage of the mixtures is attended with some risk of fire, as is indicated by the following : Two heaps of a mixture of 1 part molasses and 2 parts palm-oil cake were stored in a sugar-house. The heaps were several metres apart. After some time an odor similar to that of chicory was noticed, and upon investigation the material in both heaps was found to be carbonized. The temperature of the interior of the heaps was fully 120° C.[2]

Crawley, of the Hawaiian experiment station, states that a quantity of cane-molasses stored in a cistern in a cane-sugar house on one of the islands, boiled violently, and after twelve hours only a charred mass was left.[3]

178. Calorific Value of Molasses.—Three experiments were made with molasses, using a Mahler calorimeter, and the following results were obtained :[4]

[1] Herzfield, *Zeit. Rübenzucker-Ind.*, **43**, 1065.

[2] *Bulletin de l'Association des Chemistes de France*, **14**, 712

[3] *Journ. Am. Chem. Soc.* **19**, 238.

[4] Camille Martignon, *Bulletin de l'Association des Chemistes de France*, **14**, 366.

(1) Beet-molasses................ .... 3000 calories
(2) Cane-molasses................... 2675 "
(3) Cane-molasses from Louisiana .. 2646 "

A number of Cuban sugar-houses burn the molasses for the production of steam.

179. Fermentation.—*Ferment.*—Any substance capable of producing fermentation.

Vinous or Alcoholic Fermentation.—Liquid disturbed; rise in temperature and increase in volume; carbonic acid escapes, forming peculiar bubbles on the surface of the liquid. A temperature between 15° and 18° C. is favorable to this fermentation; between 18° and 30° the fermentation proceeds very rapidly; it is checked below 15° C., and ceases entirely below 12° C.

Acetic Fermentation. — The favorable temperatures are between 20° and 35° C. The liquid becomes turbid, and is filled with a ropy substance. Finally, the solution clears up and acetic acid is formed. Use lime to check this fermentation.

Putrid Fermentation. — This fermentation follows the acetic stage. The solution becomes turbid and viscous; ammonia is set free, and a sediment deposits. The fetid odor is repulsive.

Viscous Fermentation. — The solution becomes thick, slimy, ropy; and starchy matters and sugar are transformed into gummy substances. A mucilaginous appearance is characteristic. Small quantities of carbonic acid and hydrogen are liberated. Wash the tanks with a dilute sulphuric-acid solution to eliminate this ferment (5 per cent solution of 66° acid).

Lactic Fermentation.—This fermentation may exist in the presence of the viscous ferment. Odor acid, taste very disagreeable. This ferment is checked by acidity; hence, use sulphuric acid in washing the tanks.

Mucous Fermentation.—Sugar-beet juices are attacked by this ferment in the presence of nitrogenous bodies and the air. Mannite, gum, and carbonic acid are formed. The liquid becomes thick and ropy.

"*Frog-spawn.*"—Called "*frais de grenouilles*" by the French and *Froschlaischpilz* by the Germans. The juice assumes a jelly-like or gelatinous state; this is usually attributed to *Leuconastœ mesenteroides.* F. Glaser[1] has shown that another bacterium, which differs from the above in not flourishing in 10% neutral molasses, can produce this phenomenon. This organism is known as *Bacterium gelatinosum betæ,* and grows rapidly in sugar-beet juice gelatine. As stated, this organism does not thrive in 10% molasses, but if the slimy precipitate obtained by the addition of alcohol to the juice or its ash be added, development takes place. *Pari passu* with this gelatinous formation the sucrose is inverted and alcohol is produced. The gelatinous mass is similar to the beet gums.

This substance is insoluble in cold water, and with difficulty in cold acids, but almost completely soluble in hot acids and alkalis.

A Peculiar Fermentation of Beet-juice.—The juice occasionally becomes mucilaginous. This is due to a fermentation the products of which are dextrose, manitol, and nonvolatile organic acids. There is also an organic substance formed which is not precipitable by subacetate of lead.[2]

Soil-ferments as a Cause of the Formation of Gas in the Diffusers.—The evolution of gas in the diffusers is attributed by Neitzel to the action of soil-ferments upon the beet. The gases examined in an experiment, and drawn from the second and eighth diffusers, contained, respectively, 11.4 to 51.5% carbonic acid, 0 to 57.8% hydrogen, and 10.8 to 68% nitrogen. Free oxygen was rarely found.[3]

Fermentation of the Massecuites in the Hot-room.—According to Horsin-Déon, the "foaming" or "boiling up" of the massecuite in the hot-room tanks is due to a viscous fermentation in which the sugar is transformed into mannite, gums, carbonic acid, and water without the formation of glucose. The mannite combines with the organic acids and disappears, leaving gums which render the massecuite

---

[1] *Cent. Bl. Bacter.,* 1895, 2 abth., 1, 879; *Journ. Soc. Chem. Ind.,* 15, 200.

[2] *Anderlik. Zeit. Zücker.-Ind., Böhm.,* 18, 90

[3] Neitzel, *N. Zeit. Rübenzucker-Ind.,* 1895, 35, 22; *Journ. Soc. Chem. Ind.,* 14, 876.

viscous.[1] For further opinions relative to this matter *see*
page 215.

180. Melassigenic Salts.[2]—The following salts are
positive molasses-makers, *i.e.*, salts which promote the formation of molasses: carbonate, acetate, butyrate, and citrate of potassium.

The following have no influence on the formation of
molasses, and are classified as indifferent: sulphate, nitrate
and chloride of potassium, carbonate and chloride sodium,
calcium hydrate, valerianate, oxalate and succinate of
potassium, and oxalate, citrate, and aspartate of sodium.

The negative molasses-makers, *i.e.*, salts which promote
the crystallization of sucrose, are sulphate, nitrate, acetate,
butyrate, valerianate and succinate of sodium, sulphate
chloride and nitrate of magnesium, the chloride and nitrate
of calcium, and the aspartate of potassium.

The above classification is from the investigations of
Marschall.[2]

**181. The Chemical Composition of the Sugar-
beet.**—In addition to the carbohydrate bodies, chlorophyll
and water, the following organic substances have been
identified in the sugar-beet by various chemists :

Oxalic, formic, citric, malonic, succinic, aconitic, tricar-
ballicylic, oxycitric, malic, and tartaric acids in the juice.
The last eight were identified by von Lippmann. Other
acids formed through decompositions due to the manufacturing process are mentioned later.

The following nitrogenous bodies have been identified :
Betaïne (Scheibler), $C_5H_{11}NO_2$ or $C_5H_{13}NO_2$; Asparagine
(Scheibler), $C_4H_8N_2O_3$; glutamine (Schultze and Bosshard),
$C_5H_{10}N_2O_3$; leucine (von Lippmann), $C_6H_{13}NO_2$; legumine.
von Lippmann has identified the following nitrogenous
bodies in addition to those named. Tyrosine ($C_9H_{11}NO_3$);
Xanthine bodies,— viz.: Xanthine ($C_5H_4N_4O_2$), guanine
($C_5H_6N_5O$), hypoxanthine ($C_5H_4N_4O$), adenine ($C_5H_5N_5$),
and carnine ($C_7H_8N_4O_2$); the following decomposition-
products, which are the cause of ammonia so noticeable in

---

[1] Horsin-Déon, *Bulletin de l'Assoc. Chimistes de France*, **4**, 223.

[2] Marschall, *Z. it. Rübenzucker-Ind.*, **20**, 328, 619; **21**, 57.

beet-sugar manufacture and of the variations in the alkalinity of the juice, were also identified by von Lippmann: Arginine, guanidine, allantonine, vernine, vicine, and, in the young plant, alloxanthine.

Through the decomposition of some of the above substances, in the manufacture, the following have been identified in the molasses: Glutannic acid (Scheibler), $C_5H_9NO_4$, from glutamine, and aspartic acid, $C_4H_7NO_4$, from asparagine.

The following non-nitrogenous bodies are found in the beet: Lecithine, a " phosphorized fat;" pectose, an insoluble substance, which is converted into soluble pectine by pectase, a substance having the properties of a ferment, which is also present. Pectine in water solution is converted by heat into parapectine; acids convert it into metapectic acid, and alkalis into pectic and parapectic acids. Pectase converts pectine into pectosic acid. Of these bodies, pectine and parapectic acid are dextrorotatory, having a specific rotatory power 2.7 times that of sucrose. The pectose bodies are of great importance, both in sugar analysis and manufacture.

Pectic acid forms soluble amorphous bodies with alkalis and insoluble pectates with lime.

The cellular tissue frequently contains coniferine (E. de Lippmann); this is oxidized to vanilline, which has been found in molasses (Scheibler). Cholesterine has also been found in the molasses (Lippmann).

Among the mineral constituents of the beet are the following: Salts of potassium, sodium, rubidium, vanadium, calcium, magnesium, iron, and manganese; the bases are combined with hydrochloric, sulphuric, nitric, phosphoric, and silicic acids, also with the organic acids present in the beet.

The information relative to the composition of the beet is derived mainly from von Lippmann's[1] paper cited in the foot-note, Sidersky's, *Traité d'analyse des Matières Sucrées*, and Horsin-Déon's *Traité de la Fabrication du Sucre*.

---

[1] *Bulletin de l'Assoc. Chim. de France*, **14**, 691 and 819.

The following table showing the distribution of the nitrogen in the beet is from analyses by Ed. Urbain:[1]

	Per Cent in the Beet.	Per Cent of the Total Nitrogen.
Total nitrogen	0.198	
Nitrogen of insoluble proteïds	0.012	6.06
Albuminoid nitrogen	0.063	31.81
Nitric nitrogen	0.050	25.25
Amide and ammoniacal nitrogen.	0.069	34.84
Loss	2.04	
		100.00

There is a reducing substance present in the beet that is not a sugar. Its composition is not definitely known. It is usually termed " Bodenbender's substance," from the name of the discoverer.

182. List of Reagents Suggested for the Treatment of Beet-juice.[2]—(*von Lippmann, Zeit. Rüben-zucker-Ind.*, 1886, 621.)

Chloride of calcium, Z., II, 65.

Chloride of lime, Z., VII, 423.

Carbonate of calcium, Maumené, J. d F. S., **17**, 22.

Acetate of calcium, Durieux, Jahresber., **8**, 334.

Sulphate of calcium, Duquesne, Dingler, 196, 83.

Chloride of strontium, Kottmann, Z., **32**, 899.

Chloride of barium, Licht, Berl. Ber., **15**, 1471.

Chloride of barium with caustic soda, Plecque, D. Z. I., **2**, 51.

Barium hydroxide with sulphate of aluminum, Eisenstuck, Jahresber., **3**, 244.

Oxide of magnesium, Thénard, Z., **13**, 128.

Carbonate of magnesium, Reich, Z., **6**, 173.

---

[1] *Bulletin de l'Association des Chimistes de France*, **14**, 1095.

[2] Abbreviations : Z. = Zeitschrift des Vereins für die Rübenzucker-Industrie des Deutschen Reichs.

N. Z. = Neue Zeitschrift für die Rübenzucker-Industrie (Scheibler).

D. Z. I. = Deutsche Zuckerindustrie.

J. d. F. S. = Journal des Fabricants de Sucre.

Dingler = Dingler's Polytecnische Journal.

Berl. Ber. = Berichte der deutschen chemischen Gesellschaft.

Jahresber. = Jahresbericht.

Sulphate of magnesium, Bayvet, Z., 10, 256.

Hydrate of magnesium, Rümpler, D. Z. I., 1879, 52.

Sulphite of magnesium, Dubrenil, J. d. F. S., 13, 27.

Chloride of magnesium, Kessler, Z., 16, 760.

Dolomite, Dubrenil, Berl. Ber., 6, 155.

Sulphate of magnesium with sulphide of barium, Drummond, Dingler, 203, 325.

Acid sulphite of magnesium, Becker, Z., 1886.

Sulphate of magnesium and sulphide of calcium, Drummond, Dingler, 203, 325.

Chloride of ammonium, Licht, Jahresber., 24, 415.

Sulphate of ammonium, Beanes, Dingler, 167, 220.

Ammonia and lime, Marot, Berl. Ber. 9, 643.

Carbonate of ammonium, Stammer, Z., 9, 430.

Phosphate of magnesium, Kessler, Z., 15, 525.

Phosphate of ammonium, Kuhlmann, Z., II, 92.

Phosphate of sodium, Z., 2, 130.

Phosphate of potassium, Blanchard, Berl. Ber., 6, 153.

Double phosphate of calcium and sodium, Gwynne, Z.,
3, 292.

Tribasic phosphate of lime with phosphate of ammonium,
Leplay, Z., 12, 193.

Acid phosphate of calcium with sulphate of magnesium,
Kessler, Z., 15, 51.

Phosphate of calcium with sulphate of aluminum, Kessler, *ibid.*

Sulphite of ammonium, Beauss, Dingler, 167, 220.

Sulphite of lime, Calvert, Z., 12, 500.

Sulphite of sodium, Périer-Possoz, Z., 12, 128.

Sulphite of magnesia, Mehay, Z., 23, 27.

Bisulphite of calcium, Reynose, Z., 12, 501.

Basic sulphite of magnesium, Z., 23, 26.

Bisulphite of calcium with sulphate of aluminum, Leyde,
Z., 1, 365.

Bisulphite of iron, Becker, N. Z., 16, 6.

Hydroxide of iron with plaster, Rousseau, Z., 11, 67.

Chloride of iron, Kral, Z., 18, 317.

Ferric sulphate, Kral, *ibid.*

Ferrous sulphate, Bayvet, Z., 10, 256.

Sulphate of manganese, Massé, Z., *ibid.*

Chloride of tin, Maumené, J. d. F. S., **20**, 7.

Chloride of tin, Manoury, Z., **34**, 1275.

Stannous sulphate, Org, Z., **15**, 76.

Oxide of tin with soda, Berl. Ber., **19**, 520.

Sulphate of zinc, Kindler, Z., **3**, 556.

Nitrate of zinc, Decastro, Jahresber., **19**, 340.   •

Nitrate of zinc with alkaline sulphides, *ibid.*

Nitrate of zinc with sulphide of barium or of calcium, *ibid.*

Zinc-dust, with sulphuric acid and sulphide of barium, Crespo, Jahresber., **24**, 416.

Acetate of lead and sulphide of sodium, Maumené, in his "Traité."

Hydroxide of lead, Gwynne, Z., **3**, 292.

Acetate of lead, *ibid.*

Saccharate of lead, *ibid.*

Hydrate of aluminum, Howard, Z., **2**, 92.

Colloidal aluminum, Löwig, Z., **29**, 905.

Silicate of aluminum (Walkererde), blue clay, Fritsche, Z., **35**, 261.

Chloride of aluminum, with lime, Siemen, Jahresber., **18**, 256.

Fluoride of aluminum, Kessler, Z., **15**, 525.

Sulphate of aluminum, Kessler, Z., **15**, 51.

Alum, Kindler, Z., **3**, 556.

Phosphate of aluminum, Oxland, Z., **2**, 92.

Acid phosphate of aluminum, *ibid.*, **2**, 130.

Acetate of aluminum, Schubarth, Z., **2**, 92.

Sulphite of aluminum, Mehay, Z., **23**, 27.

Sulphite of aluminum, with hydrate of calcium, Schubarth, Z., **2**, 129.

Sulphite of aluminum with sulphate of manganese, Massé, Z., **10**, 256.

Bisulphite of aluminum, Becker, Z., 35, 924.

Hydrosulphite of aluminum, Becker, Z., 1886 (?).

Aluminates of the alkaline earths, Alicoque, D. Z. I., **2**, 51.

Aluminate of calcium, Oxland, Z., **2**, 92.

Fluosilicate of aluminum, Kessler, Z., **16**, 760.

Silica, Schubarth, Z., **2**, 92.

Silicate of sodium, Wagner, Z., **9**, 331.

to obtain dense juices, and lose only from 0.15 to 0.26 per cent of the weight of the exhausted cossettes of sucrose.

Difficulties in Diffusion-work.—Some difficulty may be experienced in the conduct of the diffusion when the supply of beets is irregular, the battery overheated, the roots imperfectly washed, or the knives improperly sharpened or set. The remedies are self-evident. The work should be so conducted that the juice may be drawn at regular intervals. The intervals should be lengthened in the event of a shortage in the supply of beets or delays in the sugar-house. A long delay in the diffusion-work or excessively slow work results in impure juices, which yield but stubbornly to subsequent treatment.

Overheating affects the purity of the juice adversely, cooks the beet-cuttings, and renders them difficult to press. Overheating is also liable to cause the cuttings to pack or mat in the diffuser and thus render the circulation of the juice slow and imperfect and the pulp difficult to remove from the cell. Overheating may also cause pectic bodies to pass into solution, which later in the manufacture result in compounds which impede the filtration of the juice.

184. "Gray" Juice.—According to Herzfeld[1] the cause of this phenomenon is somewhat obscure. Invert-sugar and similar bodies, are present in the beet, and react with the alkalis, sodium, and potassium, during the evaporation, and form apoglucic acid and humic substances, which color the juice. The oxide of iron also plays a part.

Beets exposed in mild weather to rain lose much sugar through renewed vegetation, and impart a gray color to the juice.

It is difficult to remedy this coloration. Sulphurous acid modifies it but little. Sugars from such juices lose this color if stored in the warehouse about two weeks, and subjected to frequent mixing.

185. Carbonatation.—It is the practice in many houses to pass the juice, flowing from the measuring-tank at the battery, through a heater in which its temperature is raised by the vapors from the last pan of the multiple-effect. A small quantity of lime is then added, about 5 quarts of the milk of 20° Baumé per 1000 gallons, and the juice is passed through a heater, its temperature quickly raised to about 90° C., and the lime added, preparatory to the first carbonatation. In many houses the lime is placed to the juice immediately after it leaves the measuring-tank.

In the early part of the season, the quantity of lime used for the first carbonatation is much smaller than the amount necessary later on. More lime is required when the beets are in a bad condition than when sound. The quantity used is about 15 pounds of quicklime per 100 gallons of juice, but with unsound beets this amount is often exceeded.

In France, the lime is usually slaked, and reduced to a milk of 20° Baumé with the thin juice obtained in washing the filter-cake. The practice in many houses is to place the quicklime in wire baskets and slake it directly in the juice in the carbonatation-tank. This practice is extending. The carbonatation should be effected with rich gas, i.e., containing approximately 30 per cent carbonic acid, thus not only economizing time, but producing a precipitate which is more easily removed by the filter-presses. Practice differs as to the temperature of the carbonatation, but with rich juices a temperature of approximately 85° C. appears to give the more satisfactory results. With weak juices a lower temperature is often employed.

Toward the end of the carbonatation, the juice is heated to 80° to 90° C., thus breaking up the sucrocarbonates of lime which have been formed. Practice also differs relative to this temperature.

There is an alkalinity of 1 gram to 1.6 grams, calculated as lime (CaO), per litre of juice after the first carbonatation.

Second Carbonatation or Saturation.—The quantity of lime required in the second carbonatation is from 2 to 4 pounds per 100 gallons of juice. The gas is passed into the limed juice, when working with sound beets, until a test shows that the free lime is saturated ; late in the season, or when the beets are in a bad condition, the supply of gas is cut off before the alkalinity is entirely saturated. When working without sulphur, the alkalinity is usually reduced to 0.02 gram per litre, using phenolphthalein as an indicator. In some factories, instead of leaving a slight alkalinity due to lime, a small quantity of carbonate of soda is added to the juice. This practice is of doubtful utility.

On the completion of the carbonatation, the juice is boiled two or three minutes, then filtered.

Difficulties.—A defective first carbonatation results in difficulty in the filtration of the juice, imperfect removal of the sugar from the filter-press cakes, juices of lower purity than necessary, and formation of lime salts. It may also result in difficulty in boiling the sugar to grain in the vacuum-pan. These difficulties arise from a slow carbonatation, too low a temperature (sucrocarbonates of lime in the press-cake), too little lime, excessive use of the carbonic acid, or excessive alkalinity of the juice from the second carbonatation.

186. Sulphuring.—Sulphurous acid is usually employed in the gaseous state, in the manufacture of white sugar, without bone-black.

When sulphurous acid is used the following procedure is advised : The second carbonatation is stopped when the alkalinity is reduced to .5 to .6 gram per litre, calculated as caustic lime, CaO, and the juice is boiled and filter-pressed, as usual. The filtered juice is treated with sulphurous-acid gas at a temperature of 95° C., until the alkalinity is reduced to .1 to .15 gram per litre; the juice is then boiled.

The sulphuring must be very carefully controlled, in order to avoid loss of sugar through inversion.

187. Difficulties in Filter-pressing.—With juice from sound beets, properly treated in the carbonatation process, the filtration is always easy.

If the juice have not been sufficiently heated after the first carbonatation, or the supply of gas have been cut off too soon, the juice will filter badly; also, if too little lime, or lime from stone containing too much silica or having hydraulic properties have been employed. In the U. S. Government's experiments, in carbonating sorghum-cane juices, the limestone supplied by the contractor had decided hydraulic properties. The filter-press cake soon became hard, forming an impervious slab of artificial stone. A supply of suitable lime remedied the difficulty. This illustrates the importance of a chemical examination of the limestone supplied the factory.

A German manufacturer had considerable difficulty in filter-pressing certain juices. The usual remedies were applied, without success. A sample of the filter-press cake was sent to Dr. Herzfeld, of the German Sugar Manufacturers' Association, who found, on analysis, 1.3 per cent ferric oxide and 0.3 per cent aluminic oxide in the dry sample. He explained the difficulty as follows : In the presence of iron, pectine forms a flocculent, spongy mass of ferropectine, and not calcium pectate, which is granular. It is probable that, owing to an abnormal quantity of pectine, formed by excessive heat in the battery, in the presence of iron from the lime, a large quantity of ferropectine was formed, which obstructed the cloths.

A proper adjustment of the quantity of lime added to the juice, and careful control of the diffusion and of the first carbonatation, will usually remedy difficulties in the filtration.

188. Lime-kiln.—The relative quantities of coke and limestone vary between wide limits in the practice of various sugar-houses. According to Gallois, who has made probably one of the most exhaustive studies of the lime-kiln yet published, the quantity of coke theoretically required is 6 pounds for the decomposition of 100 pounds of limestone containing 95 per cent of calcium carbonate. This is approximately 1 volume of coke per 6 volumes of limestone. Gallois advises, however, in practice, the use of 1 volume of coke per 4 to 5 volumes of limestone. These proportions of coke and limestone produce a satisfactory gas.

Some authorities recommend 3 volumes of limestone to 1 volume of coke.

The coke and stone should be well mixed, and distributed as evenly as possible in the kiln. Notes relative to the quality of the limestone are given on the following page.

The gas produced by the furnace should contain approximately 30 per cent of carbonic acid.

Difficulties.—The difficulties usually encountered in the management of the lime-kiln are as follows: A limestone containing too much silica will show a tendency to fuse, and, if overheated, will adhere firmly to the walls of kiln.[1] Stone in too small pieces, or stone and coke, or an excess of coke, will sometimes " scaffold " or bridge. The above conditions soon prevent the downward progress of the stone and lime. These difficulties are obviated by the use of suitable stone, properly mixed with the coke and evenly distributed in the kiln, and by the withdrawal of lime at regular intervals. Should the charge " scaffold " in the kiln, it can only be broken down by the withdrawal of a considerable quantity of material at the lime-doors and energetic use of an iron bar at the " peep-holes." The use of too little coke or the too rapid withdrawal of lime results in an undue proportion of underburned or raw lime. The admission of too little air to the kiln results in an imperfect combustion and an excess of carbonic oxide in the gas. This carbonic oxide not only is a loss of carbon, but, if carelessly inhaled by the workmen, may result in serious poisoning. The addition of too much air dilutes the gas. This may result from leakage in the pipes, careless charging, or from driving the gas-pump too fast.

The following table contains valuable information relative to the quality of the limestone:

" Limestone No. 3 was used in a sugar-house and caused much trouble: 'scaffolding,' difficulty in the mechanical filtration, incrustations in the triple-effect and on the vacuum-pan coils. No. 9 was substituted for this stone, and these difficulties disappeared."

---

[1] Largely based on a report by F. Dupont and J. Delavierre, *Bulletin de l'Association des Chimistes de France*, **9**, 134.

ANALYSES OF LIMESTONES AND COMMENTS ON THEIR COMPOSITION.

(Messrs. GALLOIS and DUPONT, Paris.)

Substance.	1	2	3	4	5
	%	%	%	%	%
Moisture	4.10	5.10	7.25	4.15	4.17
Sand, clay, and insoluble matter........	4.50	5.15	4.90	2.15	3.07
Organic matter	1.20	1.17	1.37	1.05	0.97
Soluble silica	2.10	1.75	3.30	1.05	0.98
Oxides of iron and alumina } (Fe_2O_3, Al_2O_3)	0.37	0.41	0.27	0.17	0.19
Carbonate of calcium ($CaCO_3$)	85.86	85.12	81.67	90.13	88.65
Carbonate of magnesium ($MgCO_3$)	0.95	0.47	0.59	0.75	0.95
Sodium and potassium (Na_2O, K_2O) ...	0.05	0.06	....	0.10	0.01
Undetermined...................... ...	0.87	0.77	0.65	0.45	1.00
	100	100	100	100	100

Substance.	6	7	8	9	10
	%	%	%	%	%
Moisture........	6.25	5.16	0.52	1.21	0.11
Sand, clay, and insoluble matter...	3.17	2.25	2.85	0.55	0.27
Organic matter.................	1.12	0.86	0.30	0.41	0.15
Soluble silica....................·.........	0.64	0.56	0.06	0.20	0.03
Oxides of iron and alumina } (Fe_2O_3. Al_2O_3)	0.15	0.20	0.32	0.23	....
Carbonate of calcium ($CaCO_3$)...	87.93	90.03	93.80	96.58	99.10
Carbonate of magnesium ($MgCO_3$).....	0.50	0.45	1.81	0.50	....
Sodium and potassium (Na_2O, K_2O).....	....	....	....	....	....
Undetermined.....................	0.24	0.39	0.34	0.32	0.34
	100	100	100	100	100

Nos. 1, 2, 3. 4 are bad, Nos. 5, 6, 7 are passable, and Nos. 8, 9, 10 are excellent."

In the examination of a limestone its physical condition as well as its chemical composition must be taken into account. The stone should be compact and hard, thus reducing the quantity of fragments and the risk of "scaffolding" in the kiln.

Excessive moisture, 5 per cent or more, in the stone reduces the temperature of the kiln when charging, involving an imperfect combustion and the production of carbonic oxide (CO); further, such stone breaks into small pieces under the influence of the heat. A small proportion of water, approximately 1 per cent, probably facilitates the decomposition of the stone, and is advantageous.

Magnesium is not objectionable, so far as the operation of the kiln is concerned, except in the presence of silicates, but introduces difficulties in the purification of the juice and forms incrustations on the heating-surfaces of the evaporating-apparatus. It forms fusible silicates at high temperatures, and thus increases the tendency to "scaffolding." The objections to sulphate of calcium are practically the same as to magnesium.

The objections to the presence of silicates are as indicated above, in the formation of fusible silicates of lime and magnesium. Part of the silica passes into the juice with the lime, retards the filtration with the presses and coats the cloths of the mechanical filters, to their detriment. Silica also forms part of the scale on the heating-surfaces. Less harm results from this substance in hard limestones, than from that in soft stone; hence, if the stone be hard and compact, a larger content of silica is admissible than in a soft stone.

When necessarily using stone of comparatively poor quality, the best obtainable coke should be employed.

189. Granulation of the Sugar in the Vacuum-pan.—An excessive alkalinity or an excess of lime salts in the sirups causes the "strike" to boil slowly or heavily. When the difficulty is due to the alkalinity, the addition of sufficient dilute hydrochloric acid to nearly neutralize the massecuite, and careful supervision of the "saturation" with carbonic acid or sulphurous acid, is usually a satisfactory remedy. Excessive alkalinity of the sirup may be corrected by the addition of superphosphate of calcium, followed by filtration, to remove the precipitate, or the sirup may be nearly neutralized with dilute hydrochloric acid. The former method is preferable. The author was once present in a French sugar-house when the pan-man reported that "the strike would not boil." The chemist determined the alkalinity of the massecuite in the pan and calculated the quantity of acid required to nearly neutralize it, then added dilute sulphuric acid, hydrochloric acid not being readily obtainable. The strike was completed without further difficulty, though the yield of sugar was diminished by the treatment.

Should the difficulty be due to excess of lime salts, they may be decomposed by the addition of a vegetable oil or carbonate of sodium. The purification of the juice should be carefully controlled, so that it may rarely be necessary to adopt one of these remedies.

Difficulty in boiling the massecuite may also arise from a large proportion of organic matter, not sugar. In this event the only remedy is to increase the proportion of lime used in the first carbonatation.

190. Second and Third Massecuites, etc.— There is occasionally a tendency in massecuites, boiled to "string-proof," to foam in the crystallizing-tanks, or, as this is usually termed by the workmen, to "boil over." This has been attributed to various causes (*see* also page 200). It is often charged to reducing the alkalinity of the juice in the first carbonatation too low. Caustic soda may be used to remedy the alkalinity. Overheating of the massecuite in the pan or in the hot-room is supposed to often be the cause of the difficulty. The usual remedies are to sprinkle water or caustic soda solution on the surface of the massecuite.

191. Gray Sugar.—A series of experiments, by Herzfeld,[1] shows that the gray or reddish-gray color, which is sometimes observed in raw sugars, is due to the solution of ferric and ferrous oxides in the juice, in the presence of which, during the saturation with sulphurous acid, the sugar is discolored. Gray sugar, as a class, is acid to phenolphthalein; this discoloration is not noticeable in the products when the liquors are kept alkaline. Certain sugars, obtained when using dry lime in the defecation, gave unsatisfactory results. The alkalinity of the dry products was not determined. That of the sirups was determined with rosolic acid as an indicator, the use of which led to great errors, since juice apparently alkaline was in reality acid, and therefore dissolved ferric and ferrous oxides. When the sugar is gray its color may be remedied by covering it with strongly alkaline sirup. The "graying" of raw sugar is attributed by Munier[2] to the formation of a double sulphate of iron and potassium. The sulphur of this double salt is chiefly derived from the decomposition of albumin.

[1] *Zeit. Rübenzucker-Ind.*, 1896, **46**, 1.
[2] *Deutsche Zucker.-Ind.*, 1895, **20**, 1744; *Journ. Soc. Chem. Ind.*, **15**, 42.

SPECIAL REAGENTS.

192. Soxhlet's Solution. — In Soxhlet's method two solutions are employed, prepared as follows :

(*A*) Dissolve 34.639 grams of copper sulphate in water and dilute to 500 cc.

(*B*) Dissolve 173 grams tartrate of soda and potash (Rochelle salt) in water, dilute to 400 cc. and mix with 100 cc. caustic-soda solution 40 per cent by volume.

193. Soldaïni's Solution. — Dissolve 40 grams of sulphate of copper and 40 grams of carbonate of sodium separately in water ; mix ; collect the precipitate on a filter and wash with cold water. Transfer the precipitate to a large flask fitted with a reflux condenser; a long glass tube will answer for this purpose. Add approximately 416 grams of bicarbonate of potassium and 1400 cc. distilled water; heat on a water-bath or a hot-plate several hours, or until the evolution of carbonic acid ceases. When no more carbonic acid is given off, filter the solution and boil the filtrate a few minutes, and dilute it to 2000 cc. The specific gravity of the solution should be approximately 1.185. Solutions to be treated with Soldaïni's reagent should be boiled in case they contain ammonia, to insure freedom from this substance. Check this solution as indicated in **175**.

194. Fehling's Solution. — The formula for Fehling's solution is as follows :

> 34.64 grams of pure crystalline copper sulphate ;
> 100.00 grams neutral potassium tartrate.

Dissolve the copper sulphate in 160 cc. distilled water ; dissolve the neutral potassic tartrate in 600 to 700 cc. caustic-soda solution, specific gravity 1.12, equivalent to approximately a 14-per-cent solution by volume ; add the copper solution to the alkali, stirring thoroughly after each addition ; make up to 1000 cc. at the temperature at which the litre flask was graduated. Check this solution as indicated in **175**.

Fehling's solution decomposes readily on exposure to strong light. The author prefers the following modification by Violette for commercial work.

195. Violette's Solution.—This solution should be prepared in small quantities at a time, since it is liable to deposit oxide of copper, even in the cold, on long exposure to light. To prepare this solution, use the following quantities of the reagents :

34.64 grams chemically pure crystallized sulphate of copper ;

187.00 grams commercially pure tartrate of soda and potash (Rochelle salt) ;

78.00 grams commercially pure caustic soda.

Dissolve the copper sulphate in 140 cc. water, and add it slowly to the solution of Rochelle salt and caustic soda, taking care to thoroughly stir the solution after each addition ; dilute to one litre.

The copper sulphate should be carefully examined for impurities. If the salt be impure it must be dissolved and recrystallized repeatedly. The crystals must be finely powdered and dried between filter-papers before weighing.

If it be desirable to make up a large quantity of Fehling or Violette solution, all risk of deposition of the copper oxide in the cold may be avoided by making a separate solution of the copper sulphate. Dissolve the alkali and dilute to one litre ; dissolve the copper and make up to exactly one litre. For the analytical work take equal volumes of the solutions. Check this solution with invert-sugar (**204**) or dextrose under the conditions adopted for the analysis. The copper in 10 cc. of this solution should be reduced by 0.05 gram invert-sugar.

196. Normal Solutions.—"Normal solutions, as a general rule, are prepared so that one litre shall contain the hydrogen equivalent of the active reagent weighed in grams (H = 1)" (Sutton). Thus, normal sulphuric acid contains 49.043 grams H_2SO_4 per litre ; normal hydrochloric acid, 36.458 grams HCl per litre, etc. Half-normal, one-fifth normal, and one-tenth normal (decinormal) solutions are frequently used, and are prepared by diluting the normal solutions. Normal, half-normal, one-fifth normal solutions, etc., are usually written as follows: N, $\dfrac{N}{2}$, $\dfrac{N}{5}$, $\dfrac{N}{10}$, etc. These solutions are prepared and checked as indicated in the following sections.

197. Standard Hydrochloric Acid.—The reagent acid has usually a specific gravity of 1.20, approximately. Acid of this specific gravity contains 40.78 per cent of hydrochloric acid (*see* table, page 272); hence, a little less than 100 grams of this acid is required to contain the 36.458 grams necessary to form a normal solution. It is advisable to dilute a somewhat larger quantity of the acid, *e.g.*, 80 cc. to 1000 cc., with distilled water, rather than to attempt to closely approximate the correct quantity. Titrate this solution with a normal alkali solution (**201**), adding the acid from a burette to 10 cc. of the alkali solution, using cochineal or other suitable indicator (**215**). The preliminary titration should, most conveniently, show the acid solution to be too strong ; for example, suppose 9.6 cc. of the acid solution is required to neutralize 10 cc. of the alkali solution, then to 9.6 $\times$ 100 = 960 cc. of the acid must be added 1000 − 960 cc. = 40 cc. of water to make one solution exactly neutralize the other. The solution should be further checked by a determination of the chlorine, preferably by the method described on page 169. This acid is a convenient one for use in preparing very accurate standard alkali and acid solutions, since its strength may be ascertained with ease and accuracy by the chlorine determination. The half-normal acid is a convenient strength, and should contain 17.725 grams of chlorine per litre.

1 cc. normal hydrochloric acid = .036458 gram HCl

$$= .03545 \quad \text{``} \quad Cl$$
$$= .02804 \quad \text{`` } \quad CaO$$
$$= .05181 \quad \text{`` } \quad SrO$$
$$= .07672 \quad \text{`` } \quad BaO.$$

198. Standard Oxalic Acid.

—This is the simplest of the normal solutions to prepare, and when strictly pure oxalic acid can be obtained it may be used in the preparation of all the standard alkali and acid solutions.

Repeatedly crystallize the purest obtainable oxalic acid, from water solution. Dry the crystals thoroughly in the air at ordinary temperatures. Reject all crystals that show indications of efflorescence. Dissolve 63.034 grams of this acid in distilled water and dilute to 1000 cc. to prepare the normal solution, or, preferably, dry the powdered acid at 100° C. to constant weight and use 45.018 grams in preparing the normal solution. It is advisable to employ weaker solutions, usually the one-tenth normal acid. This should be prepared from the normal solution as required, since the latter keeps better, provided it is not exposed to direct sunlight.

1 cc. normal oxalic acid = .06303 gram $H_2C_2O_4.2H_2O$.

199. Standard Sulphuric Acid.

—Add approximately 28 cc. of concentrated sulphuric acid to distilled water, cool the solution, and dilute to 1000 cc. Standardize by titration with normal alkali.

$$1 \text{ cc. normal sulphuric acid} = .049043 \text{ gram } H_2SO_4$$
$$= .02804 \quad `` \quad CaO$$
$$= .05181 \quad `` \quad SrO$$
$$= .07672 \quad `` \quad BaO.$$

200. Standard Sulphuric Acid for the Control of the Carbonatation.

—Add approximately 21 cc. of concentrated sulphuric acid to distilled water, cool the solution, and dilute to 1000 cc. Titrate this solution with a normal soda or potash solution, using phenolphthalein as an indicator. Dilute the acid so that 14 cc. will be required to neutralize 10 cc. of the normal alkali (201).

$$1 \text{ cc. this standard acid} = .035 \text{ gram } H_2SO_4$$
$$= .02 \quad `` \quad CaO.$$

It is usual to add the phenolphthalein to this solution before dilution to 1000 cc.

201. Standard Alkali Solutions.

— Ammonium

hydrate (NH_4HO), caustic soda ($NaHO$), and caustic potash (KHO) are used in preparing the alkali solutions. The normal soda or potash solutions are used, but the ammonia should be weaker, preferably decinormal, or, for Sidersky's method for reducing-sugars, half-normal.

Dissolve 42 grams of chemically pure caustic soda in water, in preparing the normal reagent, cool the solution, dilute to 1000 cc., and standardize by titration, against a normal acid. In preparing the potash solution, use 58 grams of chemically pure caustic potash. The table, page 274, is convenient for use in standardizing the ammonia solution. Dilute the ammonia to approximately the required strength, and standardize by titration with decinormal or half-normal acid, as may be required, using cochineal as an indicator, or for Sidersky's method for reducing-sugars, use sulphate of copper as an indicator, as directed on page 88.

$$1 \text{ cc. normal caustic-soda solution} = .0401 \text{ gram } NaOH$$
$$= .03105 \text{ '' } Na_2O$$
$$1 \text{ cc. normal caustic potash solution} = .056 \text{ '' } KHO$$
$$= .04711 \text{ '' } K_2O$$
$$1 \text{ cc. half-normal ammonia solution} = .00853 \text{ '' } NH_3$$
$$= .01754 \text{ '' } (NH_4)HO$$
$$1 \text{ cc. decinormal ammonia solution} = .00171 \text{ '' } NH_3$$
$$= .00351 \text{ '' } (NH_4)HO$$

Phenolphthalein cannot be used as an indicator with ammonia.

202. Decinormal Permanganate of Potassium.—Dissolve 3.16 grams of chemically pure and dry permanganate of potassium ($KMnO_4$) in distilled water, and dilute to 1000 cc. This solution is conveniently checked by titration with decinormal oxalic acid. To 10 cc. of decinormal oxalic acid add several volumes of water and a few cc. of dilute sulphuric acid. Warm the solution to approximately 60° C., and add the permanganate solution little by little. Discontinue the addition of the permanganate as soon as the solution acquires a faint pink- or rose-color. The temperature of the solution must be maintained at approximately 60° C., and a little time must be allowed for the reaction. In reducing-sugar determinations, check the permanganate, as indicated on page 90.

Permanganate of potassium solution should be preserved in a tightly stoppered bottle, and should be checked from time to time. The appearance of a sediment indicates a change in the solution. It is simpler to determine a factor from time to time, rather than attempt to maintain the solution strictly decinormal.

1 cc. decinormal permanganate of potash $\begin{cases} = .0316 \text{ gram } KMnO_4 \\ = .00636 \text{ " } Cu. \end{cases}$

203. Permanganate Solution for Reducing-sugar Determinations.

—This solution should be of such strength that 1 cc. is equivalent to .01 gram of copper. Dissolve 4.9763 grams of permanganate of potassium in distilled water and dilute to 1000 cc. This solution should be checked by a reducing-sugar determination in material of known composition.

204. Invert-sugar Solution.

—Bornträger[1] recommends the following method of preparing an invert-sugar solution for checking the reagents used in reducing-sugar determinations: Dissolve 2.375 grams pure sucrose in water, dilute to 100 cc., and add 10 cc. hydrochloric acid of 1.188 specific gravity. Let the mixture stand overnight in the cold. Neutralize with sodium hydrate and dilute to 1000 cc.

20 cc. of this solution contains .05 gram invert-sugar, and should reduce the copper in 10 cc. of Violette or Fehling solution.

The inversion may also be conducted under the temperature conditions given in **89** in preparing invert sugar; or pure dextrose may be substituted for it in standardizing the alkaline copper reagents.

205. Soap Solution for Clark's Test and Alkalinity Determinations.

—Courtonne recommends the following method of preparing the soap solution, which he states is quite permanent: To 28 grams or 33 cc. of olive-oil or oil of sweet almonds add 10 cc. caustic soda

---

[1] *Zeit. Angew. Chem.*, 1892, 333.

solution of 35° Baumé, and 10 cc. 90 to 95 per cent alcohol ; heat the mixture a few minutes on the boiling water-bath to saponify the oil, then add 800 to 900 cc. of 60 per cent alcohol and agitate to dissolve the soap. Filter the solution into a 1000 cc. flask, cool, and complete the volume to 1 litre with 60 per cent alcohol. The solution should be standardized as directed in **82**.

Sidersky recommends the following solution : Dissolve 50 grams of white Marseilles soap in 800 grams of 90 per cent alcohol, filter and add 500 cc. distilled water. Standardize as in **82**.

The following is Clark's method as described by Sutton : " Rub together 150 parts lead plaster (Emplast. Plumbi of the druggists) and 40 parts dry potassic carbonate. When fairly mixed add a little methylated spirit and triturate to a uniform creamy mixture. Allow to stand some hours, then throw on a filter and wash several times with methylated spirit. Dilute the strong soap solution with a mixture of one volume of distilled water and two volumes of methylated spirit (considering the soap solution as spirit) until 14.25 cc. are required to form a permanent lather with 50 cc. standard calcic chloride, the experiment being performed as in determining the hardness of water. To prepare the calcic chloride solution : Dissolve 0.2 gram pure crystallized calcite in dilute hydrochloric acid in a platinum dish. Evaporate to dryness on the water-bath, dissolve with water and again evaporate to dryness, repeating this several times. Lastly, dissolve in distilled water and complete the volume to 1000 cc."

206. Preparation of Pure Sugar.—With sugar from the cane proceed as follows : Powder the purest sugar obtainable. Wash thoroughly with 85 per cent alcohol, and, finally, once with absolute alcohol. Dry, in a thin layer, over sulphuric acid in a desiccator.

If the sugar be of beet or unknown origin, purify it by the following method recommended by Wiley: Dissolve 70 parts of sugar in 30 parts of water, then precipitate the sugar from this solution at 60° C. with an equal volume of 96 per cent alcohol. Decant the supernatant liquor while still warm, and wash the sugar with strong warm alcohol. The raffinose is removed in the alcohol solution. Finally wash the sugar with absolute alcohol and dry over sulphuric acid in a desiccator.

207. Subacetate of Lead.—*Dilute Solution.*—Heat, nearly to boiling, for about half an hour, 430 grams of neutral acetate of lead, 130 grams of litharge, and 1000 cc. of water. Cool, settle, and decant the clear solution and reduce this to 54.3° Brix with cold, recently boiled, distilled water.

This is the solution recommended for use with Pellet's aqueous methods for the direct analysis of the beet.

Concentrated Solution.—Proceed as above, except use only 250 cc. of water.

Late investigations show that highly basic solutions of subacetate of lead should not be employed.

208. Preparation of Bone-black for Decolorizing Solutions.—Powder the bone-black obtained from the sugar-house filters, or otherwise, and heat it several hours with hydrochloric or nitric acid to dissolve the mineral matter. Decant the acid and wash the bone-black with water until the washings no longer turn blue litmus-paper red. Dry the powdered char in an air-bath, at about 150° C. Preserve in a tightly stoppered bottle.

Vivien advises digesting the bone-black, reduced to a fine powder, with a large excess of acid during several days; the char is then thoroughly washed with water, and finally with dilute ammonia, and thoroughly dried. The dry char is calcined in a vessel from which the air is excluded.

209. Preparation of Hydrate of Alumina.—Hydrate of alumina, frequently termed "alumina cream," may be used instead of lead for decolorizing sugar solutions or for removing an opalescence. To a moderately concentrated solution of common alum in water add ammonia in slight excess. Wash the resulting precipitate by decantation until the wash-water no longer reacts acid with litmus-paper. This precipitate is employed in a moist state. After adding the hydrate of alumina to the solution to be examined, it should stand a few minutes, with frequent shaking. A little lead may sometimes be advantageously employed with the alumina.

210. Litmus Solution. — Powder the litmus and treat it several times with boiling-hot 80-per-cent alcohol to separate the coloring matter soluble in this reagent. Reject the alcoholic solution, boil the residue with distilled water, and filter. Divide the filtrate into two equal parts ; carefully neutralize one with sulphuric acid, then mix the two portions together. Again divide into two parts, neutralize one, and mix as before. Repeat these operations until the solution is exactly neutral; preserve in an open bottle.

211. Litmus-papers. — Take a portion of the above solution and divide into two parts. To one part add sufficient sulphuric acid to render it faintly acid ; to the other portion add caustic-soda solution to faint alkalinity. Soak strips of Swedish filter-paper in these solutions, using the acid for red paper and the alkaline for the blue. Dry the strips in a room free from acid or alkaline vapors. Preserve in an unstoppered bottle, out of contact with strong sunlight.

212. Turmeric-paper. — Treat the finely powdered turmeric first with water, to dissolve out impurities, then with alcohol, to extract the coloring matter. Soak strips of Swedish filter-paper in the alcoholic solution, and dry them out of contact with the laboratory fumes. Preserve the papers in a stoppered bottle.

213. Phenolphthalein Solution. — Dissolve 1 gram of phenolphthalein in 100 cc. of dilute alcohol. This solution is colorless when acid and red in the presence of alkalis. It should be neutralized with dilute caustic soda or potash. Phenolphthalein is not applicable in the presence of ammonia. This indicator is considered the most suitable for beet-sugar work by Herzfeld, Claassen, and Henke.[1]

214. Corallin or Rosolic Acid Solution. — Digest equal quantities of carbolic, sulphuric, and oxalic acids to-

---

[1] An extensive paper on indicators for sugar-house purposes is published by Henke in *Cent. Blatt. f. d. Zuckerind.*, 1894, Nos. 11 and 12 ; abstract in *Bulletin de l'Association des Chimistes*, **13**, 492.

gether for some time at 150° C ; dilute the mixture with water, saturate the free acid with calcium carbonate, and evaporate the mixture to dryness ; extract the color with alcohol and nearly neutralize the solution (Sutton): or, prepare a saturated solution of commercial corallin in 90% alcohol, and nearly neutralize with an alkali. This solution is more permanent than litmus, but otherwise has no advantages over the latter.

215. Cochineal Solution.—Extract 3 grams of pulverized cochineal with 50 cc. strong alcohol and 200 cc. water, with occasional agitation, for a day or two. Filter off, and neutralize the extract.

216. Phenacetolin Solution.—Dissolve 2 grams of the reagent in 1000 cc. of strong alcohol.

217. Nessler's Solution.—Dissolve 62.5 grams of potassium iodide, KI, in 250 cc. of water. Set aside about 10 cc. of this solution ; add to the larger portion a solution of mercuric chloride, $HgCl_2$, until the precipitate formed no longer redissolves. Add the 10 cc. of the iodide solution ; then continue the addition of mercuric chloride very cautiously until a slight permanent precipitate forms. Dissolve 150 grams of caustic potash in 150 cc. water, cool, and add gradually to the above solution. Dilute the mixture to 1 litre.

REAGENTS.

218. TABLE SHOWING THE IMPURITIES PRESENT IN COMMERCIAL REAGENTS; ALSO, THE STRENGTH OF SOLUTIONS, ETC., RECOMMENDED.

NAME.	SYMBOL.	IMPURITIES.	STRENGTH OF SOLUTION, ETC.
Sulphuric Acid (Oil of Vitriol).	H_2SO_4.	Pb, As, Fe, Ca, HNO_3, N_2O_4.	Concentrated and dilute. To dilute pour 1 part acid by measure into 9 parts distilled water. Use porcelain dish.
Nitric Acid.	HNO_3.	H_2SO_4, HCl.	Concentrate and dilute. To dilute add 1 part acid to 9 parts water.
Hydrochloric Acid (Muriatic Acid).	HCl.	Cl, Fe_2Cl_6, H_2SO_4, SO_2, As.	Concentrated and dilute. Dilute = 1 part acid to 9 parts water.
Nitro-hydrochloric Acid. (Aqua regia.)			Prepare when required by adding 4 parts hydrochloric to 1 part nitric acid. Use concentrated acids.
Acetic Acid.	$H_4C_2O_2$.	H_2SO_4, HCl, Cu, Pb, Fe, Ca.	Concentrated and dilute. Dilute = 1 part pure glacial acetic acid to 1 part water.
Sulphurous Acid.	H_2SO_3.		To charcoal, in a flask, add concentrated H_2SO_4. Boil, wash the gas generated by passing it through water, and finally pass it into very cold water. Preserve the solution in tightly-stoppered bottles.
Oxalic Acid.	$H_2C_2O_4$.	Fe, K, Na, Ca.	Dissolve 1 part of crystallized acid in 9 parts distilled water.
Sulphuretted Hydrogen.	H_2S.		Use in gaseous state or in water solution. Wash the gas.
Sodic Hydrate or Potassic Hydrate.	NaHO, KHO.	Al, SiO_2, phosphates, sulphates, and chlorides.	Dissolve the stick soda or potash in 20 parts water. (Soda is less expensive, and will usually answer for most purposes in place of potash.)
Ammonic Hydrate.	NH_4HO.	Sulphate, chloride, carbonate, tarry matters.	Stronger water of ammonia (.96 specific gravity) and ⅛ above strength.
Baric Hydrate.	BaO_2H_2.		Dissolve 1 part of the crystals in 20 parts water; filter, and preserve in stoppered bottle.

REAGENTS.—*Continued.*

NAME.	SYMBOL.	IMPURITIES.	STRENGTH OF SOLUTION, ETC.
Calcic Hydrate.	CaO_2H_2.		Slake lime in water, filter off the solution, and preserve out of contact with the air.
Sodic Ammonic Hydric Phosphate. (Microcosmic Salt.)	$Na(NH_4)HPO_4$.		Dry and powder the salt. It may be made as follows: Dissolve 7 parts disodic hydric phosphate (Na_2HPO_4) and 1 part ammonic chloride in 2 parts boiling water, filter, and separate the required salt by crystallization. Purify by recrystallization.
Sodic Biborate.	$Na_2B_4O_7$.		Heat to expel water of crystallization and powder.
Sodic Carbonate.	Na_2CO_3.	Chlorides, phosphates, sulphates, silicates.	Use the powdered salt or dissolve in 5 parts water.
Ammonic Sulphate.	$(NH_4)_2SO_4$.		Dissolve 1 part in 5 parts water.
Ammonic Chloride.	$(NH_4)Cl$.	Fe. Purify the commercial salt by the addition of ammonia; filter. Neutralize filtrate with HCl; concentrate and recrystallize.	Dissolve 1 part in 5 parts water.
Ammonic Nitrate.	$(NH_4)NO_3$.		Saturated solution.
Ammonic Oxalate.	$(NH_4)_2C_2O_4$.	Purify by recrystallization.	Dissolve 1 part in 20 parts water.
Ammonic Carbonate.	$(NH_4)_2CO_3$.	Pb, Fe, sulphates, chlorides.	Dissolve 1 part in 4 parts water, and add 1 part ammonia, specific gravity .880.
Ammonic molybdate.			Dissolve the salt in strong ammonia, decant the clear solution slowly into strong nitric acid, stirring thoroughly till the precipitate redissolves.
Ammonic sulphide.	$(NH_4)_2S$.		Saturate 3 parts ammonia with H_2S, then add 2 parts ammonia.
Yellow Ammonic Sulphide.	$(NH_4)_2S_2$		Prepared by dissolving sulphur in ammonic sulphide.
Potassic Sulphate.	K_2SO_4.		Dissolve 1 part in 10 parts water.
Potassic Iodide.	KI.	Iodate, carbonate.	Dissolve 1 part in 50 parts water.

REAGENTS.—*Continued.*

NAME.	SYMBOL.	IMPURITIES.	STRENGTH OF SOLUTION, ETC.
Potassic Chromate.	K_2CrO_4.	Sulphates.	Dissolve 1 part in 10 parts water.
Potassic Bichromate.	$K_2Cr_2O_7$.		Dissolve 1 part in 10 parts water.
Potassic Ferricyanide.	$K_6Fe_2Cy_{12}$.		Dissolve 1 part in 12 parts water. Better to prepare solution when required.
Potassic Ferrocyanide.	K_4FeCy_6.		Dissolve 1 part in 12 parts water, or, **for glucose** work, 1 part in 30 parts water.
Baric Chloride.	$BaCl_2$.	Purify the commercial salt by passing H_2S through it and crystallizing.	Dissolve 1 part in 10 parts water.
Baric Nitrate.	$Ba(NO_3)_2$.		Dissolve 1 part in 15 parts water.
Baric Carbonate.	$BaCO_3$.		Add water to the powdered carbonate and preserve in salt-mouthed bottle.
Calcic Chloride	$CaCl_2$.	Fe.	Dissolve 1 part in 5 parts water.
Calcic Sulphate.	$CaSO_4$.		Dissolve as much of the salt as possible in water (in the cold), filter, and preserve the filtrate.
Magnesic Sulphate.	$MgSO_4$.		Dissolve 1 part in 10 parts water.
Ferrous Sulphate.	$FeSO_4$.		Dissolve 1 part in 10 parts cold water.
Ferric Chloride	Fe_2Cl_6.		Dissolve 1 part in 10 parts water.
Cobaltous Nitrate.	$CO(NO_3)_2$.	Fe, Ni, etc.	Dissolve 1 part in 10 parts water.
Cupric Sulphate.	$CuSO_4$.	Fe, Zn.	For sugar work purify by repeated crystallizations. Even the so-called "C. P." salts cannot always be depended upon. For Fehling solution see page 216. For ordinary work dissolve 1 part in 10 parts water.
Mercuric Chloride.	$HgCl_2$.		Dissolve 1 part in 20 parts water.
Mercurous Nitrate.	$Hg_2(NO_3)_2$.		Dissolve 1 part in 20 parts **water** acidulated with 1.2 part nitric acid. Filter into a bottle containing a little metallic mercury.

REAGENTS.—*Continued.*

NAME.	SYMBOL.	IMPURITIES.	STRENGTH OF SOLUTION, ETC.
Platinic Chloride.	$PtCl_4$.		Dissolve 1 part in 10 parts water.
Argentic Nitrate.	$AgNO_3$.		Dissolve 1 part in 10 parts water.
Stannous Chloride.	$SnCl_2$.		Dissolve pure tin in strong HCl in the presence of platinum. Dilute with 4 volumes dilute HCl. Keep granulated tin in the bottle.

219. ATOMIC WEIGHTS—PARTIAL LIST.

(*The Constants of Nature*—Frank Wigglesworth Clarke.)

NAME.	SYM-BOL.	ATOMIC WT.		NAME.	SYM-BOL.	ATOMIC WT	
		$H = 1$	$0 = 16$			$H = 1$	$0 = 16$
Aluminum.	Al	26.91	27.11	Lead........	Pb	205.36	206.92
Antimony..	Sb	119.52	120.43	Magnesium.	Mg	24.10	24.28
Arsenic....	As	74.44	75.01	Manganese..	Mn	54.57	54.99
Barium....	Ba	136.39	137.43	Mercury....	Hg	198.49	200.00
Bismuth...	Bi	206.54	208.11	Nickel	Ni	58.24	58.69
Boron	B	10.86	10.95	Nitrogen....	N	13.93	14.04
Bromine...	Br	79.34	79.95	Oxygen.....	O	15.88	16.00
Calcium ...	Ca	39.76	40.07	Phosphorus	P	30.79	31.02
Carbon....	C	11.92	12.01	Platinum....	Pt	193.41	194.89
Chlorine...	Cl	35.18	35.45	Potassium..	K	38.82	39.11
Chromium.	Cr	51.74	52.14	Silicon	Si	28.18	28.40
Cobalt	Co	58.49	58.98	Silver... ...	Ag	107.11	107.92
Copper	Cu	63.12	63.60	Sodium.	Na	22.88	23.05
Fluorine...	Fl	18.91	19.06	Strontium...	Sr	86.95	87.61
Gold	Au	195.74	197.23	Sulphur	S	31.83	32.07
Hydrogen..	H	1.00	1.008	Tin	Sn	118.15	119.05
Iodine......	I	125.89	126.85	Zinc	Zn	64.91	65.41
Iron	Fe	55.60	56.02				

220. COMPARISON OF WEIGHTS AND MEASURES.

MEASURES OF WEIGHT.

	POUNDS AVOIRDUPOIS.	OUNCE AVOIRDUPOIS.	TROY GRAINS.
Milligram.........	....	....	.01543
Centigram.........	....	....	.15433
Decigram.........	....	....	1.54332
Gram	.002204	.0353	15.43316
Decagram.........	.022047	.3527	....
Hectogram........	.220474	3.5276	....
Kilogram..........	2.204737	35.2758	....

1 lb. avoirdupois = 453.593 grams.

MEASURES OF LENGTH.

	INCHES.	FEET.
Millimetre................	.03937	.003281
Centimetre...............	.39371	.032809
Decimetre................	3.93708	.328090
Metre....................	39.37079	3.280899
Decametre...............	393.70790	32.808992
Hectometre	3937.07900	328.089917
Kilometre................	39370.79000	3280.899167
Myriametre.............	393707.90000	32808.991667

1 inch = 2.53995 centimetres.   1 foot = 30.47945 centimetres.

MEASURES OF CAPACITY.

	CUBIC INCHES.	GALLONS (231 cu. in.).
Millilitre (cubic centimetre)..	.06103	.000264
Centilitre.............	.61027	.002641
Decilitre....................	6.10270	.026414
Litre (cubic decimetre)	61.02705	.26414
Decalitre....................	610.2705	2.6414
Hectolitre...................	6102.705	26.414
Kilolitre....................	61027.05	264.14
Myrialitre...	610270.5	2641.4

1 cubic inch = 16.3862 cubic centimetres.   1 cubic foot = 28.3153 litres.
1 gallon (231 cu. in.) = 3.785 litres.

MEASURES OF SURFACE.

	SQUARE FEET.	ACRES.
Centiare, square metre	10.7643	.000247
Are, 100 square metres.......	1076.4293	.024711
Hectare, 10,000 sq. metres...	107642.9342	2.471143

1 sq. inch = 6.4514 sq. centimetres.   1 sq. foot = 9.29 sq. decimetres.
1 acre = .4046 hectare.

221. RELATIVE VALUES OF DIFFERENT FUELS.—(Haswell.)

DESCRIPTION.	Lbs. steam from water at 212° F. by 1 lb. of fuel.	Relative evaporative power for equal weights.	Relative evaporative power for equal volumes.	Relative rapidities of ignition.	Relative freedom from waste.	Relative completeness of combustion.	Relative weights.
Anthracites.							
Peach Mountain, Pa.......	10.7	1	1	.505	.633	.725	.945
Beaver Meadow...........	9.88	.923	.982	.207	.748	6	1
Bituminous.							
Newcastle...............	8.66	.809	.776	.595	.887	.346	.904
Pictou..................	8.48	.792	.738	.588	.418	1	.876
Liverpool...............	7.84	.733	.663	.581	1	.333	.852
Cannelton, Ind..........	7.34	.686	.616	1	.984	.578	.842
Scotch.................	6.95	.649	.625	.521	.499	.649	.909
Pine wood, dry..........	4.69	.436	.175	....	16.417	....	....

222. Testing a Burette.—The method of testing a burette as described by Payne[1] may be applied with advantage in a sugar-house laboratory and contribute its share to the reduction of the "undetermined losses." Payne's article is given here in full, with the exception of the preliminary statements and the abridgment of the tables to an upper limit of 40° C. The author urges the adoption of Payne's suggestion relative to the standard temperature.

"Most makers choose 15° or 16° as the standard temperature, and many graduates are so marked; but we may preferably take a somewhat higher temperature, one nearer the average working temperature of our room, and in this way secure less actual deviation from the truth. Several temperatures have been proposed from 15° to 25°, and the highest of these seems to be the best.

"Having selected a standard temperature for our burette, the next point to consider is the standard unit of volume. By definition, 'The kilogram is the vacuum weight of 1000 cc. of water at its temperature of maximum density, about 4°.' Reversing this, the volume occupied by 1 kilo

---

[1] Journal of Anal. and Applied Chemistry **6**, 327.

of water at 4° (weighed in vacuo) is the volume of 1000 cc.
or 1 litre. Since we are obliged to weigh in air, and for
convenience at temperatures greater than 4°, we can only
arrive at the correct litre by knowing the conditions of our
experiment and making the proper corrections therefor.

" The true litre is independent of the expansion of water
by heat, and out of respect for the authors of the metric
system, as well as from a regard for uniformity, it may
well be retained as our actual standard.

"Our first correction depends upon the variation in weight
of 1 litre of water under a change of temperature. This
has been determined by several experimenters, and a careful comparison of their best results will give us a very
accurate table. The following has been compiled from the
latest determinations, plotted into a curve of expansion
and corrected by the method of second differences. (See
Table I.)

"At our standard temperature, 25°, the true weight of 1
litre of water is seen to be 997.27 gms. The apparent

TABLE No. 1.

Degrees C.	Density or Grams in 1 Litre.	Volume or Centimetres cu. in 1 Kilo.	Degrees C.	Density or Grams in 1 Litre.	Volume or Centimetres cu. in 1 Kilo.
0	999.86	1000.14	21	998.18	1001.82
1	999.91	1000.09	22	997.97	1002.03
2	999.95	1000.05	23	997.74	1002.26
3	999.98	1000.02	24	997.51	1002.49
4	1000.00	1000.00	25	997.27	1002.73
5	999.97	1000.03	26	997.02	1002.98
6	999.94	1000.06	27	996.76	1003.24
7	999.90	1000.10	28	996.48	1003.52
8	999.85	1000.15	29	996.19	1003.81
9	999.79	1000.21	30	995.89	1004.11
10	999.72	1000.28	31	995.58	1004.42
11	999.64	1000.36	32	995.25	1004.75
12	999.55	1000.45	33	994.92	1005.08
13	999.44	1000.56	34	994.58	1005.42
14	999.32	1000.68	35	994.23	1005.77
15	999.19	1000.81	36	993.87	1006.13
16	999.05	1000.95	37	993.50	1006.50
17	998.90	1001.10	38	993.12	1006.88
18	998.74	1001.26	39	992.73	1007.27
19	998.57	1001.43	40	992.32	1007.68
20	998.38	1001.62			

weight of 1 litre of water at 25° as weighed with brass
weights in air at the same temperature and at 760 mm.
barometric pressure would be less than this by an amount
equal to the weight of air displaced by the difference in
volume between the water and the weights. With brass at
a sp. gr. of 8, and water at 1, the difference in volume
equals ⅞ of the volume of the water or ⅞ of 1 litre. 1 litre
of air at 25° and 760 mm. B. weighs 1.1845 gms. and ⅞ of this
1.0364 gms. Hence the litre under these circumstances
weighs or at least counterbalances weights equal to 996.23
gms. This correction for loss of weight in air varies with
the barometer, but for any pressure between 730 and 780
mm. a change of less than .05 cc. per litre is occasioned,
which for our purpose may be entirely disregarded. The
temperature of the air will be approximately the same as
that of the water, a maximum difference of 5° modifying
the result by only .02 cc. per litre, and by subtracting the
correction from the previous table we get the following :

TABLE No. 2.

APPARENT WEIGHT OF 1 LITRE OF WATER AT DIFFERENT
TEMPERATURES, AS WEIGHED WITH BRASS WEIGHTS IN
AIR.

Temp. of Water, Degrees C.	Apparent Weight.	Temp. of Water, Degrees C.	Apparent Weight.
15	998.1	28	995.4
16	998.0	29	995.2
17	997.8	30	994.9
18	997.7	31	994.6
19	997.5	32	994.2
20	997.3	33	993.9
21	997.1	34	993.6
22	996.9	35	993.2
23	996.7	36	992.9
24	996.5	37	992.5
25	996.2	38	992.1
26	996.0	39	991.7
27	995.7	40	991.3

" This table at 25° gives the apparent weight of one litre of
water as measured by our burette. The expansion or contraction of the glass above or below this temperature will
modify the other figures by an amount equal to .023 cc. for
each degree, and this amount must be subtracted below 25°,

and added above 25°, to the figures of the table.   Hence we
have a final table giving the apparent weight of 1 litre of
water under ordinary circumstances as above stated.   As
most of our volumetric glassware is marked as standard at
15°, we give a table for this temperature also, although the
difference amounts to only .02 per cent.

TABLE No. 3.

APPARENT WEIGHT OF 1 LITRE OF WATER AT DIFFERENT
TEMPERATURES, AS WEIGHED WITH BRASS WEIGHTS IN
AIR.   CORRECTED FOR EXPANSION OF GLASS.

Temperature.	Apparent Weight.		Apparent Volume.	
Degrees C.	Standard at 15°.	Standard at 25°.	Standard at 15°.	Standard at 25°.
15	998.1	997.9	1001.9	1002.1
16	998.0	997.8	1002.0	1002.2
17	997.9	997.7	1002.1	1002.3
18	997.8	997.5	1002.2	1002.5
19	997.6	997.4	1002.4	1002.6
20	997.4	997.2	1002.6	1002.8
21	997.3	997.0	1002.7	1003.0
22	997.1	996.8	1002.9	1003.2
23	996.9	996.6	1003.1	1003.4
24	996.7	996.4	1003.3	1003.6
25	996.5	996.2	1003.5	1003.8
26	996.2	996.0	1003.8	1004.0
27	996.0	995.8	1004.0	1004.2
28	995.7	995.5	1004.3	1004.5
29	995.5	995.2	1004.5	1004.8
30	995.2	995.0	1004.8	1005.0
31	994.9	994.7	1005.1	1005.3
32	994.6	994.4	1005.4	1005.6
33	994.3	994.1	1005.7	1005.9
34	994.0	993.8	1006.0	1006.2
35	993.7	993.5	1006.3	1006.5
36	993.4	993.2	1006.6	1006.8
37	993.0	992.8	1007.0	1007.2
38	992.6	992.4	1007.4	1007.6
39	992.3	992.1	1007.7	1007.9
40	991.9	991.7	1008.1	1008.3

" This table is accurate to probably .1 cc. in a litre or to .01
per cent., which is about the limit of error in an ordinary
analysis.

" In testing a burette or other graduate, the conditions of
the operation should be as nearly as possible the same as
those of actual use.   The burette should be read after a lapse

of time equal to the time of an ordinary titration. We have found that in a 100 cc. burette on drawing the contents out rapidly the liquid will run down from the sides about as follows:

.1 cc. in ½ minute.
.2 cc. in 2 minutes.
.3 cc. in 5 minutes.
and .4 cc. in 15 minutes.

"Water, acid, and salt solutions about the same, but alkalies a little more slowly. As a careful titration takes usually more than 2 minutes and less than 15, we are accustomed to read the burette after 5 minutes standing.

"Select water at the same temperature as the balance-room. A convenient vessel for holding the water while weighing is a good-sized weighing-bottle or a glass-stoppered 100 cc. flask. A solution of bichromate of potash in moderately strong sulphuric acid used warm is an excellent agent for removing grease or other foreign matter from a burette-tube.

"The following example of 2 burettes purchased recently will show the method of testing and also exhibit the quality of graduated glassware to be found in the market. With two or three tested burettes and flasks in a laboratory we may readily compare others and make them equivalent.

25 CC. BURETTE. MARK B. WATER AT 25°.

	Weighings.	H_2O.	True cc.	Burette.	Difference.
Empty	25.120			0.00	
	32.931	7.811	7.84	7.82	7.82
	41.452	8.521	8.55	16.37	8.55
	49.252	7.800	7.83	24.21	7.84

24.22

SAME BURETTE AGAIN FOR TOTAL CAPACITY. WATER AT 25°.

	Weighings.	Burette.
Empty	25.084	0.00
	49.946	24.94

24.862 = 24.96 cc.   error, .02 cc.

DUPLICATE.   WATER AT 25°.

Burette.

Empty   25.086

           49.983   0.00

24.897=25.00 cc. 24.97 error, .03 cc.

"Burette readings were taken to $\frac{1}{100}$ cc., but the error of such a reading would amount to probably .03 cc. The results of the above test show the burette to be highly accurate. It will be noticed that duplicate determinations give concordant results, the variations being less than the probable error of any single reading. This fact alone will indicate the general accuracy of the method.

75 CC. BURETTE.   MARK K.   WATER AT 26°.

Burette.

	Weighings.	H$_2$O.	True cc.		Difference.	Errors. Successive.	Total.
Empty	25 065			0.00			
	32.587	7.522	7.55	7.57	7.57	+ .02	
	40.083	7.496	7.52	15.10	7.53	.00	+ .02
	47.573	7.490	7.52	22.63	7.53	+ .01	+ .03
	54.954	7.381	7.41	30.00	7.37	− .04	− .01
	62.501	7.547	7.58	37.59	7.59	+ .01	.00
	70.022	7.521	7.55	45.17	7.58	+ .03	+ .03
	77.692	7.670	7.70	52.88	7.71	+ .01	+ .04
Empty	25.129						
	32.940	7.811	7.84	60.70	7.82	− .02	+ .02
	40.461	7.521	7.55	68.25	7.55	.00	+ .02
	45.740	5.279	5.30	73.65	5.40	+ .10	+ .12
		73.58		73.65			

"The test points to a probable inaccuracy in the lower part of the burette. This fact was proven by a duplication of the weighings for the lower part of the burette, and also by a direct comparison of this burette with the 25 cc. burette marked *B*, and an error of .1 cc. was discovered between the 70 and 75 cc. marks."

In graduating apparatus to Mohr's units, instead of to true cubic centimetres, proceed as described in **233**, page 250. When the normal weight, 26.048 grams, is used with Schmidt and Haensch polariscopes, the flasks should be graduated to Mohr's units ; with the Laurent polariscope, the flasks should be graduated to true cubic centimetres.

223. TABLE SHOWING THE EVAPORATION (PERCENTAGES BY WEIGHT) NECESSARY TO REDUCE A SUGAR SOLUTION FROM A LOWER TO A HIGHER DENSITY.—(G. L. SPENCER.)

Initial Density		35° Brix 19°.6 Baumé	37° Brix 20°.7 Baumé	39° Brix 21°.8 Baumé	41° Brix 22°.9 Baumé	43° Brix 23°.95 Baumé	45° Brix 25° Baumé	47° Brix 26°.1 Baumé	49° Brix 27°.2 Baumé	51° Brix 28°.2 Baumé	53° Brix 29°.3 Baumé	55° Brix 30°.4 Baumé	57° Brix 31°.4 Baumé	59° Brix 32°.5 Baumé
Degrees Brix.	Degrees Baumé.							Per cent.						
9.	5.1	74.28	75.67	76.92	78.05	79.07	80.00	80.85	81.63	82.35	83.02	83.63	84.21	84.74
9.2	5.2	73.71	75.13	76.41	77.56	78.60	79.55	80.42	81.22	81.96	82.64	83.27	83.86	84.40
9.4	5.3	73.14	74.59	75.89	77.07	78.14	79.11	80.00	80.82	81.57	82.26	82.91	83.51	84.07
9.6	5.4	72.57	74.05	75.38	76.58	77.67	78.77	79.58	80.41	81.17	81.88	82.54	83.16	83.73
9.8	5.55	72.00	73.51	74.87	76.09	77.21	78.22	79.15	80.00	80.78	81.51	82.18	82.81	83.39
10.	5.7	71.43	73.00	74.36	75.61	76.74	77.78	78.72	79.59	80.39	81.13	81.82	82.65	83.05
10.2	5.8	70.86	72.48	73.84	75.12	76.28	77.33	78.29	79.18	80.00	80.75	81.45	82.10	82.71
10.4	5.9	70.29	71.89	73.33	74.63	75.81	76.88	77.87	78.77	79.61	80.38	81.09	81.73	82.37
10.6	6.	69.71	71.35	72.82	74.14	75.35	76.44	77.44	78.37	79.21	80.00	80.72	81.40	82.03
10.8	6.1	69.14	70.81	72.31	73.66	74.88	76.00	77.02	77.96	78.82	79.62	80.36	81.05	81.69
11.	6.2	68.57	70.27	71.79	73.17	74.42	75.55	76.59	77.55	78.43	79.25	80.00	80.70	81.35
11.2	6.3	68.00	69.73	71.28	72.68	73.96	75.11	76.17	77.14	78.04	78.87	79.63	80.35	81.01
11.4	6.5	67.43	69.19	70.77	72.19	73.49	74.66	75.74	76.74	77.65	78.49	79.27	80.00	80.68
11.6	6.6	66.86	68.65	70.25	71.70	73.02	74.22	75.32	76.33	77.25	78.11	78.91	79.65	80.34
11.8	6.7	66.29	68.11	69.74	71.22	72.56	73.77	74.89	75.92	76.86	77.73	78.55	79.29	80.00
12.	6.8	65.71	67.56	69.23	70.73	72.09	73.33	74.47	75.51	76.47	77.36	78.18	78.95	79.66
12.2	6.9	65.14	67.02	68.72	70.24	71.63	72.88	74.04	75.10	76.08	76.97	77.82	78.59	79.32
12.4	7.	64.57	66.59	68.20	69.75	71.16	72.44	73.61	74.69	75.69	76.60	77.45	78.24	78.98
12.6	7.1	64.00	66.05	67.69	69.26	70.69	72.00	73.19	74.29	75.29	76.22	77.09	77.89	78.64
12.8	7.2	63.43	65.51	67.18	68.78	70.23	71.55	72.76	73.88	74.90	75.85	76.73	77.54	78.30

EVAPORATION TABLE (PERCENTAGES BY WEIGHT).—*Continued.*

Initial Density		35° Brix 19°.6 Baumé.	37° Brix 20°.7 Baumé.	39° Brix 21°.8 Baumé.	41° Brix 22°.9 Baumé.	43° Brix 23°.95 Baumé.	45° Brix 25° Baumé.	47° Brix 26°.1 Baumé.	49° Brix 27°.2 Baumé.	51° Brix 28°.2 Baumé.	53° Brix 29°.3 Baumé.	55° Brix 30°.4 Baumé.	57° Brix 31°.4 Baumé.	59° Brix 32°.5 Baumé.
Degrees Brix	Degrees Baumé.							Per cent.						
13.	7.4	62.86	64.86	66.66	68.29	69.76	71.11	72.34	73.47	74.51	75.46	76.36	77.19	77.97
13.2	7.5	62.99	64.32	66.15	67.80	69.30	70.66	71.91	73.06	74.12	75.09	76.00	76.84	77.63
13.4	7.6	61.71	63.88	65.64	67.31	68.83	70.22	71.49	72.65	73.73	74.71	75.64	76.49	77.29
13.6	7.7	61.14	63.34	65.13	66.83	68.37	69.78	71.06	72.25	73.33	74.34	75.27	76.14	76.95
13.8	7.8	60.57	62.80	64.61	66.34	67.90	69.33	70.64	71.84	72.94	73.96	74.91	75.79	76.61
14.	7.9	60.00	62.16	64.10	65.85	67.44	68.88	70.21	71.43	72.35	73.58	74.54	75.43	76.27
14.2	8.0	59.43	61.62	63.59	65.36	66.97	68.44	69.78	71.02	72.15	73.20	74.18	75.08	75.93
14.4	8.1	58.86	61.08	63.08	64.88	66.51	68.00	69.36	70.61	71.76	72.83	73.82	74.73	75.59
14.6	8.3	58.29	60.54	62.56	64.39	66.05	67.55	68.93	70.20	71.37	72.45	73.45	74.38	75.25
14.8	8.4	57.71	60.00	62.05	63.90	65.58	67.11	68.51	69.80	70.98	72.07	73.09	74.03	74.91
15.	8.5	57.14	59.46	61.54	63.41	65.12	66.66	68.09	69.39	70.59	71.61	72.72	73.68	74.58
15.2	8.35	56.57	58.92	61.02	62.93	64.65	66.41	67.66	68.78	70.19	71.32	72.36	73.33	74.24
15.4	8.7	56.00	58.38	60.51	62.44	64.18	65.77	67.23	68.57	69.80	70.94	72.00	72.98	73.90
15.6	8.8	55.43	57.84	60.00	61.95	63.72	65.33	66.80	68.16	69.41	70.56	71.64	72.63	73.56
15.8	8.9	54.86	57.30	59.49	61.46	63.25	64.88	66.38	67.75	69.01	70.19	71.27	72.28	73.22
16.	8.0	54.29	56.75	58.97	60.97	62.79	64.44	65.96	67.35	68.62	69.81	70.91	71.93	72.88
16.2	9.2	53.71	56.21	58.46	60.48	62.32	64.00	65.53	66.94	68.23	69.43	70.54	71.58	72.54
16.4	9.3	53.14	55.77	57.95	60.00	61.86	63.55	65.17	66.53	67.84	69.06	70.18	71.23	72.20
16.6	9.4	52.57	55.17	57.44	59.51	61.39	63.11	64.68	66.12	67.45	68.68	69.82	70.88	71.90
16.8	9.5	52.00	54.59	56.93	59.02	60.93	62.66	64.25	65.71	67.06	68.30	69.45	70.52	71.52

Brix of Sirup	Initial Brix													
17.	9.6	51.43	54.05	56.41	58.53	60.46	62.22	63.83	65.30	66.66	67.92	69.09	70.17	71.18
17.2	9.7	50.86	53.51	55.90	58.05	60.00	61.77	63.40	64.90	66.27	67.54	68.73	69.82	70.85
17.4	9.9	50.29	52.97	55.39	57.58	59.53	61.33	62.98	64.49	65.88	67.17	68.36	69.47	70.51
17.6	9.9	49.71	52.43	54.88	57.07	59.07	60.88	62.51	64.08	65.49	66.79	68.00	69.12	70.17
17.8	10.0	49.14	51.89	54.36	56.58	58.60	60.44	62.13	63.67	65.10	66.41	67.63	68.77	69.83
18.	10.1	48.57	51.35	53.81	56.09	58.14	60.00	61.70	63.26	64.70	66.03	67.27	68.42	69.49
18.2	10.3	48.00	50.81	53.33	55.61	57.67	59.55	61.27	62.86	64.31	65.66	66.91	68.07	69.15
18.4	10.4	47.43	50.27	52.82	55.12	57.21	59.11	60.85	62.45	63.92	65.28	66.55	67.72	68.81
18.6	10.5	46.86	49.73	52.31	54.63	56.74	58.66	60.42	62.04	63.53	64.90	66.18	67.37	68.47
18.8	10.6	46.29	49.19	51.80	54.14	56.28	58.22	60.00	61.63	63.13	64.53	65.82	67.01	68.13
19.	10.7	45.71	48.65	51.28	53.66	55.81	57.88	59.57	61.22	62.75	64.15	65.45	66.66	67.80
19.2	10.8	45.14	48.11	50.77	53.17	55.35	57.33	59.15	60.81	62.35	63.77	65.09	66.31	67.46
19.4	10.9	44.57	47.57	50.26	52.68	54.88	56.88	58.72	60.40	61.96	63.39	64.72	65.96	67.12
19.6	11.1	44.00	47.03	49.74	52.29	54.42	56.44	58.29	60.00	61.57	63.02	64.36	65.61	66.78
19.8	11.2	43.43	46.49	49.23	51.70	53.95	56.00	57.87	59.66	61.17	62.64	64.00	65.25	66.44
20.	11.3	42.86	45.95	48.72	51.22	53.49	55.55	57.44	59.18	60.78	62.25	63.63	64.91	66.10

Formula for calculating the above table:

$$\frac{\text{Degree Brix of Sirup} - \text{Initial Degree Brix}}{\text{Degree Brix of Sirup}} \times 100 = \text{Per cent water evaporated in terms of the weight of the original solution.}$$

224. TABLE SHOWING THE EVAPORATION (PERCENTAGES BY VOLUME) NECESSARY TO REDUCE A SUGAR SOLUTION FROM A LOWER TO A HIGHER DENSITY.—(G. L. SPENCER.)

(This table was originally constructed for the old Baumé scale, hence the odd numbers.)

INITIAL DENSITY			36°.4 Brix, 20°.4 Baumé, 1.1611 sp. gr.	38°.3 Brix, 21°.4 Baumé, 1.1707 sp. gr.	40°.1 Brix, 22°.4 Baumé, 1.1799 sp. gr.	42° Brix, 23°.4 Baumé, 1.1898 sp. gr.	44° Brix, 24°.5 Baumé, 1.2003 sp. gr.	45°.8 Brix, 25°.5 Baumé, 1.2099 sp. gr.	47°.7 Brix, 26°.5 Baumé, 1.2201 sp. gr.	49°.6 Brix, 27°.5 Baumé, 1.2305 sp. gr.	51°.5 Brix, 28°.5 Baumé, 1.2411 sp. gr.	53°.5 Brix, 29°.6 Baumé, 1.2523 sp. gr.	55°.4 Brix, 30°.6 Baumé, 1.2631 sp. gr.	57°.3 Brix, 31°.6 Baumé, 1.2740 sp. gr.
Degrees Baumé	Degrees Brix	Specific Gravity	Per cent.											
5.1	9.0	1.0359	77.94	79.21	80.30	81.34	82.34	83.17	83.97	84.72	85.41	86.08	86.62	87.22
5.3	9.3	1.0372	77.18	78.49	79.62	80.69	81.73	82.59	83.42	84.19	84.90	85.60	86.21	86.78
5.5	9.7	1.0388	76.16	77.53	78.71	79.84	80.93	81.82	82.69	83.11	84.24	84.96	85.60	86.20
5.7	10.1	1.0405	75.13	76.56	77.80	78.98	80.12	81.05	81.95	82.79	83.68	84.32	84.99	85.61
5.9	10.4	1.0418	74.12	75.84	77.11	78.32	79.49	80.45	81.39	82.25	83.05	83.88	84.52	85.16
6.1	10.8	1.0434	73.34	74.87	76.19	77.46	78.68	79.77	80.65	81.54	82.38	83.19	83.91	84.57
6.3	11.1	1.0447	72.56	74.14	75.50	76.80	78.05	79.07	80.08	81.00	81.86	82.70	83.43	84.12
6.5	11.5	1.0464	71.53	73.16	74.57	75.92	77.22	78.29	79.33	80.38	81.17	82.04	82.90	83.52
6.7	11.9	1.0481	70.49	72.19	73.65	75.04	76.38	77.49	78.56	79.56	80.50	81.38	82.17	82.91
6.9	12.2	1.0493	69.71	71.45	72.95	74.38	75.76	76.89	78.00	79.02	79.97	80.89	81.70	82.46
7.1	12.6	1.0510	68.67	70.45	72.02	73.59	74.92	76.10	77.24	78.30	79.28	80.23	81.07	81.85
7.3	12.9	1.0528	67.88	69.73	71.32	72.84	74.30	75.51	76.68	77.76	78.77	79.74	80.61	81.41
7.5	13.3	1.0540	66.83	68.74	70.39	71.96	73.47	74.71	75.92	77.04	78.08	79.02	79.98	80.81
7.75	13.7	1.0557	65.76	67.73	69.44	70.97	72.54	73.82	75.08	76.24	77.31	78.33	79.27	80.13
7.9	14	1.0570	64.99	67.00	68.73	70.40	71.98	73.30	74.58	75.76	76.86	77.92	78.86	79.74
8.1	14.4	1.0587	63.93	66.00	67.78	69.50	71.14	72.49	73.81	75.02	76.16	77.25	78.32	79.12
8.4	14.8	1.0604	62.95	65.09	66.84	68.60	70.29	71.68	73.04	74.29	75.56	76.55	77.09	78.51
8.5	15.1	1.0617	62.07	64.25	66.12	67.91	69.64	71.06	72.45	73.73	74.91	76.07	77.09	78.04
8.8	15.5	1.0634	61.00	63.34	65.17	67.01	68.79	70.25	71.67	72.90	74.21	75.40	76.44	77.42
8.9	15.8	1.0647	60.19	62.49	64.49	66.34	68.14	69.64	71.09	72.43	73.68	74.89	75.96	76.96

TABLE SHOWING THE EVAPORATION, ETC.—Continued.

Initial Density			36°.4 Brix. 20°.4 Baumé. 1.1611 sp. gr.	38°.3 Brix. 21°.4 Baumé. 1.1707 sp. gr.	40°.1 Brix. 22°.4 Baumé. 1.1799 sp. gr.	42° Brix. 23°.4 Baumé. 1.1898 sp. gr.	44° Brix. 24°.5 Baumé. 1.2003 sp. gr.	45°.8 Brix. 25°.5 Baumé. 1.2099 sp. gr.	47°.7 Brix. 26°.5 Baumé. 1.2201 sp. gr.	49°.6 Brix. 27°.5 Baumé. 1.2305 sp. gr.	51°.5 Brix. 28°.5 Baumé. 1.2411 sp. gr.	53°.5 Brix. 29°.6 Baumé. 1.2523 sp. gr.	55°.4 Brix. 30°.6 Baumé. 1.2631 sp. gr.	57°.8 Brix. 31°.6 Baumé. 1.2740 sp. gr.
Degree Baumé.	Degree Brix.	Specific Gravity.						Per cent.						
9.2	16.3	1.0665	59.12	61.47	63.49	65.44	67.30	68.88	70.32	71.70	72.97	74.22	75.32	76.88
9.4	16.6	1.0682	58.04	60.45	62.52	64.50	66.42	68.00	69.53	70.90	72.26	73.53	74.66	75.71
9.6	17.	1.0700	56.96	59.44	61.55	63.59	65.56	67.17	68.74	70.20	71.54	72.85	74.01	75.08
9.8	17.4	1.0717	55.88	58.42	60.60	62.70	64.70	66.36	67.97	69.46	70.84	72.18	73.36	74.47
10.	17.7	1.0730	55.08	57.64	59.85	61.91	64.06	65.72	67.36	68.88	70.29	71.65	72.96	73.98
10.1	18	1.0744	54.24	56.87	59.14	61.32	63.40	65.11	66.77	67.59	69.76	71.15	72.37	73.52

Formula for calculating the above table:

(Initial specific gravity × initial degree Brix) ÷ (specific gravity of the sirup × degree Brix of the sirup) × 100 = per cent. sirup obtained. 100 − per cent sirup = per cent water evaporated.

225. TABLE FOR THE REDUCTION OF THE WEIGHT OR VOLUME OF A SIRUP OF A GIVEN DEGREE BRIX OR BAUMÉ TO A SIRUP OF 54.3° BRIX OR 30° BAUMÉ.—(G. L. Spencer.)

Initial Density.		Equivalent Sirup of 54 3° Brix or 30° Baumé.		Initial Density.		Equivalent Sirup of 54.3° Brix or 30° Baumé.	
Degrees Brix.	Degrees Baumé.	Per Cents by Weight.	Per Cents by Volume.	Degrees Brix.	Degrees Baumé.	Per Cents by Weight.	Per Cents by Volume
35.0	19.6	64.46	59.19	**39.0**	21.8	71.82	67.10
.1	19.65	64.64	59.38	.1	21.8	72.00	67.30
.2	19.7	64.83	59.58	.2	21.9	72.19	67.51
.3	19.8	65.01	59.77	.3	21.9	72.37	67.71
.4	19.8	65.19	59.96	.4	22.0	72.55	67.91
.5	19.9	65.38	60.16	.5	22.05	72.74	68.12
.6	19.9	65.56	60.36	.6	22.1	72.92	68.32
.7	20.0	65.74	60.56	.7	22.2	73.10	68.52
.8	20.0	65.93	60.75	.8	22.2	73.29	68.72
.9	20.1	66.11	60.94	.9	22.3	73.47	68.92
36.0	20.1	66.30	61.14	**40.0**	22.3	73.66	69.12
.1	20.2	66.48	61.33	.1	22.4	73.84	69.32
.2	20 25	66.67	61.53	.2	22.4	74.02	69.52
.3	20.3	66.85	61.72	.3	22.5	74.21	69.73
.4	20.4	67.03	61.92	.4	22.5	74.40	69.93
.5	20 4	67.22	62.12	.5	22.6	74.58	70.14
.6	20.5	67.40	62.31	.6	22.6	74.76	70.34
.7	20.5	67.59	62.50	.7	22.7	74.94	70.54
.8	20.6	67.77	62.70	.8	22.8	75.13	70.74
.9	20.6	67.95	62.91	.9	22.8	75.31	70.94
37.0	20.7	68.14	63.11	**41.0**	22.9	75.50	71.15
.1	20.7	68.32	63.31	.1	22.9	75.68	71.35
.2	20.8	68.50	63 51	.2	23.0	75.87	71.55
.3	20.9	68.69	63.70	.3	23 0	76.06	71.75
.4	20.9	68.87	63.90	.4	23.1	76.24	71.95
.5	21.0	69.06	64.10	.5	23.1	76.42	72.16
.6	21.0	69.24	64.30	.6	23.2	76.60	72.37
.7	21.1	69.42	64.49	.7	23.25	76.78	72.58
.8	21.1	69.61	64.69	.8	23.3	76.97	72.79
.9	21.2	69.79	64.89	.9	23.4	77.16	73.00
38.0	21.2	69.98	65.09	**42.0**	23.4	77.34	73.21
.1	21.3	70.16	65.29	.1	23.5	77.52	73.41
.2	21.35	70.34	65.49	.2	23.5	77.70	73.61
.3	21.4	70.53	65.69	.3	23.6	77.89	73.81
.4	21.5	70.72	65.90	.4	23.6	78.08	74.01
.5	21.5	70.90	66.10	.5	23.7	78.26	74.22
.6	21.6	71.08	66.30	.6	23.7	78.44	74.43
.7	21.6	71.26	66.50	.7	23.8	78.62	74.64
.8	21.7	71.45	66.70	.8	23.8	78.81	74.86
.9	21.7	71.63	66.90	.9	23.9	79.00	75.08

TABLE FOR THE REDUCTION OF THE WEIGHT OR VOLUME
OF A SIRUP, ETC.—*Continued.*

Initial Density.		Equivalent Sirup of 54.3° Brix or 30° Baumé.		Initial Density.		Equivalent Sirup of 54.3° Brix or 30° Baumé.	
Degrees Brix.	Degrees Baumé.	Per Cents by Weight.	Per Cents by Volume.	Degrees Brix.	Degrees Baumé.	Per Cents by Weight.	Per Cents by Volume.
43.0	23.95	79.19	75.29	47.0	26.1	86.55	83.76
.1	24.0	79.37	75.49	.1	26.2	86.73	83.97
.2	24.1	79.55	75.69	.2	26.2	86.91	84.18
.3	24.1	79.74	75.89	.3	26.3	87.10	84.39
.4	24.2	79.92	76.10	.4	26.3	87.29	84.60
.5	24.2	80.11	76.31	.5	26.4	87.47	84.82
.6	24.3	80.29	76.52	.6	26.4	87.65	85.04
.7	24.3	80.47	76.73	.7	26.5	87.83	85.26
.8	24.4	80.66	76.95	.8	26.5	88.02	85.48
.9	24.4	80.85	77.17	.9	26.6	88.21	85.70
44.0	24.5	81.03	77.38	48.0	26.6	88.39	85.92
.1	24.55	81.21	77.59	.1	26.7	88.57	86.13
.2	24.6	81.39	77.80	.2	26.75	88.75	86.35
.3	24.65	81.58	78.01	.3	26.8	88.94	86.57
.4	24.7	81.77	78.22	.4	26.9	89.13	86.79
.5	24.8	81.95	78.43	.5	26.9	89.13	87.01
.6	24.8	82.13	78.64	.6	27.0	89.49	87.23
.7	24.9	82.31	78.85	.7	27.0	89.67	87.45
.8	24.9	82.50	79.06	.8	27.1	89.86	87.67
.9	25.0	82.69	79.27	.9	27.1	90.15	87.89
45.0	25.0	82.87	79.49	49.0	27.2	90.24	88.11
.1	25.1	83.05	79.70	.1	27.2	90.42	88.33
.2	25.1	83.23	79.91	.2	27.3	90.60	88.55
.3	25.2	83.42	80.12	.3	27.3	90.78	88.77
.4	25.2	83.61	80.33	.4	27.4	90.96	88.99
.5	25.3	83.79	80.54	.5	27.4	91.16	89.21
.6	25.4	83.97	80.75	.6	27.5	91.35	89.43
.7	25.4	84.15	80.96	.7	27.6	91.54	89.65
.8	25.5	84.34	81.18	.8	27.6	91.72	89.87
.9	25.5	84.53	81.40	.9	27.7	92.90	90.09
46.0	25.6	84.71	81.61	50.0	27.7	92.08	90.31
.1	25.6	84.89	81.82	.1	27.8	92.26	90.53
.2	25.7	85.07	82.03	.2	27.8	92.45	90.75
.3	25.7	85.26	82.24	.3	27.9	92.63	90.97
.4	25.8	85.45	82.45	.4	27.9	92.82	91.19
.5	25.8	85.63	82.66	.5	28.0	93.00	91.41
.6	25.9	85.81	82.87	.6	28.0	93.19	91.63
.7	25.95	85.99	83.09	.7	28.1	93.37	91.85
.8	26.0	86.18	83.31	.8	28.1	93.55	92.07
.9	26.1	86.37	83.53	.9	28.2	93.73	92.30

TABLE FOR THE REDUCTION OF THE WEIGHT OR VOLUME
OF A SIRUP, ETC.—*Continued.*

Initial Density.		Equivalent Sirup of 54.3° Brix or 30° Baumé.		Initial Density.		Equivalent Sirup of 54.3° Brix or 30° Baumé.	
Degrees Brix.	Degrees Baumé.	Per Cents by Weight.	Per Cents by Volume.	Degrees Brix.	Degrees Baumé.	Per Cents by Weight.	Per Cents by Volume.
51.0	28.2	93.92	92.53	**55.0**	30.4	101.28	101.61
.1	28.3	94.10	92.75	.1	30.4	101.46	101.84
.2	28.35	94.29	92.97	.2	30.5	101.64	102.07
.3	28.4	94.47	93.19	.3	30.5	101.83	102.30
.4	28.5	94.65	93.41	.4	30.6	102.01	102.53
.5	28.5	94.84	93.63	.5	30.6	102.20	102.76
.6	28.6	95.02	93.85	.6	30.7	102.38	102.99
.7	28.6	95.20	94.07	.7	30.7	102.56	103.22
.8	28.7	95.39	94.30	.8	30.8	102.75	103.45
.9	28.7	95.58	94.53	.9	30.8	102.94	103.68
52.0	28.8	95.76	94.77	**56.0**	30.9	103.13	103.92
.1	28.8	95.94	94.99	.1	30.9	103.31	104.15
.2	28.9	96.13	95.21	.2	31.0	103.49	104.38
.3	28.9	96.31	95.43	.3	31.05	103.68	104.61
.4	29.0	96.50	95.65	.4	31.1	103.86	104.84
.5	29.0	96.68	95.87	.5	31.2	104.05	105.07
.6	29.1	96.87	96.09	.6	31.2	104.23	105.30
.7	29.15	97.05	96.32	.7	31.3	104.41	105.54
.8	29.2	97.23	96.55	.8	31.3	104.60	105.78
.9	29.2	97.42	96.79	.9	31.4	104.78	106.02
53.0	29.3	97.60	97.02	**57.0**	31.4	104.97	106.26
.1	29.4	97.79	97.25	.1	31.5	105.15	106.49
.2	29.4	97.98	97.48	.2	31.5	105.34	106.72
.3	29.5	98.16	97.71	.3	31.6	105.52	106.95
.4	29.5	98.34	97.94	.4	31.6	105.70	107.18
.5	29.6	98.52	98.17	.5	31.7	105.89	107.41
.6	29.6	98.70	98.40	.6	31.7	106.07	107.65
.7	29.7	98.89	98.63	.7	31.8	106.25	107.89
.8	29.7	99.07	98.86	.8	31.8	106.44	108.13
.9	29.8	99.26	99.08	.9	31.9	106.62	108.37
54.0	29.8	99.44	99.30	**58.0**	31.9	106.81	108.61
.1	29.9	99.62	99.53	.1	32.0	106.99	108.84
.2	29.9	99.81	99.76	.2	32.0	107.17	109.08
54.3	**30.0**	**100.00**	**100.00**	.3	32.1	107.35	109.32
.4	30.05	100.18	100.22	.4	32.15	107.54	109.56
.5	30.1	100.36	100.45	.5	32.2	107.73	109.80
.6	30.2	100.55	100.68	.6	32.3	107.91	110.04
.7	30.2	100.73	100.91	.7	32.3	108.09	110.28
.8	30.3	100.91	101.14	.8	32.4	108.28	110.52
.9	30.3	101.09	101.37	.9	32.4	108.47	110.76

TABLE FOR THE REDUCTION OF THE WEIGHT OR VOLUME
OF A SIRUP, ETC.—*Continued.*

Initial Density.		Equivalent Sirup of 54.3° Brix or 30° Baumé.		Initial Density.		Equivalent Sirup of 54.3° Brix or 30° Baumé.	
Degrees Brix.	Degrees Baumé.	Per Cents by Weight.	Per Cents by Volume.	Degrees Brix.	Degrees Baumé.	Per Cents by Weight.	Per Cents by Volume.
59.0	32.5	108.65	111.00	60.0	33.0	110.49	113.39
.1	32.5	108.83	111.23	.1	33.0	110.68	113.63
.2	32.6	109.02	111.47	.2	33.1	110.86	113.87
.3	32.6	109.20	111.71	.3	33.1	111.04	114.11
.4	32.7	109.38	111.95	.4	33.2	111.23	114.35
.5	32.7	109.56	112.19	.5	33.2	111.41	114.59
.6	32.8	109.75	112.43	.6	33.3	111.60	114.83
.7	32.8	109.93	112.67	.7	33.35	111.78	114.97
.8	32.9	110.12	112.91	.8	33.4	111.96	115.21
.9	32.9	110.30	113.15	.9	33.45	112.14	115.45

The above table is for use in calculating sirups within
the usual range of densities, to a standard degree Brix or
Baumé, for purposes of comparison. A convenient check
on pan and centrifugal work is a statement showing the
analysis of the sirup and the pounds of first sugar yielded
per 100 lbs., or per 100 gallons of sirup of 54.3° Brix (30°
Baumé). The volume or weight of sirup at 54.3° Brix (30°
Baumé) is obtained by multiplying the measured volume
or the weight by the number in the per cent column in the
table corresponding to the observed degree Brix or Baumé
and pointing off as in other percentage calculations.

226. TABLE SHOWING THE VOLUMES OF JUICE, IN LITRES, YIELDED IN THE DIFFUSION OF 100 KILOGRAMS OF BEETS OF VARIOUS DENSITIES. (F. DUPONT.)

See page 207.

Density of the Diffusion-juice.	[1] Density of the Normal Juice of the Beet.								
	5°	5.5°	6°	6.5°	7°	7.5°	8°	8.5°	9°
3.6°	125	137	148	161	173	184	193	206	218
3.8	118	131	143	152	163	174	185	19₅	206
4	113	124	134	144	155	165	176	186	196
4.2	......	117	127	138	147	157	167	177	187
4.4	......	112	122	132	141	150	160	170	178
4.6	......	......	116	126	135	145	153	162	170
4.8	......	......	111	121	130	138	147	156	163
5	......	......	....	116	124	132	141	149	157
5.2	......	......	......	111	120	127	135	143	151
5.4	......	......	......	..	115	122	130	138	145
5.6	......	......	......	......	111	118	126	132	140
5.8	......	......	......	......	......	114	121	127	135
6	......	......	......	......	......	110	117	124	131
6.2	......	...	......	......	.. ..	......	114	120	127
6.4	......	......	......	......	......	......	110	117	123
6.6	......	......				......	......	114	119
6.8	......	......	.. .	.. .	......	......	110	115
7	......	......				......	......	... ..	112
7.2	......	... ..	......	.. ..	.....	.....	......	....	109

Litres of juice per 100 kilos. beets.

[1] The degrees given in this table are according to the French. To convert into specific gravity, prefix 10 and move the decimal point two places to the left. Example: 3.6° = 1.036 specific gravity = 9° Brix.

227. Formulæ for Concentration and Dilution. (1) Having two solutions of known degrees Brix (B and B'), to determine the degree Brix of a mixture composed of the volumes V and V' of these solutions.

$$x = \text{degree Brix required} = \frac{VB + V'B'}{V + V'}.$$

(2) Formula for the calculation of the water required (per cent by weight) to reduce a sugar solution of a given density to any required density.

$x =$ per cent of water required; $B =$ initial degree Brix; $b =$ degree Brix after dilution; $\dfrac{B - b}{B} = E$, and $\dfrac{100E}{100 - E} = x$, the per cent required.

(3) For formulæ for the concentration of sugar solutions from stated densities to certain required densities, see pages 69, 71.

(4) To determine the volume V of a sugar solution before concentration.

$b =$ degree Brix, $s =$ the specific gravity of the solution before concentration; $B =$ degree Brix, $S =$ specific gravity after concentration to a volume of 100.

$$V = \frac{100SB}{sb}.$$

228. TABLE SHOWING A COMPARISON OF THERMOMETRIC SCALES.

(Schubarth's Handbuch der techn. Chem. III. Aufl. I. 61.)

Fahrenheit.	Centigrade.	Réaumur.	Fahrenheit.	Centigrade.	Réaumur.	Fahrenheit.	Centigrade.	Réaumur.
°	°	°	°	°	°	°	°	°
212	100	80	190	87.78	70.22	168	75.55	60.44
211	99.44	79.56	189	87.22	69.78	167	75	60
210	98.89	79.11	188	86.67	69.33	166	74.44	59.56
209	98.33	78.67	187	86.11	68.89	165	73.89	59.11
208	97.78	78.22	186	85.55	68.44	164	73.33	58.67
207	97.22	77.78	185	85	68	163	72.78	58.22
206	96.67	77.33	184	84.44	67.56	162	72.22	57.78
205	96.11	76.89	183	83.89	67.11	161	71.67	57.33
204	95.55	76.44	182	83.33	66.67	160	71.11	56.89
203	95	76	181	82.78	66.22	159	70.55	56.44
202	94.44	75.56	180	82.22	65.78	158	70	56
201	93.89	75.11	179	81.67	65.33	157	69.44	55.56
200	93.33	74.67	178	81.11	64.89	156	68.89	55.11
199	92.78	74.22	177	80.55	64.44	155	68.33	54.67
198	92.22	73.78	176	80	64	154	67.78	54.22
197	91.67	73.33	175	79.44	63.56	153	67.22	53.78
196	91.11	72.89	174	78.89	63.11	152	66.67	53.33
195	90.55	72.44	173	78.33	62.67	151	66.11	52.89
194	90	72	172	77.78	62.22	150	65.55	52.44
193	89.44	71.56	171	77.22	61.78	149	65	52
192	88.89	71.11	170	76.67	61.33	148	64.44	51.56
191	88.33	70.67	169	76.11	60.89	147	63.89	51.11

COMPARISON OF THERMOMETRIC SCALES.—*Continued.*

Fahrenheit.	Centigrade.	Réaumur.	Fahrenheit.	Centigrade.	Réaumur.	Fahrenheit.	Centigrade.	Réaumur.
°	°	°	°	°	°	°	°	°
146	63.33	50.67	83	28.33	22.67	21	−6.11	−4.89
145	62.78	50.22	82	27.78	22.22	20	−6.67	−5.33
144	62.22	49.78	81	27.22	21.78	19	−7.22	−5.78
143	61.67	49.33	80	26.67	21.33	18	−7.78	−6.22
142	61.11	48.89	79	26.11	20.89	17	−8.33	−6.67
141	60.55	48.44	78	25.55	20.44	16	−8.89	−7.11
140	60	48	77	25	20	15	−9.44	−7.56
139	59.44	47.56	76	24.44	19.56	14	−10	−8
138	58.89	47.11	75	23.89	19.11	13	−10.55	−8.44
137	58.33	46.67	74	23.33	18.67	12	−11.11	−8.89
136	57.78	46.22	73	22.78	18.22	11	−11.67	−9.33
135	57.22	45.78	72	22.22	17.78	10	−12.22	−9.78
134	56.67	45.33	71	21.67	17.33	9	−12.78	−10.22
133	56.11	44.89	70	21.11	16.89	8	−13.33	−10.67
132	55.55	44.44	69	20.55	16.44	7	−13.89	−11.11
131	55	44	68	20	16	6	−14.44	−11.56
130	54.44	43.56	67	19.44	15.56	5	−15	−12
129	53.89	43.11	66	18.89	15.11	4	−15.55	−12.44
128	53.33	42.67	65	18.33	14.67	3	−16.11	−12.89
127	52.78	42.22	64	17.78	14.22	2	−16.67	−13.33
126	52.22	41.78	63	17.22	13.78	1	−17.22	−13.78
125	51.67	41.33	62	16.67	13.33	0	−17.78	−14.22
124	51.11	40.89	61	16.11	12.89	−1	−18.33	−14.67
123	50.55	40.44	60	15.55	12.44	−2	−18.89	−15.11
122	50	40	59	15	12	−3	−19.44	−15.56
121	49.44	39.56	58	14.44	11.56	−4	−20	−16
120	48.89	39.11	57	13.89	11.11	−5	−20.55	−16.44
119	48.33	38.67	56	13.33	10.67	−6	−21.11	−16.89
118	47.78	38.22	55	12.78	10.22	−7	−21.67	−17.33
117	47.22	37.78	54	12.22	9.78	−8	−22.22	−17.78
116	46.67	37.33	53	11.67	9.33	−9	−22.78	−18.22
115	46.11	36.89	52	11.11	8.89	−10	−23.33	−18.67
114	45.55	36.44	51	10.55	8.44	−11	−23.89	−19.11
113	45	36	50	10	8	−12	−24.44	−19.56
112	44.44	35.56	49	9.44	7.56	−13	−25	−20
111	43.89	35.11	48	8.89	7.11	−14	−25.55	−20.44
110	43.33	34.67	47	8.33	6.67	−15	−26.11	−20.89
109	42.78	34.22	46	7.78	6.22	−16	−26.67	−21.33
108	42.22	33.78	45	7.22	5.78	−17	−27.22	−21.78
107	41.67	33.33	44	6.67	5.33	−18	−27.78	−22.22
106	41.11	32.89	43	6.11	4.89	−19	−28.33	−22.67
105	40.55	32.44	42	5.55	4.44	−20	−28.89	−23.11
104	40	32	41	5	4	−21	−29.44	−23.56
103	39.44	31.56	40	4.44	3.56	−22	−30	−24
102	38.89	31.11	39	3.89	3.11	−23	−30.55	−24.44
101	38.33	30.67	38	3.33	2.67	−24	−31.11	−24.89
100	37.78	30.22	37	2.78	2.22	−25	−31.67	−25.33
99	37.22	29.78	36	2.22	1.78	−26	−32.22	−25.78
98	36.67	29.33	35	1.67	1.33	−27	−32.78	−26.22
97	36.11	28.89	34	1.11	0.89	−28	−33.33	−26.67
96	35.55	28.44	33	0.55	0.44	−29	−33.89	−27.11
95	35	28	32	0.	0.	−30	−34.44	−27.56
94	34.44	27.56	31	−0.55	−0.44	−31	−35	−28
93	33.89	27.11	30	−1.11	−0.89	−32	−35.55	−28.44
92	33.33	26.67	29	−1.67	−1.33	−33	−36.11	−28.89
91	32.78	26.22	28	−2.22	−1.78	−34	−36.67	−29.33
90	32.22	25.78	27	−2.78	−2.22	−35	−37.22	−29.78
89	31.67	25.33	26	−3.33	−2.67	−36	−37.78	−30.22
88	31.11	24.89	25	−3.89	−3.11	−37	−38.33	−30.67
87	30.55	24.44	24	−4.44	−3.56	−38	−38.89	−31.11
86	30	24	23	−5	−4	−39	−39.44	−31.56
85	29.44	23.56	22	−5.55	−4.44	−40	−40	−32
84	28.89	23.11						

Formulæ for the conversion of the degrees of one thermometric scale into those of another:

$$F = \tfrac{9}{5}C + 32 = \tfrac{9}{4}R + 32.$$
$$C = \tfrac{5}{9}(F - 32) = \tfrac{5}{4}R.$$
$$R = \tfrac{4}{9}(F - 32) = \tfrac{4}{5}C.$$

Additions and subtractions are algebraic.

229. TABLE SHOWING A COMPARISON OF THERMOMETRIC SCALES.

Centi-grade.	Fahren-heit.	Réau-mur.	Centi-grade.	Fahren-heit.	Réau-mur.	Centi-grade.	Fahren-heit.	Réau-mur.
°	°	°	°	°	°	°	°	°
100	212	80	62	143.6	49.6	24	75.2	19.2
99	210.2	79.2	61	141.8	48.8	23	73.4	18.4
98	208.4	78.4	60	140	48	22	71.6	17.6
97	206.6	77.6	59	138.2	47.2	21	69.8	16.8
96	204.8	76.8	58	136.4	46.4	20	68	16
95	203	76	57	134.6	45.6	19	66.2	15.2
94	201.2	75.2	56	132.8	44.8	18	64.4	14.4
93	199.4	74.4	55	131	44	17	62.6	13.6
92	197.6	73.6	54	129.2	43.2	16	60.8	12.8
91	195.8	72.8	53	127.4	42.4	15	59	12
90	194	72	52	125.6	41.6	14	57.2	11.2
89	192.2	71.2	51	123.8	40.8	13	55.4	10.4
88	190.4	70.4	50	122	40	12	53.6	9.6
87	188.6	69.6	49	120.2	39.2	11	51.8	8.8
86	186.8	68.8	48	118.4	38.4	10	50	8
85	185	68	47	116.6	37.6	9	48.2	7.2
84	183.2	67.2	46	114.8	36.8	8	46.4	6.4
83	181.4	66.4	45	113	36	7	44.6	5.6
82	179.6	65.6	44	111.2	35.2	6	42.8	4.8
81	177.8	64.8	43	109.4	34.4	5	41	4
80	176	64	42	107.6	33.6	4	39.2	3.2
79	174.2	63.2	41	105.8	32.8	3	37.4	2.4
78	172.4	62.4	40	104	32	2	35.6	1.6
77	170.6	61.6	39	102.2	31.2	1	33.8	.8
76	168.8	60.8	38	100.4	30.4	0	32	0
75	167	60	37	98.6	29.6	−1	30.2	− .8
74	165.2	59.2	36	96.8	28.8	−2	28.4	−1.6
73	163.4	58.4	35	95	28	−3	26.6	−2.4
72	161.6	57.6	34	93.2	27.2	−4	24.8	−3.2
71	159.8	56.8	33	91.4	26.4	−5	23	−4
70	158	56	32	89.6	25.6	−6	21.2	−4.8
69	156.2	55.2	31	87.8	24.8	−7	19.4	−5.6
68	154.4	54.4	30	86	24	−8	17.6	−6.4
67	152.6	53.6	29	84.2	23.2	−9	15.8	−7.2
66	150.8	52.8	28	82.4	22.4	−10	14	−8
65	149	52	27	80.6	21.6	−11	12.2	−8.8
64	147.2	51.2	26	78.8	20.8	−12	10.4	−9.6
63	145.4	50.4	25	77	20			

230. APPROXIMATE TEMPERATURES OF IRON WHEN HEATED UNTIL IT HAS THE FOLLOWING COLORS:

	° F.	° C.		° F.	° C.
Faint red.............	977	525	Orange...............	2100	1150
Dark red.............	1292	700	White.......	2370	1300
Cherry-red	1666	908	Dazzling white,.......	2730	1500
Bright cherry-red.....	1832	1000			

231. TABLE SHOWING THE ALTERATION OF **THE VOLUME** OF GLASS VESSELS BY HEAT, THE **VOLUME** AT 15° C. BEING TAKEN AS UNITY.

(From Bailey's "Chemist's Pocket-Book.")

Temp. ° C.	Volume.	Temp. ° C.	Volume.	Temp. ° C.	Volume.
0	.99961210	15	1.00000000	30	1.00038790
1	.99963796	16	1.00002586	35	1.00051720
2	.99966382	17	1.00005172	40	1.00064650
3	.99968968	18	1.00007758	45	1.00077580
4	.99971554	19	1.00010344	50	1.00090510
5	.99974140	20	1.00012930	55	1.00103440
6	.99976726	21	1.00015516	60	1.00116370
7	.99979313	22	1.00018102	65	1.00129300
8	.99981898	23	1.00020688	70	1.00142230
9	.99984484	24	1.00023274	75	1.00155160
10	.99987070	25	1.00025860	80	1.00168090
11	.99989656	26	1.00028446	85	1.00181020
12	.99992242	27	1.00031032	90	1.00193950
13	.99994828	28	1.00033618	95	1.00206880
14	.99997414	29	1.00036204	100	1.00219810

232. COEFFICIENTS OF EXPANSION (CUBICAL) OF ORDINARY GLASS.

EXPANSION PER DEGREE FROM—

0° C. to 100° C.	0° C. to 150° C.	0° C. to 200° C.	0° C. to 250° C.	0° C. to 300° C.
.0000276	.0000284	.0000291	.0000298	.0000306

233. TABLE SHOWING THE APPARENT WEIGHT OF 1,000 MOHR'S UNITS (MOHR'S LITRE) OF WATER AT DIFFERENT TEMPERATURES AS WEIGHED WITH BRASS WEIGHTS IN THE AIR.

Corrected for expansion and contraction of the glass container, for temperatures above and below 17½° C. Based on Payne's Table, page 234.

Temp. ° C.	Apparent Weight.	Temp. ° C.	Apparent Weight.	Temp. ° C.	Apparent Weight.	Temp. ° C.	Apparent Weight.
	Grams.		Grams.		Grams.		Grams.
15	1000.3	19	999.8	24	999.8	29	997.6
16	1000.2	20	999.6	25	999.6	30	997.4
17	1000.1	21	999.4	26	998.4	31	997.1
17½	**1000.0**	22	999.2	27	998.2	32	996.8
18	999.9	23	999.0	28	997.9	33	996.5
....	....	....	....	....	....	34	996.2

The above table may be used in the graduation of sugar-flasks, burettes, etc., to Mohr's units. This unit is the volume occupied by 1 gram of water, as weighed with brass weights in the air, at 17½° C., and is frequently termed "Mohr's cc."

In checking a litre flask, it should be counterpoised on a good scale, and the number of grams of water corresponding to its temperature run into it. If the flask be correctly graduated, this quantity of water should fill it to the mark. The water should be at the temperature of the laboratory. The same principle is applied in checking other graduated ware to Mohr's units.

For methods of graduating apparatus to true cubic centimetres, see 222.

234. KOPP'S TABLE, SHOWING THE EXPANSION OF WATER FROM 0° C. TO 100° C. (32° F. TO 212° F.).

Temp. ° C.	Temp ° F.	Volume.	Temp. ° C.	Temp. ° F.	Volume.
0	32	1.000000	21	69.8	1.001776
1	33.8	.999947	22	71.6	1.001995
2	35.6	.999908	23	73.4	1.002225
3	37.4	.999885	24	75.2	1.002465
4	39.2	.999877	25	77.0	1.002715
5	41.0	.999883	30	86.0	1.004064
6	42.8	.999903	35	95.0	1.005697
7	44.6	.999938	40	104.0	1.007531
8	46.4	.999986	45	113.0	1.009541
9	48.2	1.000048	50	122.0	1.011766
10	50.0	1.000124	55	131.0	1.014100
11	51.8	1.000213	60	140.0	1.016590
12	53.6	1.000314	65	149.0	1.019302
13	55.4	1.000429	70	158.0	1.022246
14	57.2	1.000556	75	167.0	1.025440
15	59.0	1.000695	80	176.0	1.028581
16	60.8	1.000846	85	185.0	1.031894
17	62.6	1.001010	90	194.0	1.035397
18	64.4	1.001184	95	203.0	1.039094
19	66.2	1.001370	100	212.0	1.042986
20	68.0	1.001567			

235. TABLE SHOWING THE EXPANSION OF WATER AND THE WEIGHT OF A UNIT VOLUME AT DIFFERENT TEMPERATURES.

(Abridgment of F. Rossetti's Table.)

° C.	Weight. +4° C. = 1.	Volume. +4° C. = 1.	° C.	Weight. +4° C. = 1.	Volume. +4° C. = 1.
+4	1.000000	1.000000	20	0.998259	1.001744
5	0.999990	1.000010	21	0.998047	1.001957
6	0.999970	1.000030	22	0.997828	1.002177
7	0.999933	1.000067	23	0.997601	1.002405
8	0.999886	1.000114	24	0.997367	1.002641
9	0.999824	1.000176	25	0.997120	1.002888
10	0.999747	1.000253	26	0.996866	1.003144
11	0.999655	1.000354	27	0.996603	1.003408
12	0.999549	1.000451	28	0.996331	1.003682
13	0.999430	1.000570	29	0.996051	1.003965
14	0.999299	1.000701	30	0.99575	1.00425
15	0.999160	1.000841	31	0.99547	1.00455
16	0.999002	1.000999	32	0.99517	1.00486
17	0.998841	1.000116	33	0.99485	1.00518
18	0.998654	1.001348	34	0.99452	1.00551
19	0.998460	1.001542	35	0.99418	1.00586

236. TABLE SHOWING THE VOLUME OF SUGAR SOLUTIONS AT DIFFERENT TEMPERATURES.—(GERLACH.)

Temp.°C.	10 per cent.	20 per cent.	30 per cent.	40 per cent.	50 per cent.
0°	10000	10000	10000	10000	10000
5	10004.5	10007	10009	10012	10016
10	10012	10016	10021	10026	10032
15	10021	10028	10034	10042	10050
20	10033	10041	10049	10058	10069
25	10048	10057	10066	10075	10088
30	10064	10074	10084	10094	10110
35	10082	10092	10103	10114	10132
40	10101	10112	10124	10136	10156
45	10122	10134	10146	10160	10180
50	10145	10156	10170	10184	10204
55	10170	10183	10196	10210	10229
60	10197	10209	10222	10235	10253
65	10225	10236	10249	10261	10278
70	10255	10265	10277	10287	10306
75	10284	10295	10306	10316	10332
80	10316	10325	10335	10345	10360
85	10347	10355	10365	10375	10388
90	10379	10387	10395	10405	10417
95	10411	10418	10425	10435	10445
1 00	10442	10450	10456	10465	10457

237. TABLE SHOWING THE CONTRACTION OF INVERT SUGAR ON DISSOLVING IN WATER; ALSO, THE CONTRACTION OF CANE-SUGAR SOLUTIONS ON INVERSION.

(From "Manuel Agenda" Gallois and Dupont.)

Per Cent Sugar.	Volume.	Contraction.	SPECIFIC GRAVITY.	
			Cane-Sugar Solution.	Invert-Sugar Solution.
0	1.00000	0.00000	1.0000	1.0000
5	.99863	0.00137	1.0203	1.0206
10	.99744	0.00256	1.0413	1.0418
15	.99639	0.00361	1.0630	1.0631
20	.99546	0.00454	1.0854	1.0856
25	.99462	0.00538	1.1086	1.1086

238. TABLE SHOWING THE BOILING-POINT OF SUGAR SOLUTIONS.—(GERLACH.)

Strength of Solution.	Boiling-point, ° C.	Boiling-point, ° F.
Per cent.		
10	100.4	212.7
20	100.6	213.1
30	101	213.8
40	101.5	214.7
50	102	215.6
60	103	217.4
70	106.5	223.7
79	112	233.6
90.8	130	266

239. TABLE SHOWING THE SOLUBILITY OF LIME IN SOLUTIONS OF SUGAR.

Sugar in 100 parts water.	Density of Sirup.	Density after saturation with lime.	100 PARTS OF THE RESIDUE DRIED AT 120° C. CONTAIN:	
			Lime.	Sugar.
40	1.122	1.179	21	79
35	1.110	1.166	20.5	79.5
30	1.096	1.148	20.1	79.9
25	1.082	1.128	19.8	80.2
20	1.068	1.104	18.8	81.2
15	1.052	1.080	18.5	81.5
10	1.036	1.053	18.1	81.9
5	1.018	1.026	15.3	84.7

240. TABLE SHOWING THE SOLUBILITY OF SUGAR IN WATER.—(AFTER FLOURENS.)

Temp. ° C.	Sugar. Per Cent.	Degree Baumé at the observed temperature.	at 15° C.	Temp. ° C.	Sugar. Per Cent.	Degree Baumé at the observed temperature.	at 15° C.
0	64.9	35.3	34.6	55	72.8	37.5	39.3
5	65	35.35	34.9	60	74	37.9	39.9
10	65.5	35.45	35.2	65	75	38.3	40.55
15	66	35.5	35.5	70	76.1	38.6	41.1
20	66.5	35.6	35.7	75	77.2	39	41.7
25	67.2	35.8	36.25	80	78.35	39.3	42.2
30	68	36	36.7	85	79.5	39.65	42 8
35	68.8	36.2	37.1	90	80.6	39.95	43.3
40	69.75	36.4	37.5	95	81.6	40.1	43.7
45	70 8	36.75	38.1	100	82.5	40.3	44.1
50	71.8	37.1	38.7				

241. TABLE SHOWING THE SOLUBILITY OF SUGAR IN WATER. (HERZFELD.)

Temp. ° C.	Sugar. Per Cent.	Temp. ° C.	Sugar. Per Cent.	Temp. ° C.	Sugar. Per Cent.
0	64.18	35	69.55	70	76.22
5	64.87	40	70.42	75	77.27
10	65.58	45	71.32	80	78.36
15	66.53	50	72.25	85	79.46
20	67.09	55	73.20	90	80.61
25	67.89	60	74.18	95	81.77
30	67.80	65	75.88	100	82.97

The solubility is decreased by presence of a small quantity of organic or inorganic salts, but increased by a large quantity.

242. TABLE SHOWING THE SOLUBILITY OF SUGAR IN ALCOHOL.
AT 17.5° C.   (Otto Schrefeld.)

(Zeit. f. Rübenzucker-Ind., 44, 970.)

Alcohol Per Cent by Weight.	Sucrose Per Cent.	Sucrose in Grams in 100 Grams of the Mixture of Alcohol and Water Solution.
0	66.20	195.8
5*	64.25	179.7
10*	62.20	164.5
15	60.40	152.5
20*	58.55	141.2
25	56.20	128.3
30	54.05	117.8
35	51.25	105.3
40	47.75	91.3
45	43.40	76.6
50	38.55	62.7
55	32.80	48.8
60	26.70	36.4
65	19.50	24.2
70	12.25	13.9
75	7.20	7.7
80	4.05	4.2
85	2.10	2.1
90	0.95	0.09
95	0.15	0.01
Absolute	0.00	0.00

* Calculated.

243. TABLE SHOWING THE SOLUBILITY OF STRONTIA IN SUGAR SOLUTIONS.   (Sidersky.)

Per Cent Sucrose.	Strontia (SrO) Per Cent of the Solution.				Per Cent Sucrose.	Strontia (SrO) Per Cent of the Solution.			
	At 3° C.	At 15° C.	At 24° C.	At 40° C.		At 3° C.	At 15° C.	At 24° C.	At 40° C.
1	0.45	0.65	0.70	1.68	11	1.30	1.57	2.01	3.75
2	0.53	0.75	0.83	1.89	12	1.38	1.66	2.14	3.96
3	0.62	0.84	0.96	2.09	13	1.47	1.75	2.28	4.16
4	0.70	0.93	1.09	2.30	14	1.55	1.84	2.41	4.37
5	0.79	1.03	1.22	2.51	15	1.64	1.94	2.55	4.58
6	0.87	1.12	1.35	2.72	16	1.72	2.03	2.69	4.79
7	0.96	1.21	1.48	2.92	17	1.81	2.12	2.83	4.99
8	1.04	1.30	1.61	3.13	18	1.90	2.21	2.97	5.20
9	1.13	1.39	1.74	3.33	19	1.99	2.30	3.11	5.41
10	1.21	1.48	1.87	3.55	20	2.08	2.39	3.25	5.61

243a. TABLE SHOWING THE SOLUBILITY OF BARYTA IN
SUGAR SOLUTIONS.

(PELLET and SENCIER, *La fabrication du sucre*, **1**, 186.)

Sucrose per 100 cc.	Baryta (BaO) per 100 cc.	Baryta (BaO) per cent Sucrose.
2 5	4.59	18.3
5	5.46	10.9
7.5	6.56	8 7
10	7.96	7.7
12.5	9.41	7.5
15	10.00	6.6
20	10.90	5.4
25	12.90	5.1
30	14.68	4.9

243b. TABLE SHOWING THE SOLUBILITY OF CERTAIN SALTS
IN WATER IN THE PRESENCE OF SUCROSE.

(JACOBSTHAL, *Zeit. Rübenzuckerind*, **18**, 649; taken from Sidersky's
Traité d'analyse des Matières Sucrées, p. 11.)

Solution containing	5% Sucrose.	10% Sucrose.	15% Sucrose.	20% Sucrose.	25% Sucrose.
	Grams.	Grams.	Grams.	Grams.	Grams.
Sulphate of calcium.	2.095	1.946	1.593	1.539	1.333
Carb. of calcium	0.027	0.036	0.024	0.022	0.008
Oxalate of calcium .	0.033	0.047	0.012	0.008	0.001
Phosph. of calcium .	0 029	0.028	0 014	0.018	0.005
Citrate of calcium...	1.813	1.578	1.505	1.454	1.454
Carb. of magnesium	0.317	0.199	0.194	0.213	0.284

244. TABLE SHOWING THE MORE IMPORTANT CHEMICAL AND PHYSICAL PROPERTIES OF THE CARBOHYDRATES. COMPILED BY ERVIN E. EWELL.

(From von Lippmann's "Chemie der Zuckerarten," second edition; Tollens' "Handbuch der Kohlenhydrate," with reference to original papers.)

Name, Synonyms, and Formulæ.	Melting-point ° C.	Melting-point of Osazone ° C.	Source, etc.
I. MONOSACCHARIDES.			
A. Bioses.			
Glycollic aldehyde... $C_2H_4O_2 = CH_2OH.COH.$	......	169°–170°	Obtained by oxidation of glycol with nitric acid or reduction of glyoxal with zinc and acetic acid. Has not been obtained in a pure state.
B. Trioses.			
Glycerose... $C_3H_6O_3 = CH_2OH.CO.CH_2OH.$ $C_3H_6O_3 = CH_2OH.CHOH.CHO.$	Sirupy.	131°; 200°	Obtained by oxidization of glycerol with nitric acid or with bromine. A mixture of dioxyacetone and glycerinaldehyde.
C. Tetroses.			
Erythrose... $C_4H_8O_4 = CH_2OH.(CHOH)_2.CHO.$	......	¹166°–168°	Made from tetrahydric alcohol erythrol (erythrite) by oxidation with nitric acid. Probably mixed with isomeric ketone $CH_2OH.CHOH.CO.CH_2OH.$ Only its osazone has been obtained pure. Erythrol occurs free and combined in lichens and algæ.
D. Pentoses.			
1. *Aldo-pentoses.* $C_5H_{10}O_5 = CH_2OH.(CHOH)_3.CHO.$			
l-arabinose... arabinose, arabose... pectinose, pectin-sugar...	²160°; ⁴150° ⁴130°	160°	Obtained alone or mixed with various other sugars by hydrolysis of gums and other vegetable substances,
d-arabinose...	160°	159°–160°	Obtained by synthesis,
i-arabinose...		163°	Obtained by mixing equal weights of l-arabinose and d-arabinose.
xylose... wood-sugar...	⁴135°–140° ⁵141°; ⁷144° ¹²145°; ¹¹154° ⁶150°–153°	¹¹152°–155° ⁹155; ¹⁰158° ¹²160°; ⁸,⁹161° ⁶170°	Obtained alone or mixed with various other sugars by the hydrolysis of gums and other vegetable substances. Its most important source is a wood gum (xylan), extracted from various kinds of wood by means of dilute solutions of caustic alkalies.

i-xylose. Inactive xylose	......	210°–215°	This name has been given by Fischer to a synthetic sugar of which he prepared the osazone.
Ribose	Sirup.	......	Obtained by the reduction of the lactone of ribonic acid.
2. *Keto-pentoses*. $CH_2OH.(CHOH)_2.CO.CH_2OH$.	......	......	Not yet known with certainty, but their existence is evidenced by the work of Fischer.
3. *Methyl-pentoses*. $C_6H_{12}O_5 = C_5H_9O_5.CH_3$.			
Fucose	194°–105°	159°	From the seaweed *Fucus nodosus*.
Rhamnose...... Isodulcite rhamnodulcite, rhamnegin-sugar, besperidin-sugar.	180°	180°	Widely distributed in the vegetable kingdom in the form of various glucosides.
Chinovose... $C_6H_{12}O_5 = CH_3.(CHOH)_4.CHO$.	Sirup.	192°–194°	A decomposition product of chinovite, a constituent of cinchona bark.
E. HEXOSES. 1. *Aldo-hexoses*. $CH_2OH.(CHOH)_4.CHO$.			
d-glucose...... Dextrose, grape-sugar, starch-sugar.	[14]144°–146° [15]89°–90°	205°–210°	Free and combined in the form of glucosides, gums, and polysaccharides, and in mixtures with various other sugars in a host of vegetable spines and in animal secretions.
l-glucose......	141°–143°	208°	Reduction of the lactone of *l*-gluconic acid.
i-glucose...... Inactive glucose.	Sirup.	210°–212°	Obtained by mixing equal quantities of *d*- and *l*-glucoses.
d-mannose...... Mannose, dextrorotatory mannose, isomannitose, seminose.	......	[16]205°–210°	Oxidation of mannite; hydrolysis of various vegetable materials.
l-mannose...... Levorotatory mannose.	......	[16]208°	Obtained synthetically by reduction of *l*-mannonic acid.
i-mannose...... Inactive mannose.	......	[16]210°–212°	Obtained by the reduction of the lactone of *i*-mannonic acid.
d-gulose ...... Dextrorotatory gulose.	Sirup.	......	Obtained by the reduction of *d*-gulonic acid, which is in turn obtained by the reduction of glycuronic acid or *d*-saccharic acid.

[1] Fischer and Laudsteiner. [2] Scheibler, Lippmann, Conrad, and Guthzeit. [3] Ost. [4] Martin. [5] Bauer. [6] Bertrand. [7] Wheeler and Tollens. [8] Allen and Tollens. [9] Tollens. [10] Stone and Test. [11] Hébert. [12] Koch. [13] Behrend and Will. [14] Anhydride. [15] Hydrate. [16] The *d*-, *l*-, and *i*-mannoses form *d*-, *l*-, and *i*-glucosazones, respectively.

244. TABLE SHOWING THE MORE IMPORTANT PROPERTIES OF THE CARBOHYDRATES.—Continued.

Name, Synonyms, and Formulæ.	Melting-point ° C.	Melting-point of Osazone ° C.	Source, etc.
l-gulose	Sirup.	156°	Obtained synthetically by the reduction of l-gulonic acid.
i-gulose	...	157°	Obtained by the reduction of i-gulonic acid.
d-idose	...	...	Obtained by the reduction of d-idonic acid, which is obtained by heating d-gulonic acid with pyridin. Yields same osazone as d-gulose.
l-idose	...	...	Obtained from l-gulonic acid in same manner that d-gulonic acid yields d-idose.
i-idose	161°-170°	196°-197°	Existence of i-idose is reported by Fischer.
d-galactose	...	...	Rarely reported free in nature, but is widely distributed in combination. It is obtained pure or mixed by the hydrolysis of milk sugar, various glucosides and polysaccharides.
galactose, dextrorotatory galactose.	118°-120°	...	
l-galactose	162°-163°	192°-195°	Obtained synthetically.
i-galactose	140°-142°	206°	Obtained synthetically. Bears same relation to d-galactose that d-mannose bears to d-glucose. Forms d-galactosazone.
d-talose	Sirup.	...	
l-talose	...	...	Only the derivative l-talomucic acid is known.
i-talose	...	...	Only the derivative i-talite is known.
Chitose	...	...	Obtained by hydrolysis of chitin, a substance found in wing-cases of insects, etc. Found in plants.
2. $Keto$-$hexoses$, $CH_2OH(CHOH)_3.CO.CH_2OH$.			
d-fructose. fructose, levulose, fruit-sugar, chylariose.	95°-105°	...	Widely distributed both free and combined as a constituent of various polysaccharides. Forms d-glucosazone.
l-fructose	...	...	Obtained synthetically from l-glucosazone. With 2 mol. phenylhydrazine forms l-glucosazone.
i-fructose. Inactive fructose, i-levulose, α-acrose.	Sirup.	...	Obtained synthetically from acrolein. With 2 molecules phenylhydrazine forms i-glucosazone.

Sorbinose.			
Sorbose, sorbine		161°	Juice of mountain-ash berries.
Formose.		144°	Obtained by condensation of formic aldehyde. Decomposes on heating to 90°-110° C.
Methose	Sirup.	201°-206°	Obtained by condensation of formic aldehyde.
Hexoses of unknown nature and constitution, $C_6H_{12}O_6$.			
(a) Natural sugars			*Chondroglucose,* from chondrin by treatment with hydrochloric acid. *Crocose* by hydrolysis of coloring matter of saffron. *Euonlya,* from melitose by hydrolysis. *Urine-sugar,* from diabetic urine. *Hederose,* a decomposition product of a glucoside found in the ivy (*Hedera helix*). *Indiglucin,* from the hydrolysis of indican; probably identical with dextrose. *Locaose,* from the hydrolysis of Chinese green, a coloring matter from a species of Rhamnus. *Paraglucose,* formed in the vinous fermentation of cane-sugar. *Phlorose,* from the hydrolysis of phloridzin. *Scammonose,* from the hydrolysis of a glucoside of scammony. *Skimminose,* from the hydrolysis of skimmin, a constituent of a certain Japanese species of rue. *Solanose,* from hydrolysis of solanin. *Turflcose,* forms 5 to 6 per cent of the milk of the Egyptian buffalo cow. *Wine sugar,* from many natural wines.
(b) Synthetic sugars.			β-acrose only known in form of osazone. Melting-point of osazone 158°-159°. β-formose (pseudoformose, isoformose) obtained by condensation of formic aldehyde. Melting-point of osazone 148° C. Obtained synthetically.
4. *Methyl-hexoses.*			
a-rhamno-hexose. $CH_3(CHOH)_4COH.$	180°	200°	Obtained synthetically.
β-rhamnus-hexose.		200°	Hydrolysis of the glucoside digitalin.
Digitalose. $(_7H_{14}O_6)$	Sirup.		Obtained synthetically.
F. *Heptoses.*			
a-gluco-heptose. $CH_2OH(CHOH)_5CHO.$	180°-190°	195°	Obtained synthetically.
β-gluco-heptose $CH_2OH(CHOH)_5CHO.$	Sirup.		Obtained synthetically. Forms a-glucoheptosazone.

¹ Anhydride. ² Hydrate.

244. TABLE SHOWING THE MORE IMPORTANT PROPERTIES OF THE CARBOHYDRATES.—Continued.

Name, Synonyms, and Formulæ.	Melting-point ° C.	Melting-point of Osazone ° C.	Source, etc.
d-manno-heptose........	134°	260°	Formula, $C_7H_{14}O_7$.
l-manno-heptose........	Sirup.	203°	Formula, $C_7H_{14}O_7$.
i-manno-heptose........	Sirup.	210°	Formula, $C_7H_{14}O_7$.
a-gala-heptose........		220°	
β-gala-heptose........			
Rhamno-heptose.... $CH_3.(CHOH)_5.CHO.$	Sirup.	200°	A methyl-heptose.
G. OCTOSES. $C_8H_{16}O_8.$			
a-gluco-octose........		210°–212°	Hydrated $= C_8H_{16}O_8 + 2H_2O.$ Melting-point 93°.
β-gluco-octose........	Sirup.		
d-manno-octose........		223°	
Gala-octose........			
Rhamno-octose........		216°	A methyl-octose.
H. NONOSES. $(C_9H_{18}O_9.)$			
a-gluco-nonost........	Sirup.	230°–233°	
d-gluco-nonose........	130°	217°	
II. DISACCHARIDES.			
A. Derivatives of Pentoses.			
Arabinon (di-arabiuose)........ $C_{10}H_{18}O_9.$			Arabinic acid (metapectic acid).
B. Derivatives of Hexoses.			
Sucrose........ Saccharose, cane-sugar.	160°		Very widely distributed in the vegetable kingdom. Sucrose, when free from rafinose, is not browned at 120°–125°.
$C_{12}H_{22}O_{11}.$ Trehalose........ Mycose, trehalbiose.	200°		From ergot and other fungi. The hydrate, $C_{12}H_{22}O_{11} + H_2O$ melts at 130°, becomes anhydrous, solidifies, and again melts at 200°. Forms no osazone.
Lactose........ Milk sugar, lactobiose.		200°	From the milk of various mammals. The hydrate, $C_{12}H_{22}O_{11} + H_2O$ can be dried at 100° and at 145°–150° begins to lose water and decomposes.

Maltose...... Maltobiose, ptyalose, cerealose.		206°	From starch by the action of diastase and by the action of ptyalin. The hydrate, $C_{12}H_{22}O_{11}.H_2O$, loses its water at 100°–105° in vacuo, at 100° to 110° in the air with browning.
Isomaltose......		150°–153°	From malted grain. Begins to decompose at 200°, and melts with marked decomposition at 200°.
Melibiose......		176°–178°	Formed, together with d-fructose, by careful inversion of melitriose. [melecitose.
Turanose......		215°–220°	Formed, together with d-glucose, by careful inversion of
Cyclamose......			From cyclamen tubers.
Agavose......			From the juice of the agave.
Para-saccharose......			Formed by the action of a certain yeast plant on cane-sugar in the presence of ammonium phosphate.
			Obtained synthetically from a milk-sugar derivative.
C. DERIVATIVES OF HEPTOSES...... $C_{13}H_{24}O_{12}$.			
III. TRISACCHARIDES. $C_{18}H_{32}O_{16}$.			
Raffinose...... Melitriose, gossypose, cotton-sugar......			Beet-molasses, cotton-seed, eucalyptus manna, etc. Recently reported in cane-molasses. The hydrate $(C_{18}H_{32}O_{16} + 5H_2O)$ loses its water on careful heating to 80°, then to 100°–105°; heated quickly it melts below 100°, losing part of its water of crystallization; decomposition begins at 125°–130°.
Melecitose...... Melicitriose.	148°–150°		Manna of *Pinus larix*. The hydrate $(C_{18}H_{32}O_{16} + 2H_2O)$ loses its water on gradual heating; the anhydrous residue melts at 148°–150°.
Stachyose......			Tubers of *Stachys tuberifera*.
Gentianose......	210°		From the roots of *Gentiana lutea*.
Lactosinose...... Lactosin.			From the roots of *Silene vulgaris* and other members of the order Caryophyllaceae.
Secalose			From unripe rye.
A sugar of the formula $C_{18}H_{32}O_{16}$......		150°–153°	Formed together with d-glucose by the action of diastase on [starch at a high temperature.
IV. POLYSACCHARIDES. Starch $(C_6H_{10}O_5)_x$.			Polysaccharides include soluble starch, dextrin, inulin, the galactans, and a number of plant constituents that are transformed into mono-saccharides by hydrolysis with dilute acids. Very widely distributed in the vegetable kingdom.

244. TABLE SHOWING THE MORE IMPORTANT PROPERTIES OF THE CARBOHYDRATES.—*Continued.*

Name.	Specific Rotatory Power, (α)D.	Behavior with Alkaline Copper Sol.	Behavior with Yeast.	Oxidation Products.	Reduction Products.
I. MONOSACCHARIDES.					
A. BIOSES.					
Glycollic aldehyde.		*		First glycollic then oxalic acid.	
B. TRIOSES.					
Glycerose.		R.R.T.		First glyceric then tartaric acid.	
C. TETROSES.					
Erythrose.	Osazone inactive.	S.R.	Readily fermented.	First erythritic then tartaric acid.	
D. PENTOSES.					
1. *Aldo-pentoses.*					
l-arabinose.	[1]104.4° at 18° [2]105.4° at 20°	R.	Not fermented by yeast. Fermentable by *Bacillus æthaceticus.*	By gentle oxidization, arabonic acid; more vigorous oxid., l-trioxyglutaric ac.	Arabite.
d-arabinose.	[3]105.1°	R.		d-trioxyglutaric acid.	
i-arabinose.	Inactive.				
Xylose	See foot-note 4.	R.	Not fermented.	Xylo-trioxyglutaric and trioxybutyric acid.	Xylite.
i-xylose.	Inactive.				
Ribose.					Adonite.
2. *Keto-pentoses.*					
3. *Methyl-pentoses.*					
Fucose.		R.			
Rhamnose.	—77°	R.	Unfermentable	l-trioxyglutaric, carbonic, formic, and oxalic acids.	Rhamnite.
Chinovose.	88°.9°	S.R.			
E. HEXOSES.					
1. *Aldo-hexoses.*					
d-glucose.	See foot-note 6.	R.	Fermentable,	d-gluconic acid by action of d-sorbite. Br or Cl; by HNO₃, saccharic acid.	d-sorbite.
l-glucose.	—51.4° in 4% sol.		Unfermentable.		

	Rotation		Fermentation	Acid products	Alcohol products
i-glucose.....	Inactive.	......	Half fermented, the l-glucose remaining.	First i-gluconic then i-saccharic acid.	
d-mannose.......... ..	12.89°–14.36° at 29°	R.	Fermentable.	d-mannonic acid by Br or HNO₃; further with HNO₃, d-manno-saccharic acid.	d-mannite (mannitol).
l-mannose......... ...	Levorotatory.	R.	Unfermentable.	....	l-mannite.
i-mannose............	Inactive.	R.	Is half fermented, the l-mannose remaining.	l-mannonic acid, with Br.	i-mannite.
d-gulose............	Weakly dextro-rotatory.	......	Unfermentable.	d-saccharic acid.	d-sorbite.
l-gulose............	....	......	Unfermentable.	First l-gulonic then l-saccharic acid.	l-sorbite.
l-idose........	....	......	....	First l-idonic then ido-saccharic acid.	
d-galactose............	See foot–note 7.	R.	Fermentable.	With Cl or Br, d-galactonic acid; with HNO₃, mucic acid.	Dulcite with other alcohols.
l-galactose...... ...	–73.6° to –74.2°	......	Unfermentable.	First l-galactonic then mucic acid.	Dulcite.
i-galactose..	Inactive.	......	Is half fermented, l-galactose remaining.	....	
d-talose....	....	......	Unfermentable.	First d-talonic then d-talo-mucic acid.	d-talite.
Chitose	Dextrorotatory.	R.	Unfermentable.	....	

* R.R.T. = reduces at room temperature; R. = reduces; S.R. = strongly reducing; N.R. = no reduction.

1 Scheibler.   2 Lippmann.   3 Kiliani.

4 Rotatory power increases with the concentration; $(\alpha)_D=18.425°$ when c=9; $(\alpha)_D=23.702°$ when c=61.7 (Tollens and Schulze).

5 Fischer and Piloty.

6 According to Tollens: 10% sol. of hydrate, $(\alpha)\frac{20}{D}=47.92°$; 10% sol. of anhydride, $(\alpha)\frac{20}{D}=52.74°$. The rotatory power increases with the concentration, and is indicated by the following formulæ: For hydrate, $(\alpha)\frac{20}{D}=47.73+.01534p+.0003883p^2$;

For anhydride, $(\alpha)\frac{20}{D}=52.5+.018796p+.0005163p^2$;

p = per cent d-glucose by weight of the solution.

7 According to Rindell and Meissl, when t = 10° to 30° and p = 4.89% to 35.36%, $(\alpha)_D=83.888+.0785p-.209t.$ t = temp.

244. TABLE SHOWING THE MORE IMPORTANT PROPERTIES OF THE CARBOHYDRATES.—*Continued*.

Name.	Specific Rotatory Power, $(\alpha)_D$	Behavior with Alkaline Copper Sol.	Behavior with Yeast.	Oxidation Products.	Reduction Products.
2. *Keto-hexoses*.					
d-fructose........	See foot-note 1.	R.	Fermentable.	A mixture of formic, glycollic and trioxybutyric acids.	d-mannite and d-sorbite.
l-fructose........	Dextrorotatory,	......	Unfermentable. Is half fermented, the l-fructose remaining.	......	
i-fructose........	Inactive.	R.	Unfermentable.	......	i-mannite (acrite).
Sorbinose........	−43.4°	R.	Unfermentable.	Aposorbonic and other acids.	d-sorbite.
Formose..........	Inactive.	R.	Unfermentable.		
Methose..........	See foot-note 2.	R.	Fermentable.		
3. *Hexoses* of unknown nature and constitution					
(a) Natural sugars,					
(b) Synthetic sugars,					
β-formose........	Inactive.	R.	Unfermentable.		
4. *Methyl-hexoses*.					
α-rhamino-hexose....				Mucic acid.	α-rhamnohexite.
β-rhamno-hexose....	−61.4° at 20°			l-talomucic acid.	
Digitalose........	−79.4° at 28°		Unfermentable.		
F. *Heptoses*.					
α-gluco-heptose...	−19.7° at 20°, 10 gms. in 100 cc.		Not readily fermented.		α-gluco-heptite.
d-manno-heptose.....	68.64° at 20°		Unfermentable.	Inactive pentoxypimelinic acid.	d-manno-heptite.
l-manno-heptose.....			Unfermentable.		l-manno heptite.
i-manno-lieptose...					i-manno-heptite.
α-gala-heptose.....				Active pentoxypimelinic acid.	
β-gala-heptose.....					

				Hydrolysis Products.
Rhamno-heptose ..	8.4° at 20°			
G. OCTOSES.				
a-glueo-octose......	3=43.9° at 20°, 4=50.5° at 30°		Unfermentable.	α-glueo-octite.
d-manno-octose.......	$-3.3°(ca)$ at 20°	R.	Unfermentable.	d-manno-octite.
Rhamno-heptose				
H. NONOSES.				
a-glueo-nonose....	Weakly dextrorotatory.		Unfermentable.	α-glueo-nonite.
d-manno-nonose.......	50°(ca) at 20°		Fermentable.	
II. DISACCHARIDES.				
A. DERIVATIVES OF PENTOSES.				
Arabinon	198.8°	R.		l-arabinose.
B. DERIVATIVES OF HEXOSES.				
Sucrose...............	See foot-note 5.	N.R.	Fermentable.	Various acids according to conditions. $\begin{cases} d\text{-glucose.} \\ d\text{-fructose.} \end{cases}$

[1] $(a)\frac{20}{D}$ (25 grams in 100 cc. of solution) $= -91.8°$. The specific rotatory power diminishes as the temperature increases, and increases with the concentration.

[2] Chondroglucose, $(a)_j = -45.8°$; crocose, dextrorotatory; eucalyn, $(a)_j = 65°$; urine-sugar, $(a)_j = -96.07°$; hederose, $(a)_D = 98.88°$; locaose, inactive; paraglucose, $(a)_j = 40°$; phlorose, $(a)_D = 38°$ to $41°$; scammonose, $(a)\frac{19}{D} = 17.78°$; skimminose, $(a)_D = 24°$; solanose, $(a)\frac{20}{D} = 28.5°$; tewflkose, $(a)_D = 48.7°$. All reduce copper except skimminose (?), and wine-sugar. Eucalyn, hederose, and wine-sugar are unfermentable, and scammonose is fermentable. On oxidation eucalyn yields oxalic acid.

[3] Hydrate.   [4] Anhydride.   [5] Sucrose in 4% solution, $(a)\frac{20}{D} = 66.8$; 70% solution, $(a)\frac{20}{D} = 65.5°$ (Tollens).

244. TABLE SHOWING THE MORE IMPORTANT PROPERTIES OF THE CARBOHYDRATES.—Continued.

Name.	Specific Rotatory Power, $(\alpha)_D$	Behavior with Alkaline Copper Sol.	Behavior with Yeast.	Oxidation Products.	Hydrolysis Products.
Trehalose	197.28°.	....	Unfermentable.	Forms oxalic acid on heating with dilute HNO_3.	d-glucose.
Lactose	152.53° at 20° for hydrate.	R.	No true alcoholic fermentation.	Mucic and other acids according to conditions.	d-glucose and other d-galactose.
Maltose	137° at 20° in 11% sol.	R.	Fermentable.	d-saccharic acid on treatment with HNO_3.	d-glucose.
Isomaltose	139°–140°	R.	Hydrolysis followed by fermentation.	....	d-glucose.
Meliblose	127.3°–139°	R.	....	....	d-glucose and d-galactose.
Turanose	65°–68°	R.	Ferments with difficulty or not at all.	....	d-glucose.
Cyclamose	−15.15°. Inactive.	R.	Unfermentable.	...	Not determined.
Agavose	$(\alpha)j = 108°$	R.		...	Not determined.
Para-saccharose					
C. DERIVATIVES OF HEPTOSES.					
Sugar of the formula $C_{18}H_{74}O_{13}$.	....	....	....	....	d-galactose and d-glucoheptose.
III. TRISACCHARIDES.					
Raffinose	105°–105.7°.	N.R.	Fermentable.	With HNO_3, oxalic, saccharic, and mucic acids.	First d-fructose and melibiose; further action decomposes the latter into d glucose and d-galactose.

Melecitose..........	88.6°–88.8° for anhydride.	N.R.	Unfermentable.	....	First d-glucose and turanose; further action decomposes the latter into d-glucose.
Stachyose...	147.9°–148.1° for anhydride.	N.R.	....	Mucic and other acids.	d-galactose, d-glucose, and d-fructose.
Gentianose........	65.7°				d-galactose and two other sugars.
Lactosinose...	211.7° at 10°			Mucic acid.	d-fructose.
Secalose .... A sugar of the formula $C_{18}H_{32}O_{16}$ . .	−28.6° to −28.9°	R.	Ferments with difficulty or not at all.	....	d-glucose.
IV. POLYSACCHARIDES.					
Starch............	....				d-glucose, maltose, dextrine, isomaltose, according to conditions.

¹ Schmoeger.

245. FREEZING MIXTURES.—(Walker's List.)

Parts	Centigrade.	Fahrenheit.	Réaumur.
Ammonium Nitrate... 1 Water.................... 1	From + 4°.4 to − 15°.5	From + 40° to + 4°	From + 3°.5 to − 12°.4
Ammonium Chloride.. 5 Potassium Nitrate .... 5 Water 16	From + 10° to − 12°.2	From + 50° to + 10°	From + 8° to − 9°.8
Ammonium Chloride.. 5 Potassium Nitrate .... 5 Sodium Sulphate...... 8 Water. 16	From + 10° to − 15°.5	From + 50° to + 4°	From + 8° to − 12°.4
Sodium Nitrate .... 3 Nitric Acid, diluted.... 2	From + 10° to − 19°.4	From + 50° to − 3°	From + 8° to − 15°.5
Ammonium Nitrate... 1 Sodium Carbonate.... 1 Water................... 1	From + 10° to − 21°.7	From + 50° to − 7°	From + 8° to − 17°.3
Sodium Phosphate.... 9 Nitric Acid, diluted... 4	From + 10° to − 24°.4	From + 50° to − 12°	From + 8° to − 19°.5
Sodium Sulphate...... 5 Sulphuric Acid, dilut.. 4	From + 10° to − 16°.1	From + 50° to + 3°	From + 8° to − 12°.9
Sodium Sulphate...... 6 Ammonium Chloride.. 4 Potassium Nitrate.... 2 Nitric Acid, diluted.... 4	From + 10° to − 23°.3	From + 50° to − 10°	From + 8° to − 18°.6
Sodium Sulphate...... 6 Ammonium Nitrate... 5 Nitric Acid, diluted... 4	From + 10° to − 40°	From + 50° to − 40°	From + 8° to − 32°
Snow or pounded ice.. 2 Sodium Chloride (common salt) 1	to − 20°.5	to − 5°	to − 16°.4
Snow or pounded ice.. 5 Sodium Chloride (common salt) .... 2 Ammonium Chloride.. 1	to − 24°.4	to − 12°	to − 19°.5
Snow or pounded ice..24 Sodium Chloride (common salt) ....10 Ammonium Chloride.. 5 Potassium Nitrate.... 5	to − 27°.7	to − 18°	to − 22°.2
Snow or pounded ice..12 Sodium Chloride (common salt)... ...... 5 Ammonium Nitrate... 5	to − 31°.6	to − 25°	to − 25°.3
Snow................. 3 Sulphuric Acid, dilu'd 2	From 0° to − 30°.5	From + 32° to − 23°	From 0° to − 24°.4
Snow .... 8 Hydrochloric Acid.... 5	From 0° to − 32°.8	From + 32° to − 27°	From 0° to − 26°.2
Snow............... 7 Nitric Acid, diluted... 4	From 0° to − 34°.4	From + 32° to − 30°	From 0° to − 27°.5
Snow................. 4 Calcium Chloride (Chloride of Lime).. 5	From 0° to − 40°	From + 32° to − 40°	From 0° to − 32°
Snow............... 2 Calcium Chloride, crystallized 3	From 0° to − 45°.5	From + 32° to − 50°	From 0° to − 36°.4
Snow................. 3 Potash 4	From 0° to − 46°.1	From + 32° to − 51°	From 0° to − 36°.9

246. TABLE SHOWING THE STRENGTH OF SULPHURIC ACID (OIL OF VITRIOL) OF DIFFERENT DENSITIES, AT 15° CENTIGRADE.—(OTTO'S TABLE.)

Per Cent of H_2SO_4.	Specific Gravity.	Per Cent of SO_3.	Per Cent of H_2SO_4.	Specific Gravity.	Per Cent of SO_3.
100	1.8426	81.63	50	1.3980	40.81
99	1.8420	80.81	49	1.3866	40.00
98	1.8406	80.00	48	1.3790	39.18
97	1.8400	79.18	47	1.3700	38.36
96	1.8384	78.36	46	1.3610	37.55
95	1.8376	77.55	45	1.3510	36.73
94	1.8356	76.73	44	1.3420	35.82
93	1.8340	75.91	43	1.3330	35.10
92	1.8310	75.10	42	1.3240	34.28
91	1.8270	74.28	41	1.3150	33.47
90	1.8220	73.47	40	1.3060	32.65
89	1.8100	72.65	39	1.2976	31.83
88	1.8090	71.83	38	1.2890	31.02
87	1.8020	71.02	37	1.2810	30.20
86	1.7940	70.10	36	1.2720	29.38
85	1.7800	69.38	35	1.2640	28.57
84	1.7770	68.57	34	1.2560	27.75
83	1.7670	67.75	33	1.2476	26.94
82	1.7560	66.94	32	1.2390	26.12
81	1.7450	66.12	31	1.2310	25.30
80	1.7340	65.30	30	1.2230	24.49
79	1.7220	64.48	29	1.2150	23.67
78	1.7100	63.67	28	1.2066	22.85
77	1.6980	62.85	27	1.1980	22.03
76	1.6860	62.04	26	1.1900	21.22
75	1.6750	61.22	25	1.1820	20.40
74	1.6630	60.40	24	1.1740	19.58
73	1.6510	59.59	23	1.1670	18.77
72	1.6390	58.77	22	1.1590	17.95
71	1.6270	57.95	21	1.1516	17.14
70	1.6150	57.14	20	1.1440	16.32
69	1.6040	56.32	19	1.1360	15.51
68	1.5920	55.59	18	1.1290	14.69
67	1.5800	54.69	17	1.1210	13.87
66	1.5860	53.87	16	1.1136	13.06
65	1.5570	53.05	15	1.1060	12.24
64	1.5450	52.22	14	1.0980	11.42
63	1.5340	51.42	13	1.0910	10.61
62	1.5230	50.61	12	1.0830	9.79
61	1.5120	49.79	11	1.0756	8.98
60	1.5010	48.98	10	1.0680	8.16
59	1.4900	48.16	9	1.0610	7.34
58	1.4800	47.34	8	1.0536	6.53
57	1.4690	46.53	7	1.0464	5.71
56	1.4586	45.71	6	1.0390	4.89
55	1.4480	44.89	5	1.0320	4.08
54	1.4380	44.07	4	1.0256	3.26
53	1.4280	43.26	3	1.0190	2.44
52	1.4180	42.45	2	1.0130	1.63
51	1.4080	41.63	1	1.0064	0.81

247. ANTHON'S TABLE FOR THE DILUTION OF SULPHURIC ACID.

To 100 parts of Water at 15° to 20° C. add...parts of Sulphuric Acid of 1.84 Specific Gravity.	Specific Gravity of diluted Acid.	To 100 parts of Water at 15° to 20° C. add...parts of Sulphuric Acid of 1.84 Specific Gravity.	Specific Gravity of diluted Acid.	To 100 parts of Water at 15° to 20° C. add...parts of Sulphuric Acid of 1.84 Specific Gravity.	Specific Gravity of diluted Acid.
1	1.009	130	1.456	370	1.723
2	1.015	140	1.473	380	1.727
5	1.035	150	1.490	390	1.730
10	1.060	160	1.510	400	1.733
15	1.090	170	1.530	410	1.737
20	1.113	180	1.543	420	1.740
25	1.140	190	1.556	430	1.743
30	1.165	200	1.568	440	1.746
35	1.187	210	1.580	450	1.750
40	1.210	220	1.593	460	1.754
45	1.229	230	1.606	470	1.757
50	1.248	240	1.620	480	1.760
55	1.265	250	1.630	490	1.763
60	1.280	260	1.640	500	1.766
65	1.297	270	1.648	510	1.768
70	1.312	280	1.654	520	1.770
75	1.326	290	1.667	530	1.772
80	1.340	300	1.678	540	1.774
85	1.357	310	1.689	550	1.776
90	1.372	320	1.700	560	1.777
95	1.396	330	1.705	580	1.778
100	1.398	340	1.710	590	1.780
110	1.420	350	1.714	600	1.782
120	1.438	360	1.719		

248. TABLE SHOWING THE STRENGTH OF NITRIC ACID (HNO₃) BY SPECIFIC GRAVITY. HYDRATED AND ANHYDRIDE.

TEMPERATURE 15°.

(Fresenius, Zeitschrift f. analyt. Chemie. 5. 449.)

Sp. Gr. at 15° C.	100 PARTS CONTAIN—		Sp. Gr. at 15° C.	100 PARTS CONTAIN—	
	N_2O_3	NO_3H		N_2O_5	NO_3H
1.530	85.71	100.00	1.488	75.43	88.00
1.530	85.57	99.84	1.486	74.95	87.45
1.530	85.47	99.72	1.482	73.86	86.17
1.529	85.30	99.52	1.478	72.86	85.00
1.523	83.90	97.89	1.474	72.00	84.00
1.520	83.14	97.00	1.470	71.14	83.00
1.516	82.28	96.00	1.467	70.28	82.00
1.514	81.66	95.27	1.463	69.39	80.96
1.509	80.57	94.00	1.460	68.57	80.00
1.506	79.72	93.01	1.456	67.71	79.00
1.503	78.85	92.00	1.451	66.56	77.66
1.499	78.00	91.00	1.445	65.14	76.00
1.495	77.15	90.00	1.442	64.28	75.00
1.494	76.77	89.56	1.438	63.44	74.01

TABLE SHOWING THE STRENGTH OF NITRIC ACID.—*Continued*.

Sp. Gr. at 15° C.	100 PARTS CONTAIN—		Sp. Gr. at 15° C.	100 PARTS CONTAIN—	
	N_2O_5	NO_3H		N_2O_5	NO_3H
1.435	62.57	73.00	1.295	39.97	46.64
1.432	62.05	72.39	1.284	38.57	45.00
1.429	61.06	71.24	1.274	37.31	43.53
1.423	60.00	69.96*	1.264	36.00	42.00
1.419	59.31	69.20	1.257	35.14	41.00
1.414	58.29	68.00	1.251	34.28	40.00
1.410	57.43	67.00	1.244	33.43	39.00
1.405	56.57	66.00	1.237	32.58	37.95
1.400	55.77	65.07	1.225	30.86	36.00
1.395	54.85	64.00	1.218	29.29	35.00
1.393	54.50	63.59	1.211	29.02	33.86
1.386	53.14	62.00	1.198	27.43	32.00
1.381	52.46	61.21	1.192	26.57	31.00
1.374	51.43	60.00	1.185	25.71	30.00
1.372	51.08	59.59	1.179	24.85	29.00
1.368	50.47	58.88	1.172	24.00	28.00
1.363	49.71	58.00	1.166	23.14	27.00
1.358	48.86	57.00	1.157	22.04	25.71
1.353	48.08	56.10	1.138	19.71	23.00
1.346	47.14	55.00	1.120	17.14	20.00
1.341	46.29	54.00	1.105	14.97	17.47
1.339	46.12	53.81†	1.089	12.85	15.00
1.335	45.40	53.00	1.077	11.14	13.00
1.331	44.85	52.33	1.067	9.77	11.41
1.323	43.70	50.99	1.045	6.62	7.22
1.317	42.83	49.97	1.022	3.42	4.00
1.312	42.00	49.00	1.010	1.71	2.00
1.304	41.14	48.00	0.999	0.00	0.00
1.298	40.44	47.18			

* Formula : $NO_3H + 1\frac{1}{2}H_2O$.   † Formula : $NO_3H + 3H_2O$.

249. TABLE SHOWING THE AMOUNT OF CaO IN MILK OF LIME OF VARIOUS DENSITIES AT 15° C.

(FROM BLATNER'S TABLE.)

Deg. Brix.	Degree Baumé.	Weight of one litre, Milk of Lime.	CaO per litre.	Per Cent CaO.	Deg. Brix.	Degree Baumé.	Weight of one litre, Milk of Lime.	CaO per litre.	Per Cent CaO.
		Grams.	Grams.				Grams.	Grams.	
1.8	1	1007	7.5	0.745	29	16	1125	159	14.13
3.6	2	1014	16.5	1.64	30.8	17	1134	170	15
5.4	3	1022	26	2.54	32.7	18	1142	181	15.85
7.2	4	1029	36	3.5	34.6	19	1152	193	16.75
9	5	1037	46	4.43	36.4	20	1162	206	17.72
10.8	6	1045	56	5.36	38.3	21	1171	218	18.61
12.6	7	1052	65	6.18	40.1	22	1180	229	19.4
14.4	8	1060	75	7.08	42	23	1190	242	20.34
16.2	9	1067	84	7.87	43.9	24	1200	255	21.25
18	10	1075	94	8.74	45.8	25	1210	268	22.15
19.8	11	1083	104	9.6	47.7	26	1220	281	23.03
21.7	12	1091	115	10.54	49.6	27	1231	295	23.96
23.5	13	1100	126	11.45	51.5	28	1241	309	24.9
25.3	14	1108	137	12.35	53.5	29	1252	324	25.87
27.2	15	1116	148	13.26	55.4	30	1263	339	26.84

250. TABLE SHOWING THE STRENGTH OF HYDROCHLORIC
ACID (MURIATIC ACID) SOLUTIONS.
TEMPERATURE, 15° C.
(Graham-Otto's Lehrb. d. Chem. 3 Aufl. II. Bd. 1. Abth. p. 382.)

Sp. Gr	HCl.	Cl.	Sp. Gr.	HCl.	Cl.	Sp. Gr.	HCl.	Cl.
1.2000	40.777	39.675	1.1328	26.913	26.186	1.0657	13.456	13.094
1.1982	40.369	39.278	1.1308	26.505	25.789	1.0637	13.049	12.697
1.1964	39.961	38.882	1.1287	26.098	25.392	1.0617	12.641	12.300
1.1946	39.554	38.485	1.1267	25.690	24.996	1.0597	12.233	11.903
1.1928	39.146	38.089	1.1247	25.282	24.599	1.0577	11.825	11.506
1.1910	38.738	37.692	1.1226	24.874	24.202	1.0557	11.418	11.109
1.1893	38.330	37.296	1.1206	24.466	23.805	1.0537	11.010	10.712
1.1875	37.923	36.900	1.1185	24.058	23.408	1.0517	10.602	10.316
1.1857	37.516	36.503	1.1164	23.650	23.012	1.0497	10.194	9.919
1.1846	37.108	36.107	1.1143	23.242	22.615	1.0477	9.786	9.522
1.1822	36.700	35.707	1.1123	22.834	22.218	1.0457	9.379	9.126
1.1802	36.292	35.310	1.1102	22.426	21.822	1.0437	8.971	8.729
1.1782	35.884	34.913	1.1082	22.019	21.425	1.0417	8.563	8.332
1.1762	35.476	34.517	1.1061	21.611	21.028	1.0397	8.155	7.935
1.1741	35.068	34.121	1.1041	21.203	20.632	1.0377	7.747	7.538
1.1721	34.660	33.724	1.1020	20.796	20.235	1.0357	7.340	7.141
1.1701	34.252	33.328	1.1000	20.388	19.887	1.0337	6.932	6.745
1.1681	33.845	32.931	1.0980	19.980	19.440	1.0318	6.524	6.348
1.1661	33.437	32.535	1.0960	19.572	19.044	1.0298	6.116	5.951
1.1641	33.029	32.136	1.0939	19.165	18.647	1.0279	5.709	5.554
1.1620	32.621	31.746	1.0919	18.757	18.250	1.0259	5.301	5.158
1.1599	32.213	31.343	1.0899	18.349	17.854	1.0239	4.893	4.762
1.1578	31.805	30.946	1.0879	17.941	17.457	1.0220	4.486	4.365
1.1557	31.398	30.550	1.0859	17.534	17.060	1.0200	4.078	3.968
1.1537	30.990	30.153	1.0838	17.126	16.664	1.0180	3.670	3.571
1.1515	30.582	29.757	1.0818	16.718	16.267	1.0160	3.262	3.174
1.1494	30.174	29.361	1.0798	16.310	15.870	1.0140	2.854	2.778
1.1473	29.767	28.964	1.0778	15.902	15.474	1.0120	2.447	2.381
1.1452	29.359	28.567	1.0758	15.494	15.077	1.0100	2.039	1.984
1.1431	28.951	28.171	1.0738	15.087	14.680	1.0080	1.631	1.588
1.1410	28.544	27.772	1.0718	14.679	14.284	1.0060	1.124	1.191
1.1389	28.136	27.376	1.0697	14.271	13.887	1.0040	0.816	0.795
1.1369	27.728	26.979	1.0677	13.863	13.490	1.0020	0.408	0.397
1.1349	27.321	26.583						

251. TABLE SHOWING THE AMOUNT OF CaO IN MILK OF
LIME OF VARIOUS DENSITIES.—(MATEGCZEK.)

Degree Brix.	Degree Baumé.	1 kilo CaO per .. litres Milk of Lime.	Degree Brix.	Degree Baumé.	1 kilo CaO per .. litres Milk of Lime.
18	10	7.50	38.3	21	4.28
20	11	7.10	40.2	22	4.16
21.7	12	6.70	42.0	23	4.05
23.5	13	6.30	43.9	24	3.95
25.3	14	5.88	45.8	25	3.87
27.2	15	5.50	47.7	26	3.81
29	16	5.25	49.6	29	3.75
30.9	17	5.01	51.6	28	3.70
32.7	18	4.80	53.5	29	3.65
34.6	19	4.68	55.5	30	3.60
36.5	20	4.42			

252. TABLE SHOWING THE QUANTITY OF SODIUM OXIDE IN SOLUTIONS OF VARIOUS DENSITIES.

(Fresenius Anl. z. quant. Analyse. V. Aufl. f. 730.)

According to Dalton.		According to Tünnermann at 15° C.					
Sp. Gr.	Per Cent Na_2O.	Sp. Gr.	Per Cent Na_2O.	Sp. Gr.	Per Cent Na_2O.	Sp. Gr.	Per Cent Na_2O.
2.00	77.8	1.4285	30.220	1.2982	20.550	1.1528	10.275
1.85	63.6	1.4193	29.616	1.2912	19.945	1.1428	9.670
1.72	53.8	1.4101	29.011	1.2843	19.341	1.1330	9.066
1.63	46.6	1.4011	28.407	1.2775	18.730	1.1233	8.462
1.56	41.2	1.3923	27.802	1.2708	18.132	1.1137	7.857
1.50	36.8	1.3836	27.200	1.2642	17.528	1.1042	7.253
1.47	34.0	1.3751	26.594	1.2578	16.923	1.0948	6.648
1.44	31.0	1.3668	25.989	1.2515	16.319	1.0855	6.044
1.40	29.0	1.3586	25.385	1.2453	15.714	1.0764	5.440
1.36	26.0	1.3505	24.780	1.2392	15.110	1.0675	4.835
1.32	23.0	1.3426	24.176	1.2280	14.506	1.0587	4.231
1.29	19.0	1.3349	23.572	1.2178	13.901	1.0500	3.626
1.23	16.0	1.3273	22.967	1.2058	13.297	1.0414	3.022
1.18	13.0	1.3198	22.363	1.1948	12.692	1.0330	2.418
1.12	9.0	1.3143	21.894	1.1841	12.088	1.0246	1.813
1.06	4.7	1.3125	21.758	1.1734	11.484	1.0163	1.209
		1.3053	21.154	1.1630	10.879	1.0081	0.604

253. TABLE SHOWING THE QUANTITY OF POTASSIC OXIDE IN SOLUTIONS OF VARIOUS DENSITIES.

(Fresenius Anl. z. quant. Analyse. V. Aufl. f. 730.)

According to Dalton.		According to Tünnermann at 15° C.			
Sp. Gr.	K_2O. Per Cent.	Sp. Gr.	K_2O. Per Cent.	Sp. Gr.	K_2O. Per Cent.
1.68	51.2	1.3300	28.290	1.1437	14.145
1.60	47.7	1.3131	27.158	1.1308	13.013
1.52	42.9	1.2966	26.027	1.1182	11.882
1.47	39.9	1.2803	24.895	1.1059	10.750
1.44	36.8	1.2648	23.764	1.0938	9.619
1.42	34.4	1.2493	22.632	1.0819	8.487
1.39	32.4	1.2342	21.500	1.0703	7.355
1.36	29.4	1.2268	20.935	1.0589	6.224
1.32	26.3	1.2122	19.803	1.0478	5.092
1.28	23.4	1.1979	18.671	1.0369	3.961
1.23	19.5	1.1839	17.540	1.0260	2.829
1.19	16.2	1.1702	16.408	1.0153	1.697
1.15	13.0	1.1568	15.277	1.0050	0.5658
1.11	9.5				
1.06	4.7				

254. TABLE SHOWING THE STRENGTH OF SOLUTIONS OF AMMONIA BY SPECIFIC GRAVITY AT 14° C.—(ABRIDGED FROM CARIUS' TABLE.)

Per Cent Ammonia (NH_3).	Specific Gravity.	Per Cent Ammonia (NH_3).	Specific Gravity.	Per Cent Ammonia (NH_3).	Specific Gravity.
1.	0.9959	13.	0.9484	25.	0.9106
1.4	0.9941	13.4	0.9470	25.4	0.9094
2.	0.9915	14.	0.9449	26.	0.9078
2.4	0.9899	14.4	0.9434	26.4	0.9068
3.	0.9873	15.	0.9414	27.	0.9052
3.4	0.9855	15.4	0.9400	27.4	0.9041
4.	0.9831	16.	0.9380	28.	0.9026
4.4	0.9815	16.4	0.9366	28.4	0.9016
5.	0.9790	17.	0.9347	29.	0.9001
5.4	0.9773	17.4	0.9333	29.4	0.8991
6.	0.9749	18.	0.9314	30.	0.8976
6.4	0.9733	18.4	0.9302	30.4	0.8967
7.	0.9709	19.	0.9283	31.	0.8953
7.4	0.9693	19.4	0.9271	31.4	0.8943
8.	0.9670	20.	0.9251	32.	0.8929
8.4	0.9654	20.4	0.9239	32.4	0.8920
9.	0.9631	21.	0.9221	33.	0.8907
9.4	0.9616	21.4	0.9209	33.4	0.8898
10.	0.9593	22.	0.9191	34.	0.8885
10.4	0.9578	22.4	0.9180	34.4	0.8877
11.	0.9556	23.	0.9162	35.	0.8864
11.4	0.9542	23.4	0.9150	35.4	0.8856
12.	0.9520	24.	0.9133	36.	0.8844
12.4	0.9505	24.4	0.9122		

255. TABLE SHOWING THE PERCENTAGE OF ACETATE OF LEAD IN SOLUTIONS OF THE SALT, OF DIFFERENT DENSITIES, AT 15° C.—(GERLACH.)

Specific Gravity.	Per Cent of the Salt.	Specific Gravity.	Per Cent of the Salt.	Specific Gravity.	Per Cent of the Salt.
1.0127	2	1.1384	20	1.2768	36
1.0255	4	1.1544	22	1.2966	38
1.0386	6	1.1704	24	1.3163	40
1.0520	8	1.1869	26	1.3376	42
1.0654	10	1.2040	28	1.3588	44
1.0796	12	1.2211	30	1.3810	46
1.0739	14	1.2395	32	1.4041	48
1.1084	16	1.2578	34	1.4271	50
1.1234	18				

256. TABLE SHOWING A COMPARISON OF THE DEGREES BRIX AND BAUMÉ, AND OF THE SPECIFIC GRAVITY OF SUGAR SOLUTIONS AT 17½° C.—(STAMMER.)

Degree Brix (Per Cent Sugar).	Degree Baumé (corrected).	Specific Gravity.	Degree Brix (Per Cent Sugar).	Degree Baumé (corrected).	Specific Gravity.	Degree Brix (Per Cent Sugar).	Degree Baumé (corrected).	Specific Gravity.
0.0	0.0	1.00000	3.0	1.7	1.01173	6.0	3.4	1.02373
.1	0.1	1.00038	.1	1.8	1.01213	.1	3.5	1.02413
.2	0.1	1.00077	.2	1.8	1.01252	.2	3.5	1.02454
.3	0.2	1.00116	.3	1.9	1.01292	.3	3.6	1.02494
.4	0.2	1.00155	.4	1.9	1.01332	.4	3.6	1.02535
.5	0.3	1.00195	.5	2.0	1.01371	.5	3.7	1.02575
.6	0.3	1.00232	.6	2.0	1.01411	.6	3.7	1.02616
.7	0.4	1.00271	.7	2.1	1.01451	.7	3.8	1.02657
.8	0.45	1.00310	.8	2.2	1.01491	.8	3.9	1.02694
.9	0.5	1.00349	.9	2.2	1.01531	.9	3.9	1.02738
1.0	0.6	1.00388	4.0	2.3	1.01570	7.0	4.0	1.02779
.1	0.6	1.00427	.1	2.3	1.01610	.1	4.0	1.02819
.2	0.7	1.00466	.2	2.4	1.01650	.2	4.1	1.02860
.3	0.7	1.00505	.3	2.4	1.01690	.3	4.1	1.02901
.4	0.8	1.00544	.4	2.5	1.01730	.4	4.2	1.02942
.5	0.85	1.00583	.5	2.55	1.01770	.5	4.25	1.02983
.6	0.9	1.00622	.6	2.6	1.01810	.6	4.3	1.03024
.7	1.0	1.00662	.7	2.7	1.01850	.7	4.4	1.03064
.8	1.0	1.00701	.8	2.7	1.01890	.8	4.4	1.03105
.9	1.1	1.00740	.9	2.8	1.01930	.9	4.5	1.03146
2.0	1.1	1.00779	5.0	2.8	1.01970	8.0	4.5	1.03187
.1	1.2	1.00818	.1	2.9	1.02010	.1	4.6	1.03228
.2	1.2	1.00858	.2	2.95	1.02051	.2	4.6	1.03270
.3	1.3	1.00897	.3	3.0	1.02091	.3	4.7	1.03311
.4	1.4	1.00936	.4	3.1	1.02131	.4	4.8	1.03352
.5	1.4	1.00976	.5	3.1	1.02171	.5	4.8	1.03393
.6	1.5	1.01015	.6	3.2	1.02211	.6	4.9	1.03434
.7	1.5	1.01055	.7	3.2	1.02252	.7	4.9	1.03475
.8	1.6	1.01094	.8	3.3	1.02292	.8	5.0	1.03517
.9	1.6	1.01134	.9	3.35	1.02333	.9	5.0	1.03558

CORRECTION FOR TEMPERATURE, BRIX SPINDLE.—(F. SACHS.)

Temp. °C.	Temp. °F.	APPROXIMATE DEGREE BRIX AND CORRECTION.			
		0	5	10	15
13	55.4	.14	.18	.19	.21
14	57.2	.12	.15	.16	.17
15	59.	.09	.11	.12	.14
16	60.8	.06	.07	.08	.09
17	62.6	.02	.02	.03	.03
18	64.4	.02	.03	.03	.03
19	66.2	.06	.08	.08	.09
20	68.	.11	.14	.15	.17
21	69.8	.16	.20	.22	.24
22	71.6	.21	.26	.29	.31
23	73.4	.27	.32	.35	.37
24	75.2	.32	.38	.41	43
25	77.	.37	44	.47	49

Subtract. NOTE.—For temperatures above 17½° C. add the correction to the reading at the observed temperature; below 17½° subtract.

Add. Obtain Baumé corrections from the corresponding degree Brix.

TABLE SHOWING A COMPARISON OF THE DEGREES BRIX AND BAUMÉ, ETC., OF SUGAR SOLUTIONS.—*Continued.*

Degree Brix (Per Cent Sugar).	Degree Baumé (corrected).	Specific Gravity.	Degree Brix (Per Cent Sugar).	Degree Baumé (corrected).	Specific Gravity.	Degree Brix (Per Cent Sugar).	Degree Baumé (corrected).	Specific Gravity.
9.0	5.1	1.03599	12.0	6.8	1.04852	15.0	8.5	1.06133
.1	5.2	1.03640	.1	6.8	1.04894	.1	8.5	1.06176
.2	5.2	1.03682	.2	6.9	1.04937	.2	8.55	1.06219
.3	5.3	1.03723	.3	7.0	1.04979	.3	8.6	1.06262
.4	5.3	1.03765	.4	7.0	1.05021	.4	8.7	1.06306
.5	5.4	1.03806	.5	7.1	1.05064	.5	8.8	1.06349
.6	5.4	1.03848	.6	7.1	1.05106	.6	8.8	1.06392
.7	5.5	1.03889	.7	7.2	1.05149	.7	8.9	1.06436
.8	5.55	1.03931	.8	7.2	1.05191	.8	8.9	1.06479
.9	5.6	1.03972	.9	7.3	1.05233	.9	9.0	1.06522
10.0	5.7	1.04014	13.0	7.4	1.05276	16.0	9.0	1.06566
.1	5.7	1.04055	.1	7.4	1.05318	.1	9.1	1.06609
.2	5.8	1.04097	.2	7.5	1.05361	.2	9.2	1.06653
.3	5.8	1.04139	.3	7.5	1.05404	.3	9.2	1.06696
.4	5.9	1.04180	.4	7.6	1.05446	.4	9.3	1.06740
.5	5.9	1.04222	.5	7.6	1.05489	.5	9.3	1.06783
.6	6.0	1.04264	.6	7.7	1.05532	.6	9.4	1.06827
.7	6.1	1.04306	.7	7.75	1.05574	.7	9.4	1.06871
.8	6.1	1.04348	.8	7.8	1.05617	.8	9.5	1.06914
.9	6.2	1.04390	.9	7.9	1.05660	.9	9.5	1.06958
11.0	6.2	1.04431	14.0	7.9	1.05703	17.0	9.6	1.07002
.1	6.3	1.04473	.1	8.0	1.05746	.1	9.7	1.07046
.2	6.3	1.04515	.2	8.0	1.05789	.2	9.7	1.07090
.3	6.4	1.04557	.3	8.1	1.05831	.3	9.8	1.07133
.4	6.5	1.04599	.4	8.1	1.05874	.4	9.8	1.07177
.5	6.5	1.04641	.5	8.2	1.05917	.5	9.9	1.07221
.6	6.6	1.04683	.6	8.3	1.05960	.6	9.9	1.07265
.7	6.6	1.04726	.7	8.3	1.06003	.7	10.0	1.07309
.8	6.7	1.04768	.8	8.4	1.06047	.8	10.0	1.07353
.9	6.7	1.04810	.9	8.4	1.06090	.9	10.1	1.07397

CORRECTION FOR TEMPERATURE, BRIX SPINDLE.—(F. SACHS.)

Temp. °C.	Temp. °F.	Approximate Degree Brix and Correction.				
		15	20	25	30	
13	55.4	.21	.22	.24	.26	*Subtract*
14	57.2	.17	.18	.19	.21	
15	59.	.14	.14	.15	.16	
16	60.8	.09	.10	.10	.11	
17	62.6	.03	.03	.04	.04	
18	64.4	03	.03	.03	.03	*Add*
19	66.2	.09	.09	.10	.10	
20	68.	.17	.17	.18	.18	
21	69.8	.24	.24	.25	.25	
22	71.6	.31	.31	.32	.32	
23	73.4	.37	.38	.39	.39	
24	75.2	.43	.44	.46	.46	
25	77.	.49	.51	.53	.54	

NOTE.—For temperatures above 17½° C. add the correction to the reading at the observed temperature; below 17½° C. subtract.

Obtain Baumé corrections from corresponding degree Brix.

TABLE SHOWING A COMPARISON OF THE DEGREES BRIX AND BAUMÉ, ETC.—Continued.

Degree Brix (Per Cent Sugar).	Degree Baumé (corrected).	Specific Gravity.	Degree Brix (Per Cent Sugar).	Degree Baumé (corrected).	Specific Gravity.	Degree Brix (Per Cent Sugar).	Degree Baumé (corrected).	Specific Gravity.
18.0	10.1	1.07441	23.0	13.0	1.09686	28.0	15.7	1.12013
.1	10.2	1.07485	.1	13.0	1.09732	.1	15.8	1.12060
.2	10.3	1.07530	.2	13.1	1.09777	.2	15.8	1.12107
.3	10.3	1.07574	.3	13.1	1.09823	.3	15.9	1.12155
.4	10.4	1.07618	.4	13.2	1.09869	.4	16.0	1.12202
.5	10.4	1.07662	.5	13.2	1.09915	.5	16.0	1.12250
.6	10.5	1.07706	.6	13.3	1.09961	.6	16.1	1.12297
.7	10.5	1.07751	.7	13.3	1.10007	.7	16.1	1.12345
.8	10.6	1.07795	.8	13.4	1.10053	.8	16.2	1.12393
.9	10.6	1.07839	.9	13.5	1.10099	.9	16.2	1.12440
19.0	10.7	1.07884	24.0	13.5	1.10145	29.0	16.3	1.12488
.1	10.8	1.07928	.1	13.6	1.10191	.1	16.3	1.12536
.2	10.8	1.07973	.2	13.6	1.10237	.2	16.4	1.12583
.3	10.9	1.08017	.3	13.7	1.10283	.3	16.5	1.12631
.4	10.9	1.08062	.4	13.7	1.10329	.4	16.5	1.12679
.5	11.0	1.08106	.5	13.8	1.10375	.5	16.6	1.12727
.6	11.1	1.08151	.6	13.8	1.10421	.6	16.6	1.12775
.7	11.1	1.08196	.7	13.9	1.10468	.7	16.7	1.12823
.8	11.2	1.08240	.8	14.0	1.10514	.8	16.7	1.12871
.9	11.2	1.08285	.9	14.0	1.10560	.9	16.8	1.12919
20.0	11.3	1.08329	25.0	14.1	1.10607	30.0	16.8	1.12967
.1	11.3	1.08374	.1	14.1	1.10653	.1	16.9	1.13015
.2	11.4	1.08419	.2	14.2	1.10700	.2	16.95	1.13063
.3	11.5	1.08464	.3	14.2	1.10746	.3	17.0	1.13111
.4	11.5	1.08509	.4	14.3	1.10793	.4	17.1	1.13159
.5	11.6	1.08553	.5	14.3	1.10839	.5	17.1	1.13207
.6	11.6	1.08599	.6	14.4	1.10886	.6	17.2	1.13255
.7	11.7	1.08643	.7	14.5	1.10932	.7	17.2	1.13304
.8	11.7	1.08688	.8	14.5	1.10979	.8	17.3	1.13352
.9	11.8	1.08733	.9	14.6	1.11026	.9	17.3	1.13400
21.0	11.8	1.08778	26.0	14.6	1.11072	31.0	17.4	1.13449
.1	11.9	1.08824	.1	14.7	1.11119	.1	17.4	1.13497
.2	11.95	1.08869	.2	14.7	1.11166	.2	17.5	1.13545
.3	12.0	1.08914	.3	14.8	1.11213	.3	17.6	1.13594
.4	12.0	1.08959	.4	14.85	1.11259	.4	17.6	1.13642
.5	12.1	1.09004	.5	14.9	1.11306	.5	17.7	1.13691
.6	12.1	1.09049	.6	15.0	1.11353	.6	17.7	1.13740
.7	12.2	1.09095	.7	15.0	1.11400	.7	17.8	1.13788
.8	12.3	1.09140	.8	15.1	1.11447	.8	17.8	1.13837
.9	12.3	1.09185	.9	15.1	1.11494	.9	17.9	1.13885
22.0	12.4	1.09231	27.0	15.2	1.11541	32.0	17.95	1.13934
.1	12.5	1.09276	.1	15.2	1.11588	.1	18.0	1.13983
.2	12.5	1.09321	.2	15.3	1.11635	.2	18.0	1.14032
.3	12.6	1.09367	.3	15.3	1.11682	.3	18.1	1.14081
.4	12.6	1.09412	.4	15.4	1.11729	.4	18.2	1.14129
.5	12.7	1.09458	.5	15.5	1.11776	.5	18.2	1.14178
.6	12.7	1.09503	.6	15.5	1.11824	.6	18.3	1.14227
.7	12.8	1.09549	.7	15.6	1.11871	.7	18.3	1.14276
.8	12.85	1.09595	.8	15.6	1.11918	.8	18.4	1.14325
.9	12.9	1.09640	.9	15.7	1.11965	.9	18.4	1.14374

TABLE SHOWING A COMPARISON OF THE DEGREES BRIX
AND BAUMÉ, ETC.—*Continued.*

Degree Brix (Per Cent Sugar).	Degree Baumé (corrected).	Specific Gravity.	Degree Brix (Per Cent Sugar).	Degree Baumé (corrected).	Specific Gravity.	Degree Brix (Per Cent Sugar).	Degree Baumé (corrected).	Specific Gravity.
33.0	18.5	1.14423	38.0	21.2	1.16920	43.0	23.95	1.19505
.1	18.55	1.14472	.1	21.3	1.16971	.1	24.0	1.19558
.2	18.6	1.14521	.2	21.35	1.17022	.2	24.1	1.19611
.3	18.7	1.14570	.3	21.4	1.17072	.3	24.1	1.19653
.4	18.7	1.14620	.4	21.5	1.17123	.4	24.2	1.19716
.5	18.8	1.14669	.5	21.5	1.17174	.5	24.2	1.19769
.6	18.8	1.14718	.6	21.6	1.17225	.6	24.3	1.19822
.7	18.9	1.14767	.7	21.6	1.17276	.7	24.3	1.19875
.8	18.9	1.14817	.8	21.7	1.17327	.8	24.4	1.19927
.9	19.0	1.14866	.9	21.7	1.17379	.9	24.4	1.19980
34.0	19.05	1.14915	39.0	21.8	1.17430	44.0	24.5	1.20033
.1	19.1	1.14965	.1	21.8	1.17481	.1	24.55	1.20086
.2	19.2	1.15014	.2	21.9	1.17532	.2	24.6	1.20139
.3	19.2	1.15064	.3	21.9	1.17583	.3	24.65	1.20192
.4	19.3	1.15113	.4	22.0	1.17635	.4	24.7	1.20245
.5	19.3	1.15163	.5	22.05	1.17686	.5	24.8	1.20299
.6	19.4	1.15213	.6	22.1	1.17737	.6	24.8	1.20352
.7	19.4	1.15262	.7	22.2	1.17789	.7	24.9	1.20405
.8	19.5	1.15312	.8	22.2	1.17840	.8	24.9	1.20458
.9	19.5	1.15362	.9	22.3	1.17892	.9	25.0	1.20512
35.0	19.6	1.15411	40.0	22.3	1.17943	45.0	25.0	1.20565
.1	19.65	1.15461	.1	22.4	1.17995	.1	25.1	1.20618
.2	19.7	1.15511	.2	22.4	1.18046	.2	25.1	1.20672
.3	19.8	1.15561	.3	22.5	1.18098	.3	25.2	1.20725
.4	19.8	1.15611	.4	22.5	1.18150	.4	25.2	1.20779
.5	19.9	1.15661	.5	22.6	1.18201	.5	25.3	1.20832
.6	19.9	1.15710	.6	22.6	1.18253	.6	25.4	1.20886
.7	20.0	1.15760	.7	22.7	1.18305	.7	25.4	1.20939
.8	20.0	1.15810	.8	22.8	1.18357	.8	25.5	1.20993
.9	20.1	1.15861	.9	22.8	1.18408	.9	25.5	1.21046
36.0	20.1	1.15911	41.0	22.9	1.18460	46.0	25.6	1.21100
.1	20.2	1.15961	.1	22.9	1.18512	.1	25.6	1.21154
.2	20.25	1.16011	.2	23.0	1.18564	.2	25.7	1.21208
.3	20.3	1.16061	.3	23.0	1.18616	.3	25.7	1.21261
.4	20.4	1.16111	.4	23.1	1.18668	.4	25.8	1.21315
.5	20.4	1.16162	.5	23.1	1.18720	.5	25.8	1.21369
.6	20.5	1.16212	.6	23.2	1.18772	.6	25.9	1.21423
.7	20.5	1.16262	.7	23.25	1.18824	.7	25.95	1.21477
.8	20.6	1.16313	.8	23.3	1.18877	.8	26.0	1.21531
.9	20.6	1.16363	.9	23.4	1.18929	.9	26.1	1.21585
37.0	20.7	1.16413	42.0	23.4	1.18981	47.0	26.1	1.21639
.1	20.7	1.16464	.1	23.5	1.19033	.1	26.2	1.21693
.2	20.8	1.16514	.2	23.5	1.19086	.2	26.2	1.21747
.3	20.9	1.16565	.3	23.6	1.19138	.3	26.3	1.21802
.4	20.9	1.16616	.4	23.6	1.19190	.4	26.3	1.21856
.5	21.0	1.16666	.5	23.7	1.19243	.5	26.4	1.21910
.6	21.0	1.16717	.6	23.7	1.19295	.6	26.4	1.21964
.7	21.1	1.16768	.7	23.8	1.19348	.7	26.5	1.22019
.8	21.1	1.16818	.8	23.8	1.19400	.8	26.5	1.22073
.9	21.2	1.16869	.9	23.9	1.19453	.9	26.6	1.22127

TABLE SHOWING A COMPARISON OF THE DEGREES BRIX AND BAUMÉ, ETC.—*Continued.*

Degree Brix (Per Cent Sugar).	Degree Baumé (corrected).	Specific Gravity.	Degree Brix (Per Cent Sugar).	Degree Baumé (corrected).	Specific Gravity.	Degree Brix (Per Cent Sugar).	Degree Baumé (corrected).	Specific Gravity.
48.0	26.6	1.22182	**53.0**	29.3	1.24951	**58.0**	31.9	1.27816
.1	26.7	1.22236	.1	29.4	1.25008	.1	32.0	1.27874
.2	26.75	1.22291	.2	29.4	1.25064	.2	32.0	1.27932
.3	26.8	1.22345	.3	29.5	1.25120	.3	32.1	1.27991
.4	26.9	1.22400	.4	29.5	1.25177	.4	32.15	1.28049
.5	26.9	1.22455	.5	29.6	1.25233	.5	32.2	1.28107
.6	27.0	1.22509	.6	29.6	1.25290	.6	32.3	1.28166
.7	27.0	1.22564	.7	29.7	1.25347	.7	32.3	1.28224
.8	27.1	1.22619	.8	29.7	1.25403	.8	32.4	1.28283
.9	27.1	1.22673	.9	29.8	1.25460	.9	32.4	1.28342
49.0	27.2	1.22728	**54.0**	29.8	1.25517	**59.0**	32.5	1.28400
.1	27.2	1.22783	.1	29.9	1.25573	.1	32.5	1.28459
.2	27.3	1.22838	.2	29.9	1.25630	.2	32.6	1.28518
.3	27.3	1.22893	.3	30.0	1.25687	.3	32.6	1.28576
.4	27.4	1.22948	.4	30.05	1.25747	.4	32.7	1.28635
.5	27.4	1.23003	.5	30.1	1.25801	.5	32.7	1.28694
.6	27.5	1.23058	.6	30.2	1.25857	.6	32.8	1.28753
.7	27.6	1.23113	.7	30.2	1.25914	.7	32.8	1.28812
.8	27.6	1.23168	.8	30.3	1.25971	.8	32.9	1.28871
.9	27.7	1.23223	.9	30.3	1.26028	.9	32.9	1.28930
50.0	27.7	1.23278	**55.0**	30.4	1.26086	**60.0**	33.0	1.28989
.1	27.8	1.23334	.1	30.4	1.26143	.1	33.0	1.29048
.2	27.8	1.23389	.2	30.5	1.26200	.2	33.1	1.29107
.3	27.9	1.23444	.3	30.5	1.26257	.3	33.1	1.29166
.4	27.9	1.23499	.4	30.6	1.26314	.4	33.2	1.29225
.5	28.0	1.23555	.5	30.6	1.26372	.5	33.2	1.29284
.6	28.0	1.23610	.6	30.7	1.26429	.6	33.3	1.29343
.7	28.1	1.23666	.7	30.7	1.26486	.7	33.35	1.29403
.8	28.1	1.23721	.8	30.8	1.26544	.8	33.4	1.29462
.9	28.2	1.23777	.9	30.8	1.26601	.9	33.45	1.29521
51.0	28.2	1.23832	**56.0**	30.9	1.26658	**61.0**	33.5	1.29581
.1	28.3	1.23888	.1	30.9	1.26716	.1	33.6	1.29640
.2	28.35	1.23943	.2	31.0	1.26773	.2	33.6	1.29700
.3	28.4	1.23999	.3	31.05	1.26831	.3	33.7	1.29759
.4	28.5	1.24055	.4	31.1	1.26889	.4	33.7	1.29819
.5	28.5	1.24111	.5	31.2	1.26946	.5	33.8	1.29878
.6	28.6	1.24166	.6	31.2	1.27004	.6	33.8	1.29938
.7	28.6	1.24222	.7	31.3	1.27062	.7	33.9	1.29998
.8	28.7	1.24278	.8	31.3	1.27120	.8	33.9	1.30057
.9	28.7	1.24334	.9	31.4	1.27177	.9	34.0	1.30117
52.0	28.8	1.24390	**57.0**	31.4	1.27235	**62.0**	34.0	1.30177
.1	28.8	1.24446	.1	31.5	1.27293	.1	34.1	1.30237
.2	28.9	1.24502	.2	31.5	1.27351	.2	34.1	1.30297
.3	28.9	1.24558	.3	31.6	1.27409	.3	34.2	1.30356
.4	29.0	1.24614	.4	31.6	1.27464	.4	34.2	1.30416
.5	29.0	1.24670	.5	31.7	1.27525	.5	34.3	1.30476
.6	29.1	1.24726	.6	31.7	1.27583	.6	34.3	1.30535
.7	29.15	1.24782	.7	31.8	1.27641	.7	34.4	1.30596
.8	29.2	1.24839	.8	31.8	1.27699	.8	34.4	1.30657
.9	29.2	1.24895	.9	31.9	1.27758	.9	34.5	1.30717

TABLE SHOWING A COMPARISON OF THE DEGREES BRIX AND BAUMÉ, ETC.—*Continued*.

Degree Brix (Per Cent Sugar).	Degree Baumé (corrected).	Specific Gravity.	Degree Brix (Per Cent Sugar).	Degree Baumé (corrected).	Specific Gravity.	Degree Brix (Per Cent Sugar).	Degree Baumé (corrected).	Specific Gravity.
63.0	34.5	1.30777	68.0	37.1	1.33836	73.0	39.6	1.36995
.1	34.6	1.30837	.1	37.1	1.33899	.1	39.7	1.37059
.2	34.6	1.30897	.2	37.2	1.33961	.2	39.7	1.37124
.3	34.7	1.30958	.3	37.3	1.34023	.3	39.8	1.37188
.4	34.7	1.31018	.4	37.3	1.34085	.4	39.8	1.37252
.5	34.8	1.31078	.5	37.4	1.34148	.5	39.9	1.37317
.6	34.85	1.31139	.6	37.4	1.34210	.6	39.9	1.37381
.7	34.9	1.31199	.7	37.5	1.34273	.7	40.0	1.37446
.8	34.95	1.31260	.8	37.5	1.34335	.8	40.0	1.37510
.9	35.0	1.31320	.9	37.6	1.34398	.9	40.1	1.37575
64.0	35.1	1.31381	69.0	37.6	1.34460	74.0	40.1	1.37639
.1	35.1	1.31442	.1	37.7	1.34523	.1	40.2	1.37704
.2	35.2	1.31502	.2	37.7	1.34585	.2	40.2	1.37768
.3	35.2	1.31563	.3	37.8	1.34648	.3	40.3	1.37833
.4	35.3	1.31624	.4	37.8	1.34711	.4	40.3	1.37898
.5	35.3	1.31684	.5	37.9	1.34774	.5	40.4	1.37962
.6	35.4	1.31745	.6	37.9	1.34836	.6	40.4	1.38027
.7	35.4	1.31806	.7	38.0	1.34899	.7	40.5	1.38092
.8	35.5	1.31867	.8	38.0	1.34962	.8	40.5	1.38157
.9	35.5	1.31928	.9	38.1	1.35025	.9	40.6	1.38222
65.0	35.6	1.31989	70.0	38.1	1.35088	75.0	40.6	1.38287
.1	35.6	1.32050	.1	38.2	1.35151	.1	40.7	1.38352
.2	35.7	1.32111	.2	38.2	1.35214	.2	40.7	1.38417
.3	35.7	1.32172	.3	38.3	1.35277	.3	40.8	1.38482
.4	35.8	1.32233	.4	38.3	1.35340	.4	40.8	1.38547
.5	35.8	1.32294	.5	38.4	1.35403	.5	40.9	1.38612
.6	35.9	1.32355	.6	38.4	1.35466	.6	40.9	1.38677
.7	35.9	1.32417	.7	38.5	1.35530	.7	41.0	1.38743
.8	36.0	1.32478	.8	38.5	1.35593	.8	41.0	1.38808
.9	36.0	1.32539	.9	38.6	1.35656	.9	41.1	1.38873
66.0	36.1	1.32601	71.0	38.6	1.35720	76.0	41.1	1.38939
.1	36.1	1.32662	.1	38.7	1.35783	.1	41.2	1.39004
.2	36.2	1.32724	.2	38.7	1.35847	.2	41.2	1.39070
.3	36.2	1.32785	.3	38.8	1.35910	.3	41.3	1.39135
.4	36.3	1.32847	.4	38.8	1.35974	.4	41.3	1.39201
.5	36.3	1.32908	.5	38.9	1.36037	.5	41.4	1.39266
.6	36.4	1.32970	.6	38.9	1.36101	.6	41.4	1.39332
.7	36.4	1.33031	.7	39.0	1.36164	.7	41.5	1.39397
.8	36.5	1.33093	.8	39.0	1.36228	.8	41.5	1.39463
.9	36.5	1.33155	.9	39.1	1.36292	.9	41.6	1.39529
67.0	36.6	1.33217	72.0	39.1	1.36355	77.0	41.6	1.39595
.1	36.6	1.33278	.1	39.2	1.36419	.1	41.7	1.39660
.2	36.7	1.33340	.2	39.2	1.36483	.2	41.7	1.39726
.3	36.75	1.33402	.3	39.3	1.36547	.3	41.8	1.39792
.4	36.8	1.33464	.4	39.3	1.36611	.4	41.8	1.39858
.5	36.85	1.33526	.5	39.4	1.36675	.5	41.9	1.39924
.6	36.9	1.33588	.6	39.4	1.36739	.6	41.9	1.39990
.7	36.95	1.33650	.7	39.5	1.36803	.7	42.0	1.40056
.8	37.0	1.33712	.8	39.5	1.36867	.8	42.0	1.40122
.9	37.0	1.33774	.9	39.6	1.36931	.9	42.1	1.40188

TABLE SHOWING A COMPARISON OF THE DEGREES BRIX AND BAUMÉ, ETC.—*Continued.*

Degree Brix (Per Cent Sugar)	Degree Baumé (corrected)	Specific Gravity	Degree Brix (Per Cent Sugar)	Degree Baumé (corrected)	Specific Gravity	Degree Brix (Per Cent Sugar)	Degree Baumé (corrected)	Specific Gravity
78.0	42.1	1.40254	83.0	44.6	1.43614	88.0	47.0	1.47074
.1	42.2	1.40321	.1	44.6	1.43682	.1	47.0	1.47145
.2	42.2	1.40387	.2	44.7	1.43750	.2	47.1	1.47215
.3	42.3	1.40453	.3	44.7	1.43819	.3	47.1	1.47285
.4	42.3	1.40520	.4	44.8	1.43887	.4	47.2	1.47356
.5	42.4	1.40586	.5	44.8	1.43955	.5	47.2	1.47426
.6	42.4	1.40652	.6	44.9	1.44024	.6	47.3	1.47496
.7	42.5	1.40719	.7	44.9	1.44092	.7	47.3	1.47567
.8	42.5	1.40785	.8	45.0	1.44161	.8	47.4	1.47637
.9	42.6	1.40852	.9	45.0	1.44229	.9	47.4	1.47708
79.0	42.6	1.40918	84.0	45.1	1.44298	89.0	47.45	1.47778
.1	42.7	1.40985	.1	45.1	1.44367	.1	47.5	1.47849
.2	42.7	1.41052	.2	45.15	1.44435	.2	47.55	1.47920
.3	42.8	1.41118	.3	45.2	1.44504	.3	47.6	1.47991
.4	42.8	1.41185	.4	45.25	1.44573	.4	47.6	1.48061
.5	42.9	1.41252	.5	45.3	1.44641	.5	47.7	1.48132
.6	42.9	1.41318	.6	45.35	1.44710	.6	47.7	1.48203
.7	43.0	1.41385	.7	45.4	1.44779	.7	47.8	1.48274
.8	43.0	1.41452	.8	45.4	1.44848	.8	47.8	1.48345
.9	43.1	1.41519	.9	45.5	1.44917	.9	47.9	1.48416
80.0	43.1	1.41586	85.0	45.5	1.44986	90.0	47.9	1.48486
.1	43.2	1.41653	.1	45.6	1.45055	.1	48.0	1.48558
.2	43.2	1.41720	.2	45.6	1.45124	.2	48.0	1.48629
.3	43.2	1.41787	.3	45.7	1.45193	.3	48.1	1.48700
.4	43.3	1.41854	.4	45.7	1.45262	.4	48.1	1.48771
.5	43.3	1.41921	.5	45.8	1.45331	.5	48.2	1.48842
.6	43.4	1.41989	.6	45.8	1.45401	.6	48.2	1.48913
.7	43.45	1.42056	.7	45.9	1.45470	.7	48.3	1.48985
.8	43.5	1.42123	.8	45.9	1.45539	.8	48.3	1.49056
.9	43.55	1.42190	.9	46.0	1.45609	.9	48.35	1.49127
81.0	43.6	1.42258	86.0	46.0	1.45678	91.0	48.4	1.49199
.1	43.65	1.42325	.1	46.1	1.45748	.1	48.45	1.49270
.2	43.7	1.42393	.2	46.1	1.45817	.2	48.5	1.49342
.3	43.7	1.42460	.3	46.2	1.45887	.3	48.5	1.49413
.4	43.8	1.42528	.4	46.2	1.45956	.4	48.6	1.49485
.5	43.8	1.42595	.5	46.3	1.46026	.5	48.6	1.49556
.6	43.9	1.42663	.6	46.3	1.46095	.6	48.7	1.49628
.7	43.9	1.42731	.7	46.35	1.46165	.7	48.7	1.49700
.8	44.0	1.42798	.8	46.4	1.46235	.8	48.8	1.49771
.9	44.0	1.42866	.9	46.45	1.46304	.9	48.8	1.49843
82.0	44.1	1.42934	87.0	46.5	1.46374	92.0	48.9	1.49915
.1	44.1	1.43002	.1	46.55	1.46444	.1	48.9	1.49987
.2	44.2	1.43070	.2	46.6	1.46514	.2	49.0	1.50058
.3	44.2	1.43137	.3	46.65	1.46584	.3	49.0	1.50130
.4	44.3	1.43205	.4	46.7	1.46654	.4	49.05	1.50202
.5	44.3	1.43273	.5	46.7	1.46724	.5	49.1	1.50274
.6	44.4	1.43341	.6	46.8	1.46794	.6	49.15	1.50346
.7	44.4	1.43409	.7	46.8	1.46864	.7	49.2	1.50419
.8	44.5	1.43478	.8	46.9	1.46934	.8	49.2	1.50491
.9	44.5	1.43546	.9	46.9	1.47004	.9	49.3	1.50563

TABLE SHOWING A COMPARISON OF THE DEGREES BRIX
AND BAUMÉ, ETC.—*Continued.*

Degree Brix (Per Cent Sugar).	Degree Baumé (corrected).	Specific Gravity.	Degree Brix (Per Cent Sugar).	Degree Baumé (corrected).	Specific Gravity.
93.0	49.3	1.50635	94.0	49.8	1.51359
.1	49.4	1.50707	.1	49.85	1.51431
.2	49.4	1.50779	.2	49.9	1.51504
.3	49.5	1.50852	.3	49.9	1.51577
.4	49.5	1.50924	.4	50.0	1.51649
.5	49.6	1.50996	.5	50.0	1.51722
.6	49.6	1.51069	.6	50.1	1.51795
.7	49.7	1.51141	.7	50.1	1.51868
.8	49.7	1.51214	.8	50.2	1.51941
.9	49.8	1.51286	.9	50.2	1.52014
			95.0	50.3	1.52087

257. TABLE FOR THE CORRECTION OF READINGS ON THE
BRIX SCALE FOR VARIATIONS IN TEMPERATURE FROM THE
STANDARD, 17½° C. (63½° F.).—(F. SACHS.)

Temp. ° C.	Temp. ° F.	APPROXIMATE DEGREE BRIX AND CORRECTION.												
		0	5	10	15	20	25	30	35	40	50	60	70	75
0	32	.27	.30	.41	.52	.62	.72	.82	.92	.98	1.11	1.22	1.25	1.29
5	41	.23	.30	.37	.44	.52	.59	.65	.72	.75	.80	.88	.91	.94
10	50	.20	.26	.29	.33	.36	.39	.42	.45	.48	.50	.54	.58	.61
11	51.8	.18	.23	.26	.28	.31	.34	.36	.39	.41	.43	.47	.50	.53
12	53.6	.16	.20	.22	.24	.26	.29	.31	.33	.34	.36	.40	.42	.46
13	55.4	.14	.18	.19	.21	.22	.24	.26	.27	.28	.29	.33	.35	.39
14	57.2	.12	.15	.16	.17	.18	.19	.21	.22	.22	.23	.26	.28	.32
15	59	.09	.11	.12	.14	.14	.15	.16	.17	.16	.17	.19	.21	.25
16	60.8	.06	.07	.08	.09	.10	.10	.11	.12	.12	.12	.14	.16	.18
17	62.6	.02	.02	.03	.03	.03	.04	.04	.04	.04	.04	.05	.05	.06

Add the correction to readings above 17½° C. (63½ F.) and subtract
the correction from those below this temperature.

Temp. ° C.	Temp. ° F.	0	5	10	15	20	25	30	35	40	50	60	70	75
18	64.4	.02	.03	.03	.03	.03	.03	.03	.03	.03	.03	.03	.03	.02
19	66.2	.06	.08	.08	.09	.09	.10	.10	.10	.10	.10	.10	.08	.06
20	68	.11	.14	.15	.17	.17	.18	.18	.18	.19	.19	.18	.15	.11
21	69.8	.16	.20	.22	.24	.24	.25	.25	.25	.26	.26	.25	.22	.18
22	71.6	.21	.26	.29	.31	.31	.32	.32	.32	.33	.34	.32	.29	.25
23	73.4	.27	.32	.35	.37	.38	.39	.39	.39	.40	.42	.39	.36	.33
24	75.2	.32	.38	.41	.43	.44	.46	.46	.47	.47	.50	.46	.43	.40
25	77	.37	.44	.47	.49	.51	.53	.54	.55	.55	.58	.54	.51	.48
26	78.8	.43	.50	.54	.56	.58	.60	.61	.62	.62	.66	.62	.58	.55
27	80.6	.49	.57	.61	.63	.65	.68	.68	.69	.70	.74	.70	.65	.62
28	82.4	.56	.64	.68	.70	.72	.76	.76	.78	.78	.82	.78	.72	.70
29	84.2	.63	.71	.75	.78	.79	.84	.84	.86	.86	.90	.86	.80	.78
30	86	.70	.78	.82	.87	.87	.92	.92	.94	.94	.98	.94	.88	.86
35	95	1.10	1.17	1.22	1.24	1.30	1.32	1.33	1.35	1.36	1.39	1.34	1.27	1.25
40	104	1.50	1.61	1.67	1.71	1.73	1.79	1.79	1.80	1.82	1.83	1.78	1.69	1.65
50	122		2.65	2.71	2.74	2.78	2.80	2.80	2.80	2.80	2.79	2.70	2.56	2.51
60	140		3.87	3.88	3.88	3.88	3.88	3.88	3.90	3.82	3.70	3.43	3.41
70	158			5.18	5.20	5.14	5.13	5.10	5.08	5.06	4.90	4.72	4.47	4.35
80	176			6.62	6.59	6.54	6.46	6.38	6.30	6.26	6.06	5.82	5.50	5.33

258. TABLE SHOWING THE WEIGHT PER CUBIC FOOT, AND U. S. GALLON (231 Cu. In.) OF SUGAR SOLUTIONS AT 17½° C.

(Calculated from Stammer's Table of Specific Gravities.)

Degree Brix	Degree Baumé (corrected)	Weight of 1 cubic ft.	Weight of 1 gallon (231 cu. in.)	Degree Brix	Degree Baumé (corrected)	Weight of 1 cubic ft.	Weight of 1 gallon (231 cu. in.)	Degree Brix	Degree Baumé (corrected)	Weight of 1 cubic ft.	Weight of 1 gallon (231 cu. in.)
		Lbs.	Lbs.			Lbs.	Lbs.			Lbs.	Lbs.
1	0:6	62.59	8.36	28	15.7	69.84	9.33	55	30.4	78.62	10.51
1.5	0.85	62.72	8.38	28.5	16.0	69.99	9.35	55.5	30.6	78.79	10.53
2	1.1	62.84	8.39	29	16.3	70.14	9.38	56	30.9	78.97	10.55
2.5	1.4	62.96	8.40	29.5	16.6	70.29	9.39	56.5	31.2	79.15	10.57
3	1.7	63.08	8.42	30	16.8	70.44	9.41	57	31.4	79.33	10.60
3.5	2.0	63.20	8.44	30.5	17.1	70.59	9.43	57.5	31.7	79.51	10.62
4	2.3	63.32	8.46	31	17.4	70.74	9.45	58	31.9	79.70	10.65
4.5	2.55	63.44	8.48	31.5	17.7	70.89	9.47	58.5	32.2	79.87	10.67
5	2.8	63.57	8.50	32	17.95	71.04	9.49	59	32.5	80.05	10.70
5.5	3.1	63.70	8.52	32.5	18.2	71.19	9.51	59.5	32.7	80.24	10.72
6	3.4	63.83	8.53	33	18.5	71.35	9.53	60	33.0	80.43	10.75
6.5	3.7	63.95	8.55	33.5	18.8	71.50	9.55	60.5	33.2	80.62	10.77
7	4.0	64.08	8.57	34	19.05	71.65	9.58	61	33.5	80.80	10.80
7.5	4.25	64.21	8.59	34.5	19.3	71.80	9.60	61.5	33.8	80.98	10.82
8	4.5	64.34	8.60	35	19.6	71.96	9.62	62	34.0	81.17	10.85
8.5	4.8	64.47	8.61	35.5	19.9	72.11	9 64	62.5	34.3	81.35	10.87
9	5.1	64.60	8.63	36	20.1	72.27	9.66	63	34.5	81 54	10.90
9.5	5.4	64.72	8.65	36.5	20.4	72.43	9 68	63.5	34.8	81 73	10.92
10	5.7	64.84	8.67	37	20.7	72.59	9.70	64	35.1	81.92	10.95
10.5	5.9	64.97	8.69	37.5	21.0	72.74	9.72	64.5	35.3	82.11	10.97
11	6.2	65.11	8.71	38	21.2	72.90	9.74	65	35.6	82.30	11.00
11.5	6 5	65.24	8.72	38.5	21.5	73.06	9.76	65.5	35.8	82.49	11.02
12	6.8	65.38	8.74	39	21.8	73.22	9.78	66	36.1	82.68	11.05
12.5	7.1	65.51	8.76	39.5	22.05	73.38	9.80	66.5	36.3	82.87	11.07
13	7.4	65.64	8.78	40	22.3	73.54	9.83	67	36.6	83.06	11.10
13.5	7.6	65.77	8.79	40.5	22.6	73.70	9.85	67.5	36.85	83.25	11.12
14	7.9	65.91	8.81	41	22.9	73.86	9.87	68	37.1	83.45	11.15
14.5	8.2	66.04	8.82	41.5	23.1	74.02	9.89	68.5	37.4	83.64	11.17
15	8.5	66.18	8.84	42	23.4	74.18	9.91	69	37.6	83.84	11.20
15.5	8.8	66.31	8.86	42.5	23.7	74.34	9.93	69.5	37.9	84.03	11.23
16	9.0	66.44	8.88	43	23.95	74.51	9.96	70	38.1	84.23	11.26
16.5	9.3	66 58	8.90	43.5	24.2	74.67	9.98	70.5	38.4	84.42	11.28
17	9 6	66 72	8.92	44	24.5	74.84	10.00	71	38.6	84.62	11.31
17.5	9.9	66.85	8.93	44.5	24 8	75.00	10.02	71.5	38 9	84.82	11.33
18	10.1	66 99	8.95	45	25.0	75.17	10 05	72	39.1	85.02	11.36
18.5	10.4	67.13	8.97	45.5	25.3	75.34	10.07	72.5	39.4	85.21	11.39
19	10.7	67.27	8.99	46	25.6	75.51	10.09	73	39.6	85.41	11.42
19.5	11.0	67.41	9.01	46.5	25.8	75.67	10.11	73.5	39.9	85.61	11.44
20	11.3	67.55	9.03	47	26.1	75.84	10.13	74	40.1	85 81	11.47
20.5	11.6	67.69	9.04	47.5	26.4	76.01	10.15	74.5	40.4	86.01	11.49
21	11.8	67.83	9.06	48	26.6	76.18	10.18	75	40.6	86.22	11.52
21.5	12.1	67.97	9.08	48.5	26.9	76.35	10.20	75.5	40.9	86.42	11.55
22	12.4	68.11	9.10	49	27.2	76.52	10.23	76	41.1	86.63	11.58
22 5	12.7	68.25	9.13	49.5	27.4	76.69	10.25	76.5	41.4	86.83	11.60
23	13.0	68.39	9.16	50	27.7	76.87	10.27	77	41.6	87.04	11.63
23.5	13.2	68.54	9.17	50.5	28.0	77.04	10 29	77.5	41.9	87 24	11.66
24	13.5	68.68	9.18	51	28.2	77.21	10.32	78	42.1	87.45	11 69
24.5	13.8	68.82	9.20	51.5	28.5	77.38	10 34	78.5	42.4	87.65	11.71
25	14.1	68.96	9.22	52	28.8	77.56	10.36	79	42.6	87.86	11.74
25.5	14.3	69.11	9.24	52.5	29.0	77.73	10.38	79.5	42.9	88.07	11.77
26	14.6	69.26	9.26	53	29.3	77.91	10.41	80	43.1	88 28	11.80
26.5	14.9	69 41	9.27	53.5	29.6	78.08	10.43	80.5	43.3	88.49	11.82
27	15 2	69.55	9.29	54	29.8	78.26	10.46	81	43.6	88.70	11.85
27.5	15.5	69.69	9.31	54.5	30.1	78.44	10.48	81.5	43.8	88.91	11.88

TABLE SHOWING THE WEIGHT PER CUBIC FOOT AND U. S. GALLON (231 Cu. In.) OF SUGAR SOLUTIONS.—*Continued.*

Degree Brix.	Degree Baumé (corrected).	Weight of 1 cubic ft.	Weight of 1 gallon (231 cu. in.).	Degree Brix.	Degree Baumé (corrected).	Weight of 1 cubic ft.	Weight of 1 gallon (231 cu. in.).	Degree Brix.	Degree Baumé (corrected).	Weight of 1 cubic ft.	Weight of 1 gallon (231 cu. in.).
		Lbs.	Lbs.			Lbs	Lbs.			Lbs.	Lbs.
82	44.1	89.13	11.91	86.5	46.3	91.04	12.17	91	48.4	93.02	12.43
82.5	44.3	89.34	11.94	87	46.5	91.26	12.20	91.5	48.6	93.24	12.46
83	44.6	89.55	11.97	87.5	46.7	91.48	12.23	92	48.9	93.47	12.49
83.5	44.8	89.76	11.99	88	47.0	91.70	12.26	92.5	49.1	93.69	12.52
84	45.1	89.97	12.02	88.5	47.2	91.92	12.28	93	49.3	93.92	12.55
84.5	45.3	90.18	12.05	89	47.45	92.14	12.31	93.5	49.6	94.14	12.58
85	45.5	90.40	12.08	89.5	47.7	92.36	12.34	94	49.8	94.37	12.61
85.5	45.8	90.61	12.11	90	47.9	92.58	12.37	94.5	50.0	94.60	12.64
86	46.0	90.83	12.14	90.5	48.2	92.80	12.40	95	50.3	94.83	12.67

259. SCHMITZ' TABLE FOR THE CALCULATION OF PER CENTS SUCROSE, ALLOWANCE BEING MADE FOR VARIATIONS IN THE SPECIFIC ROTATORY POWER OF CANE-SUGAR.

(CORRECTED FOR AN INCREASE IN VOLUME OF 1-10.) FOR INSTRUMENTS WHOSE FACTOR IS 26.048 GRAMS.

DEGREE BRIX.

Polariscopic Reading	0.5	1.0	1.5	2.0	2.5	3.0	3.5	4.0	4.5	5.0	5.5	6.0	6.5	7.0	7.5	8.0	8.5	9.0	9.5	10.0
1	0.29	0.29	0.29	0.28	0.28	0.28	0.28	0.28	0.28	0.28	0.28	0.28	0.28	0.28	0.28	0.28	0.28	0.28	0.28	0.28
2		0.57	0.57	0.57	0.57	0.56	0.56	0.56	0.56	0.56	0.56	0.56	0.56	0.56	0.55	0.55	0.55	0.55	0.55	0.55
3		0.85	0.85	0.85	0.85	0.85	0.85	0.84	0.84	0.84	0.84	0.84	0.84	0.83	0.83	0.83	0.83	0.83	0.83	0.82
4			1.14	1.13	1.13	1.13	1.13	1.13	1.12	1.12	1.12	1.12	1.11	1.11	1.11	1.11	1.11	1.10	1.10	1.10
5			1.42	1.42	1.41	1.41	1.41	1.41	1.40	1.40	1.40	1.40	1.39	1.39	1.39	1.38	1.38	1.38	1.38	1.37
6				1.70	1.70	1.69	1.69	1.69	1.68	1.68	1.68	1.67	1.67	1.67	1.66	1.66	1.66	1.66	1.65	1.65
7				1.98	1.98	1.98	1.97	1.97	1.96	1.96	1.96	1.95	1.95	1.95	1.94	1.94	1.93	1.93	1.93	1.92
8					2.36	2.26	2.26	2.25	2.25	2.24	2.24	2.23	2.23	2.22	2.22	2.22	2.21	2.21	2.20	2.20
9						2.54	2.54	2.53	2.53	2.52	2.52	2.51	2.51	2.50	2.50	2.49	2.49	2.48	2.48	2.47
10						2.82	2.82	2.81	2.81	2.80	2.80	2.79	2.79	2.78	2.78	2.77	2.76	2.76	2.75	2.75
11							3.10	3.09	3.09	3.08	3.08	3.07	3.06	3.06	3.05	3.05	3.04	3.03	3.03	3.02
12							3.38	3.38	3.37	3.36	3.36	3.35	3.34	3.34	3.33	3.32	3.32	3.31	3.30	3.30
13								3.66	3.65	3.64	3.64	3.63	3.62	3.61	3.61	3.60	3.59	3.59	3.58	3.57
14								3.94	3.93	3.92	3.92	3.91	3.90	3.89	3.89	3.88	3.87	3.86	3.85	3.85
15									4.21	4.20	4.19	4.19	4.18	4.17	4.16	4.15	4.15	4.14	4.13	4.12
16									4.49	4.48	4.47	4.47	4.46	4.45	4.44	4.43	4.42	4.41	4.40	4.40
17										4.77	4.76	4.75	4.74	4.73	4.72	4.71	4.70	4.69	4.68	4.67
18											5.03	5.02	5.01	5.00	5.00	4.99	4.97	4.97	4.96	4.95
19											5.32	5.31	5.29	5.28	5.27	5.26	5.25	5.24	5.23	5.22
20												5.58	5.57	5.56	5.55	5.54	5.53	5.52	5.51	5.50
21												5.86	5.85	5.84	5.84	5.82	5.81	5.79	5.78	5.77
22													6.13	6.12	6.11	6.09	6.08	6.07	6.06	6.05
23													6.41	6.40	6.38	6.37	6.36	6.35	6.33	6.32

7.0	7.5	8.0	8.5	9.0	9.5	10.0	
6.67	6.66	6.65	6.64	6.62	6.61	6.60	24
....	6.94	6.93	6.91	6.90	6.89	6.87	25
....	7.22	7.20	7.19	7.17	7.16	7.15	26
....	....	7.48	7.46	7.45	7.44	7.42	27
....	....	7.76	7.74	7.73	7.71	7.70	28
....	....	....	8.02	8.00	7.99	7.97	29
....	....	....	....	8.28	8.26	8.25	30
....	....	....	....	8.55	8.54	8.52	31
....	....	....	....	8.83	8.81	8.80	32
....	....	....	....	....	9.09	9.07	33
....	....	....	....	....	....	9.35	34
....	....	....	....	....	....	9.68	35
....	....	....	....	....	....	....	36
....	....	....	....	....	....	....	37

DIRECTIONS FOR USING SCHMITZ' TABLE.

Note the degree Brix of the solution. Measure out 100 cc., add the lead, and dilute to 110 cc. Polarize as usual. Take the number in the table opposite the integral part of the polariscopic reading and under the degree Brix nearest that observed, and add to it the number corresponding to the tenths as shown in the small table. The sum so obtained is the per cent sucrose in the solution.

DEGREE BRIX FROM 0.5 TO 12.0	
Tenths of the Polariscopic Reading.	Per Cent Sucrose.
0.1	0.03
0.2	0.06
0.3	0.08
0.4	0.11
0.5	0.14
0.6	0.17
0.7	0.19
0.8	0.22
0.9	0.25

Degree Brix.

Polariscopic Reading.	20.0	19.5	19.0	18.5	18.0	17.5	17.0	16.5	16.0	15.5	15.0	14.5	14.0	13.5	13.0	12.5	12.0	11.5	11.0	10.5	Polariscopic Reading.
1	0.26	0.27	0.27	0.27	0.27	0.27	0.27	0.27	0.27	0.27	0.27	0.27	0.27	0.27	0.27	0.27	0.27	0.27	0.27	0.26	1
2	0.53	0.53	0.53	0.53	0.53	0.53	0.53	0.54	0.54	0.54	0.54	0.54	0.54	0.54	0.54	0.54	0.55	0.55	0.55	0.55	2
3	0.79	0.79	0.79	0.80	0.80	0.80	0.80	0.80	0.80	0.81	0.81	0.81	0.81	0.81	0.81	0.82	0.82	0.82	0.82	0.82	3
4	1.06	1.06	1.06	1.06	1.06	1.07	1.07	1.07	1.07	1.08	1.08	1.08	1.08	1.08	1.09	1.09	1.09	1.09	1.10	1.10	4
5	1.32	1.32	1.32	1.33	1.33	1.33	1.34	1.34	1.34	1.34	1.35	1.35	1.35	1.35	1.36	1.36	1.36	1.36	1.37	1.37	5
6	1.59	1.59	1.59	1.59	1.60	1.60	1.60	1.61	1.61	1.61	1.62	1.62	1.62	1.62	1.63	1.63	1.64	1.64	1.64	1.64	6
7	1.85	1.85	1.85	1.86	1.86	1.86	1.87	1.87	1.88	1.88	1.88	1.89	1.89	1.89	1.90	1.90	1.91	1.91	1.91	1.92	7
8	2.11	2.12	2.12	2.12	2.13	2.13	2.14	2.14	2.15	2.15	2.15	2.16	2.16	2.17	2.17	2.18	2.18	2.18	2.19	2.19	8
9	2.37	2.38	2.38	2.39	2.39	2.40	2.40	2.41	2.41	2.42	2.42	2.43	2.43	2.44	2.44	2.45	2.45	2.46	2.46	2.47	9
10	2.64	2.64	2.65	2.65	2.66	2.67	2.67	2.68	2.68	2.69	2.69	2.70	2.70	2.71	2.71	2.72	2.73	2.73	2.74	2.74	10
11	2.90	2.91	2.91	2.92	2.92	2.93	2.94	2.94	2.95	2.95	2.96	2.97	2.97	2.98	2.99	2.99	3.00	3.00	3.01	3.02	11
12	3.17	3.17	3.54	3.55	3.39	3.56	3.20	3.21	3.22	3.22	3.23	3.24	3.24	3.25	3.26	3.26	3.27	3.28	3.28	3.29	12
13	3.43	3.54	3.54	3.55	3.56	3.56	3.57	3.58	3.59	3.59	3.50	3.51	3.51	3.52	3.53	3.54	3.54	3.55	3.56	3.56	13
14	3.69	3.90	3.81	3.98	3.99	4.00	3.84	3.85	3.81	3.86	3.87	3.88	3.88	3.89	3.80	3.81	3.82	3.83	3.86	3.84	14
15	3.96	4.03	3.97	4.35	4.36	4.36	4.01	4.02	4.02	4.03	4.04	4.05	4.06	4.06	4.07	4.06	4.09	4.10	4.11	4.11	15
16	4.22	4.23	4.24	4.51	4.52	4.26	4.27	4.28	4.29	4.31	4.31	4.32	4.33	4.33	4.34	4.35	4.36	4.37	4.78	4.39	16
17	4.49	4.49	4.50	4.51	4.52	4.80	4.54	4.55	4.56	4.57	4.58	4.59	4.60	4.61	4.62	4.62	4.64	4.64	4.65	4.66	17
18	4.75	4.76	4.77	4.78	4.79	4.80	4.81	4.82	4.83	4.84	4.85	4.86	4.87	4.88	4.89	4.90	4.91	4.91	4.93	4.93	18
19	5.01	5.12	5.13	5.14	5.15	5.16	5.18	5.19	5.10	5.11	5.12	5.13	5.14	5.15	5.16	5.12	5.18	5.19	5.20	5.21	19
20	5.29	5.29	5.30	5.31	5.32	5.33	5.34	5.35	5.36	5.39	5.39	5.40	5.41	5.42	5.43	5.44	5.45	5.46	5.47	5.49	20
21	5.54	5.55	5.56	5.58	5.59	5.60	5.61	5.62	5.63	5.65	5.66	5.67	5.68	5.69	5.70	5.71	5.73	5.74	5.73	5.76	21
22	5.80	5.82	5.83	5.84	5.85	5.87	5.88	5.89	5.90	5.91	5.93	5.94	5.95	5.96	5.97	5.99	6.00	6.01	6.02	6.08	22
23	6.07	6.08	6.09	6.11	6.12	6.13	6.14	6.16	6.17	6.18	6.20	6.21	6.22	6.23	6.34	6.36	6.27	6.28	6.30	6.31	23
24	6.33	6.35	6.36	6.37	6.39	6.40	6.41	6.43	6.44	6.45	6.46	6.48	6.49	6.50	6.52	6.53	6.54	6.56	6.57	6.58	24
25	6.60	6.61	6.63	6.64	6.65	6.67	6.68	6.69	6.71	6.72	6.73	6.75	6.76	6.78	6.79	6.80	6.82	6.83	6.84	6.86	25
26	6.86	6.88	6.89	6.90	6.92	6.93	6.95	6.96	6.97	6.99	7.00	7.02	7.03	7.05	7.06	7.07	7.09	7.10	7.12	7.13	26

SCHMITZ' TABLE FOR THE CALCULATION OF PER CENTS SUCROSE.—*Continued.*

	10.5	11.0	11.5	12.0	12.5	13.0	13.5	14.0	14.5	15.0	15.5	16.0	16.5	17.0	17.5	18.0	18.5	19.0	19.5	20.0	
27																					27
28																					28
29																					29
30																					30
31																					31
32																					32
33																					33
34																					34
35																					35
36																					36
37																					37
38																					38
39																					39

DEGREE BRIX FROM 0.5 TO 12.0.

Tenths of the Polariscopic Reading.	Per Cent Sucrose.
0.1	0.03
0.2	0.06
0.3	0.08
0.4	0.11
0.5	0.14
0.6	0.17
0.7	0.19
0.8	0.22
0.9	0.25

DEGREE BRIX FROM 12.5 TO 20.0.

Tenths of the Polariscopic Reading.	Per Cent Sucrose.
0.1	0.03
0.2	0.05
0.3	0.08
0.4	0.11
0.5	0.13
0.6	0.16
0.7	0.19
0.8	0.21
0.9	0.24

SCHMITZ' TABLE FOR THE CALCULATION OF PER CENTS SUCROSE.—*Continued.*

Degree Brix.

Polari-scopic Reading	11.5	12.0	12.5	13.0	13.5	14.0	14.5	15.0	15.5	16.0	16.5	17.0	17.5	18.0	18.5	19.0	19.5
40	10.93	10.91	10.89	10.86	10.84	10.82	10.80	10.78	10.76	10.73	10.71	10.69	10.67	10.64	10.62	10.60	10.58
41		11.18	11.16	11.14	11.12	11.09	11.07	11.05	11.03	11.00	10.98	10.96	10.94	10.91	10.89	10.87	10.85
42		11.46	11.43	11.41	11.39	11.36	11.34	11.32	11.30	11.27	11.25	11.23	11.20	11.18	11.16	11.13	11.11
43			11.71	11.68	11.66	11.64	11.61	11.59	11.56	11.54	11.52	11.49	11.47	11.45	11.42	11.40	11.38
44			11.98	11.95	11.93	11.91	11.88	11.86	11.83	11.81	11.79	11.76	11.74	11.71	11.69	11.66	11.64
45			12.25	12.23	12.20	12.18	12.15	12.13	12.10	12.08	12.05	12.03	12.01	11.98	11.96	11.93	11.91
46				12.50	12.47	12.45	12.42	12.40	12.37	12.35	12.32	12.30	12.27	12.25	12.22	12.20	12.17
47					12.74	12.72	12.69	12.67	12.64	12.61	12.59	12.56	12.54	12.51	12.49	12.46	12.44
48					13.02	12.99	12.97	12.94	12.91	12.88	12.86	12.83	12.81	12.78	12.75	12.73	12.70
49						13.26	13.24	13.21	13.18	13.15	13.13	13.10	13.07	13.05	13.02	12.99	12.97
50							13.50	13.48	13.45	13.42	13.40	13.37	13.34	13.31	13.29	13.26	13.23
51							13.78	13.75	13.72	13.69	13.66	13.64	13.61	13.58	13.55	13.52	13.50
52								14.02	13.99	13.96	13.93	13.90	13.88	13.85	13.82	13.79	13.76
53								14.29	14.26	14.23	14.20	14.17	14.14	14.11	14.08	14.05	14.03
54									14.53	14.50	14.47	14.44	14.41	14.38	14.35	14.32	14.29
55									14.80	14.77	14.74	14.71	14.68	14.65	14.62	14.59	14.56
56										15.03	15.00	14.97	14.94	14.91	14.88	14.85	14.82
57										15.30	15.27	15.24	15.21	15.18	15.15	15.12	15.09
58										15.57	15.54	15.51	15.48	15.45	15.42	15.38	15.35
59											15.81	15.78	15.75	15.71	15.68	15.65	15.62
60												16.05	16.01	15.98	15.95	15.92	15.88
61												16.31	16.28	16.25	16.21	16.18	16.15
62													16.55	16.52	16.48	16.45	16.41
63													16.82	16.78	16.75	16.71	16.68
64														17.05	17.01	16.98	16.94
65														17.32	17.28	17.24	17.21
66															17.55	17.51	17.47
67															17.81	17.78	17.74
68																18.04	18.00
69																18.31	18.27
70																	18.53

Degree Brix from 11.5 to 22.5.

Tenths of the Polari-scopic Reading.	Per Cent Sucrose.
0.1	0.03
0.2	0.05
0.3	0.08
0.4	0.11
0.5	0.13
0.6	0.16
0.7	0.19
0.8	0.21
0.9	0.24

SCHMITZ' TABLE FOR THE CALCULATION OF PER CENTS SUCROSE.—*Continued.*

Polariscopic Reading.	Degree Brix.									Polariscopic Reading.
	20.0	20.5	21.0	21.5	22.0	22.5	23.0	23.5	24.0	
40	10.56	10.54	10.52	10.49	10.47	10.45	10.43	10.41	10.38	40
41	10.82	10.80	10.78	10.76	10.74	10.71	10.69	10.67	10.65	41
42	11.09	11.07	11.04	11.02	11.00	10.97	10.95	10.93	10.90	42
43	11.35	11.33	11.31	11.28	11.26	11.24	11.21	11.19	11.17	43
44	11.62	11.59	11.57	11.55	11.52	11.50	11.47	11.45	11.42	44
45	11.88	11.86	11.83	11.81	11.78	11.76	11.73	11.71	11.69	45
46	12.15	12.12	12.09	12.07	12.05	12.02	12.00	11.97	11.94	46
47	12.41	12.39	12.36	12.33	12.31	12.28	12.26	12.23	12.21	47
48	12.67	12.65	12.62	12.60	12.57	12.54	12.52	12.49	12.47	48
49	12.94	12.91	12.88	12.86	12.83	12.81	12.78	12.75	12.73	49
50	13.20	13.18	13.15	13.12	13.09	13.07	13.04	13.01	12.99	50
51	13.47	13.44	13.41	13.39	13.36	13.33	13.30	13.27	13.25	51
52	13.73	13.70	13.68	13.65	13.62	13.59	13.56	13.53	13.51	52
53	14.00	13.97	13.94	13.91	13.88	13.85	13.82	13.79	13.77	53
54	14.26	14.23	14.20	14.17	14.14	14.11	14.08	14.06	14.02	54
55	14.53	14.50	14.47	14.44	14.41	14.38	14.35	14.32	14.29	55
56	14.79	14.76	14.73	14.70	14.67	14.64	14.61	14.58	14.55	56
57	15.06	15.02	14.99	14.96	14.93	14.90	14.87	14.84	14.81	57
58	15.32	15.29	15.26	15.23	15.19	15.16	15.13	15.10	15.07	58
59	15.58	15.55	15.52	15.49	15.46	15.42	15.39	15.36	15.33	59
60	15.85	15.82	15.78	15.75	15.72	15.69	15.65	15.62	15.59	60
61	16.11	16.08	16.05	16.01	15.98	15.95	15.91	15.88	15.85	61
62	16.38	16.35	16.31	16.28	16.24	16.21	16.18	16.14	16.11	62
63	16.64	16.61	16.57	16.54	16.51	16.47	16.44	16.40	16.37	63
64	16.91	16.87	16.84	16.80	16.77	16.73	16.70	16.66	16.63	64
65	17.17	17.14	17.10	17.07	17.03	17.00	16.96	16.92	16.89	65
66	17.44	17.40	17.37	17.33	17.29	17.26	17.22	17.19	17.15	66
67	17.70	17.67	17.63	17.59	17.56	17.52	17.48	17.45	17.41	67
68	17.97	17.93	17.89	17.86	17.82	17.78	17.74	17.71	17.67	68
69	18.23	18.19	18.16	18.12	18.08	18.04	18.00	17.97	17.93	69
70	18.50	18.46	18.42	18.38	18.35	18.31	18.27	18.23	18.19	70
71	18.76	18.72	18.68	18.65	18.61	18.57	18.53	18.49	18.45	71
72	19.03	18.99	18.95	18.91	18.87	18.83	18.79	18.75	18.71	72
73	....	19.25	19.21	19.17	19.13	19.09	19.05	19.01	18.97	73
74	....	19.52	19.48	19.44	19.40	19.35	19.31	19.27	19.23	74
75	....	19.78	19.74	19.70	19.66	19.62	19.57	19.53	19.49	75
76	....	....	20.00	19.96	19.92	19.88	19.84	19.80	19.75	76
77	....	....	20.27	20.22	20.18	20.14	20.10	20.06	20.01	77
78	....	....	....	20.49	20.45	20.40	20.36	20.32	20.27	78
79	....	....	....	20.75	20.71	20.66	20.62	20.58	20.54	79
80	....	....	....	....	20.97	20.93	20.88	20.84	20.80	80

Degree Brix from 23 to 24.

Tenths of the Polariscopic Reading.	Per Cent Sucrose.	Tenths of the Polariscopic Reading.	Per Cent Sucrose.
0.1	0.03	0.6	0.16
0.2	0.05	0.7	0.18
0.3	0.08	0.8	0.21
0.4	0.10	0.9	0.23
0.5	0.13		

260. TABLE SHOWING THE VOLUME OF JUICE REQUIRED TO GIVE TWO OR THREE TIMES THE CORRECT POLARISCOPIC READING.

(Divide the Reading by 2 for instruments whose factor is 26.048 grams, and by 3 for those whose factor is 16.19.)

Degree. Brix.	Factor 26.048 gr. Required— cc.	Degree Brix.	Factor 26.048 gr. Required— cc.	Degree. Brix.	Factor 16.19 gr. Required— cc.	Degree Brix.	Factor 16.19 gr. Required— cc.
5	51.1	12.9	49.5	5	47.6	12.7	46.2
5.4	51	13.4	49.4	5.7	47.5	13.3	46.1
5.9	50.9	13.9	49.3	6.3	47.4	13.8	46
6.4	50.8	14.4	49.2	6.8	47.3	14.3	45.9
6.9	50.7	14.9	49.1	7.3	47.2	14.8	45.8
7.4	50.6	15.4	49	7.8	47.1	15.3	45.7
7.9	50.5	15.9	48.9	8.3	47	15.9	45.6
8.4	50.4	16.4	48.8	8.9	46.9	16.4	45.5
8.9	50.3	16.9	48.7	9.5	46.8	17	45.4
9.4	50.2	17.4	48.6	10	46.7	17.5	45.3
9.9	50.1	17.9	48.5	10.5	46.6	18	45.2
10.4	50	18.4	48.4	11	46.5	18.6	45.1
10.9	49.9	18.9	48.3	11.6	46.4	19.1	45
11.4	49.8	19.4	48.2	12.1	46.3		
11.9	49.7	19.9	48.1				
12.4	49.6						

261. TABLE FOR THE ESTIMATION OF THE APPROXIMATE PER CENT TOTAL SOLIDS IN MASSECUITE, MOLASSES, ETC. (F. E. COOMBS.)

(Dilution of sample = 100 grams to 500 cc.)

Degrees Brix of Diluted Sample. (Corrected for Temperature.)	Per Cent Solids in Original Sample.	Degrees Brix of Diluted Sample. (Corrected for Temperature.)	Per Cent Solids in Original Sample.
14.0	73.99	16.0	85.25
.1	74.55	.1	85.82
.2	75.11	.2	86.39
.3	75.65	.3	86.96
.4	76.22	.4	87.53
.5	76.79	.5	88.10
.6	77.35	.6	88.67
.7	77.91	.7	89.24
.8	78.47	.8	89.81
.9	79.04	.9	90.38
15.0	79.60	17.0	90.95
.1	80.16	.1	91.52
.2	80.72	.2	92.10
.3	81.29	.3	92.67
.4	81.86	.4	93.22
.5	82.42	.5	93.82
.6	82.99	.6	94.39
.7	83.55	.7	94.97
.8	84.12	.8	95.54
.9	84.68	.9	96.12

262. Table for the Calculation of the Per Cent Sucrose in Molasses, Massecuite, etc. (F. E. Coombs).—A portion of the solution used in estimating the approximate total solids, equivalent to 10 grams of the material (*see* **261**), is transferred to a 100 cc. sugar-flask, clarified, and polarized as usual. To calculate the sucrose, find the integral part of the polariscopic reading in the first column, follow the line to the right to the number under the tenths of the reading, and enter this number as the per cent sucrose in the material.

Integral Part of Polariscope Reading.	Fractional Part of Polariscope Reading.									
	.0	.1	.2	.3	.4	.5	.6	.7	.8	.9
8.0	20.84	21.10	21.36	21.62	21.88	22.14	22.40	22.66	22.92	23.18
9.0	23.44	23.70	23.96	24.22	24.48	24.74	25.01	25.27	25.53	25.79
10.0	26.04	26.30	26.56	26.82	27.08	27.34	27.61	27.87	28.13	28.39
11.0	28.65	28.91	29.17	29.43	29.69	29.95	30.22	30.48	30.74	31.00
12.0	31.26	31.52	31.78	32.04	32.30	32.56	32.82	33.08	33.34	33.60
13.0	33.86	34.12	34.38	34.64	34.90	35.16	35.43	35.69	35.95	36.21
14.0	36.47	36.73	36.99	37.25	37.51	37.77	38.03	38.29	38.55	38.81
15.0	39.07	39.33	39.59	39.85	40.11	40.37	40.63	40.89	41.15	41.41
16.0	41.68	41.94	42.20	42.46	42.72	42.98	43.24	43.50	43.76	44.02
17.0	44.28	44.54	44.80	45.06	45.32	45.58	45.84	46.10	46.36	46.62
18.0	46.89	47.15	47.41	47.67	47.93	48.19	48.45	48.71	48.97	49.23
19.0	49.49	49.75	50.01	50.27	50.53	50.79	51.05	51.31	51.57	51.83
20.0	52.10	52.36	52.62	52.88	53.14	53.40	53.66	53.92	54.18	54.44

263. Formulæ[1] for the Calculation of Inversion in the Diffusion-battery.—The author is indebted to Lieut. A. B. Clements, U.S.N., for the following formulæ, unless otherwise indicated :

(1) $\qquad x = b\dfrac{F_2 - F_1}{1 + \dfrac{100}{95} F_2} =$ inversion in the battery per cent

diffusion-juice;

$F_1 = \dfrac{\text{per cent sucrose in the diffusion-juice}}{\text{per cent glucose in the diffusion-juice}};$

$F_2 = \dfrac{\text{per cent sucrose in the normal juice}}{\text{per cent glucose in the normal juice}};$

$b =$ per cent glucose in the diffusion-juice;

$\dfrac{100}{95} = 1.05263.$

(2) $\qquad x = a\dfrac{r_1 - r_2}{r_2 + \dfrac{10000}{95}} =$ inversion in the battery per cent

diffusion-juice;

$a =$ per cent sucrose in the diffusion-juice;

$r_1 = \dfrac{\text{per cent glucose in the diffusion-juice}}{\text{per cent sucrose in the diffusion-juice}} \times 100;$

$r_2 = \dfrac{\text{per cent glucose in the normal juice}}{\text{per cent sucrose in the normal juice}} \times 100;$

$\dfrac{10000}{95} = 105.263.$

(3) $[p - (100 - e)P]\ .95 = x =$ inversion in the battery per cent diffusion-juice. $p =$ per cent glucose in diffusion-juice; $P =$ per cent glucose in the normal juice $\div 100$; $e =$ evaporation necessary to concentrate the diffusion-juice to the same percentage of sugars as in the normal juice. To obtain e subtract the sum of the sugars in the diffusion-juice from that in the normal juice and divide the remainder by the sum of the sugars in the normal juice. Multiply the quotient by 100.

This formula only gives approximate results. The error amounts to less than 15 lbs. sucrose per 1,000,000 lbs. of juice when the inversion does not exceed 1 per cent (G. L. Spencer).

---

[1] Based upon the formula of Dr. Stubbs of the Louisiana Experiment Station.

264. [1] RECIPROCALS OF NUMBERS FROM 11 TO 36, ADVANCING BY TENTHS.

Number.	Reciprocal.	Number.	Reciprocal.	Number.	Reciprocal.	Number.	Reciprocal.	Number.	Reciprocal.
11.0	.0909	16.0	.0625	21.0	.0476	26.0	.0385	31.0	.0322
11.1	.0900	16.1	.0621	21.1	.0474	26.1	.0383	31.1	.0321
11.2	.0893	16.2	.0617	21.2	.0472	26.2	.0381	31.2	.0320
11.3	.0885	16.3	.0613	21.3	.0469	26.3	.0380	31.3	.0319
11.4	.0877	16.4	.0610	21.4	.0467	26.4	.0379	31.4	.0318
11.5	.0869	16.5	.0606	21.5	.0465	26.5	.0377	31.5	.0317
11.6	.0862	16.6	.0602	21.6	.0463	26.6	.0376	31.6	.0316
11.7	.0855	16.7	.0599	21.7	.0461	26.7	.0374	31.7	.0315
11.8	.0847	16.8	.0595	21.8	.0459	26.8	.0373	31.8	.0314
11.9	.0840	16.9	.0592	21.9	.0457	26.9	.0372	31.9	.0313
12.0	.0833	17.0	.0588	22.0	.0454	27.0	.0370	32.0	.0312
12.1	.0826	17.1	.0585	22.1	.0452	27.1	.0369	32.1	.0311
12.2	.0820	17.2	.0581	22.2	.0450	27.2	.0368	32.2	.0310
12.3	.0813	17.3	.0578	22.3	.0448	27.3	.0366	32.3	.0309
12.4	.0806	17.4	.0575	22.4	.0446	27.4	.0365	32.4	.0308
12.5	.0800	17.5	.0571	22.5	.0444	27.5	.0364	32.5	.0308
12.6	.0794	17.6	.0568	22.6	.0442	27.6	.0362	32.6	.0307
12.7	.0787	17.7	.0565	22.7	.0440	27.7	.0361	32.7	.0305
12.8	.0781	17.8	.0562	22.8	.0438	27.8	.0360	32.8	.0305
12.9	.0775	17.9	.0559	22.9	.0437	27.9	.0358	32.9	.0304
13.0	.0769	18.0	.0555	23.0	.0435	28.0	.0357	33.0	.0303
13.1	.0763	18.1	.0552	23.1	.0432	28.1	.0356	33.1	.0302
13.2	.0757	18.2	.0549	23.2	.0431	28.2	.0355	33.2	.0301
13.3	.0752	18.3	.0546	23.3	.0429	28.3	.0353	33.3	.0300
13.4	.0746	18.4	.0543	23.4	.0427	28.4	.0352	33.4	.0299
13.5	.0741	18.5	.0540	23.5	.0425	28.5	.0351	33.5	.0298
13.6	.0735	18.6	.0538	23.6	.0424	28.6	.0350	33.6	.0297
13.7	.0730	18.7	.0535	23.7	.0422	28.7	.0348	33.7	.0296
13.8	.0725	18.8	.0532	23.8	.0420	28.8	.0347	33.8	.0295
13.9	.0719	18.9	.0529	23.9	.0418	28.9	.0346	33.9	.0295
14.0	.0714	19.0	.0526	24.0	.0417	29.0	.0345	34.0	.0294
14.1	.0709	19.1	.0523	24.1	.0415	29.1	.0344	34.1	.0293
14.2	.0704	19.2	.0521	24.2	.0413	29.2	.0342	34.2	.0292
14.3	.0699	19.3	.0518	24.3	.0411	29.3	.0341	34.3	.0291
14.4	.0694	19.4	.0515	24.4	.0409	29.4	.0340	34.4	.0290
14.5	.0690	19.5	.0513	24.5	.0408	29.5	.0339	34.5	.0289
14.6	.0685	19.6	.0510	24.6	.0406	29.6	.0338	34.6	.0289
14.7	.0680	19.7	.0508	24.7	.0405	29.7	.0337	34.7	.0288
14.8	.0676	19.8	.0505	24.8	.0403	29.8	.0335	34.8	.0287
14.9	.0671	19.9	.0502	24.9	.0402	29.9	.0334	34.9	.0286
15.0	.0667	20.0	.0500	25.0	.0400	30.0	.0333	35.0	.0285
15.1	.0662	20.1	.0497	25.1	.0398	30.1	.0332	35.1	.0284
15.2	.0658	20.2	.0495	25.2	.0397	30.2	.0331	35.2	.0284
15.3	.0654	20.3	.0493	25.3	.0395	30.3	.0330	35.3	.0283
15.4	.0649	20.4	.0490	25.4	.0394	30.4	.0329	35.4	.0282
15.5	.0645	20.5	.0488	25.5	.0392	30.5	.0328	35.5	.0282
15.6	.0641	20.6	.0485	25.6	.0391	30.6	.0327	35.6	.0281
15.7	.0637	20.7	.0483	25.7	.0389	30.7	.0326	35.7	.0280
15.8	.0633	20.8	.0481	25.8	.0388	30.8	.0325	35.8	.0279
15.9	.0629	20.9	.0478	25.9	.0386	30.9	.0324	35.9	.0278

[1] See page 87 for suggestions relative to the use of this table.

265. TABLE FOR THE DETERMINATION OF COEFFICIENTS OF PURITY.—(G. KOTTMANN.)

Per Cent Sucrose.	Per Cent of Non-Sucrose = Degree Brix minus Per Cent Sucrose.									Per Cent Sucrose.
	1.0	1.1	1.2	1.3	1.4	1.5	1.6	1.7	1.8	
8.0	88.9	87.9	87.0	86.0	85.1	84.2	83.3	82.5	81.6	8.0
8.2	89.1	88.2	87.2	86.3	85.4	84.5	83.7	82.8	82.0	8.2
8.4	89.4	88.4	87.5	86.6	85.7	84.8	84.0	83.2	82.3	8.4
8.6	89.6	88.7	87.8	86.9	86.0	85.1	84.3	83.5	82.7	8.6
8.8	89.8	88.9	88.0	87.1	86.3	85.4	84.6	83.8	83.0	8.8
9.0	90.0	89.1	88.2	87.4	86.5	85.7	84.9	84.1	83.3	9.0
9.2	90.2	89.3	88.5	87.6	86.8	86.0	85.2	84.4	83.6	9.2
9.4	90.4	89.5	88.7	87.8	87.0	86.2	85.5	84.7	83.9	9.4
9.6	90.6	89.7	88.9	88.1	87.3	86.5	85.7	85.0	84.2	9.6
9.8	90.7	89.9	89.1	88.3	87.5	86.7	86.0	85.2	84.5	9.8
10.0	90.9	90.1	89.3	88.5	87.7	87.0	86.2	85.5	84.7	10.0
10.2	91.1	90.3	89.5	88.7	87.9	87.2	86.4	85.7	85.0	10.2
10.4	91.2	90.4	89.7	88.9	88.1	87.4	86.7	86.0	85.2	10.4
10.6	91.4	90.6	89.8	89.1	88.3	87.6	86.9	86.2	85.5	10.6
10.8	91.5	90.8	90.0	89.3	88.5	87.8	87.1	86.4	85.7	10.8
11.0	91.7	90.9	90.2	89.4	88.7	88.0	87.3	86.6	85.9	11.0
11.2	91.8	91.1	90.3	89.6	88.9	88.2	87.5	86.8	86.2	11.2
11.4	91.9	91.2	90.5	89.8	89.1	88.4	87.7	87.0	86.4	11.4
11.6	92.1	91.3	90.6	89.9	89.2	88.5	87.9	87.2	86.6	11.6
11.8	92.2	91.5	90.8	90.1	89.4	88.7	88.1	87.4	86.8	11.8
12.0	92.3	91.6	90.9	90.2	89.6	88.9	88.2	87.6	87.0	12.0
12.2	92.4	91.7	91.0	90.4	89.7	89.1	88.4	87.8	87.1	12.2
12.4	92.5	91.9	91.2	90.5	89.9	89.2	88.6	87.9	87.3	12.4
12.6	92.6	92.0	91.3	90.6	90.0	89.4	88.7	88.1	87.5	12.6
12.8	92.8	92.1	91.4	90.8	90.1	89.5	88.9	88.3	87.7	12.8
13.0	92.9	92.2	91.5	90.9	90.3	89.7	89.0	88.4	87.8	13.0
13.2	93.0	92.3	91.7	91.0	90.4	89.8	89.2	88.6	88.0	13.2
13.4	93.1	92.4	91.8	91.2	90.5	89.9	89.3	88.7	88.2	13.4
13.6	93.2	92.5	91.9	91.3	90.7	90.1	89.5	88.9	88.3	13.6
13.8	93.2	92.6	92.0	91.4	90.8	90.2	89.6	89.0	88.5	13.8
14.0	93.3	92.7	92.1	91.5	90.9	90.3	89.7	89.2	88.6	14.0
14.2	93.4	92.8	92.2	91.6	91.0	90.4	89.9	89.3	88.8	14.2
14.4	93.5	92.9	92.3	91.7	91.1	90.6	90.0	89.4	88.9	14.4
14.6	93.6	93.0	92.4	91.8	91.3	90.7	90.1	89.6	89.0	14.6
14.8	93.7	93.1	92.5	91.9	91.4	90.8	90.2	89.7	89.2	14.8
15.0	93.7	93.2	92.6	92.0	91.5	90.9	90.4	89.8	89.3	15.0
15.2	93.8	93.3	92.7	92.1	91.6	91.0	90.5	89.9	89.4	15.2
15.4	93.9	93.3	92.8	92.2	91.7	91.1	90.6	90.1	89.5	15.4
15.6	94.0	93.4	92.8	92.3	91.8	91.2	90.7	90.2	89.7	15.6
15.8	94.1	93.5	92.9	92.4	91.9	91.3	90.8	90.3	89.8	15.8
16.0	94.1	93.6	93.0	92.5	92.0	91.4	90.9	90.4	89.9	16.0
16.2	94.2	93.7	93.1	92.6	92.0	91.5	91.0	90.5	90.0	16.2
16.4	94.3	93.7	93.2	92.6	92.1	91.6	91.1	90.6	90.1	16.4
16.6	94.3	93.8	93.3	92.7	92.2	91.7	91.2	90.7	90.2	16.6
16.8	94.4	93.9	93.3	92.8	92.3	91.8	91.3	90.8	90.3	16.8
17.0	94.4	93.9	93.4	92.9	92.4	91.9	91.4	90.9	90.4	17.0

TABLE FOR THE DETERMINATION OF COEFFICIENTS OF
PURITY.—Continued.

Per Cent Sucrose	Per Cent of Non-Sucrose = Degree Brix minus Per Cent Sucrose.									Per Cent Sucrose
	1.9	2.0	2.1	2.2	2.3	2.4	2.5	2.6	2.7	
8.0	80.8	80.0	79.2	78.4	77.7	76.9	76.2	75.5	74.8	8.0
8.2	81.2	80.4	79.6	78.8	78.1	77.4	76.6	75.9	75.2	8.2
8.4	81.5	80.8	80.0	79.2	78.5	77.8	77.1	76.4	75.7	8.4
8.6	81.9	81.1	80.4	79.6	78.9	78.2	77.5	76.8	76.1	8.6
8.8	82.2	81.5	80.7	80.0	79.3	78.6	77.9	77.2	76.5	8.8
9.0	82.6	81.8	81.1	80.4	79.6	78.9	78.3	77.6	76.9	9.0
9.2	82.9	82.1	81.4	80.7	80.0	79.3	78.6	77.9	77.3	9.2
9.4	83.2	82.5	81.7	81.0	80.3	79.7	79.0	78.3	77.7	9.4
9.6	83.5	82.8	82.1	81.4	80.7	80.0	79.3	78.7	78.0	9.6
9.8	83.8	83.1	82.4	81.7	81.0	80.3	79.7	79.0	78.4	9.8
10.0	84.0	83.3	82.6	82.0	81.3	80.6	80.0	79.4	78.7	10.0
10.2	84.3	83.6	82.9	82.3	81.6	81.0	80.3	79.7	79.1	10.2
10.4	84.6	83.9	83.2	82.5	81.9	81.2	80.6	80.0	79.4	10.4
10.6	84.8	84.1	83.5	82.8	82.2	81.5	80.9	80.3	79.7	10.6
10.8	85.0	84.4	83.7	83.1	82.4	81.8	81.2	80.6	80.0	10.8
11.0	85.3	84.6	84.0	83.3	82.7	82.1	81.5	80.9	80.3	11.0
11.2	85.5	84.8	84.2	83.6	83.0	82.4	81.8	81.2	80.6	11.2
11.4	85.7	85.1	84.4	83.8	83.2	82.6	82.0	81.4	80.9	11.4
11.6	85.9	85.3	84.7	84.1	83.5	82.9	82.3	81.7	81.1	11.6
11.8	86.1	85.5	84.9	84.3	83.7	83.1	82.5	81.9	81.4	11.8
12.0	86.3	85.7	85.1	84.5	83.9	83.3	82.8	82.2	81.6	12.0
12.2	86.5	85.9	85.3	84.7	84.1	83.6	83.0	82.4	81.9	12.2
12.4	86.7	86.1	85.5	84.9	84.4	83.8	83.2	82.7	82.1	12.4
12.6	86.9	86.3	85.7	85.1	84.6	84.0	83.4	82.9	82.4	12.6
12.8	87.1	86.5	85.9	85.3	84.8	84.2	83.7	83.1	82.6	12.8
13.0	87.2	86.7	86.1	85.5	85.0	84.4	83.9	83.3	82.8	13.0
13.2	87.4	86.8	86.3	85.7	85.2	84.6	84.1	83.5	83.0	13.2
13.4	87.6	87.0	86.5	85.9	85.4	84.8	84.3	83.7	83.2	13.4
13.6	87.7	87.2	86.6	86.1	85.5	85.0	84.5	83.9	83.4	13.6
13.8	87.9	87.3	86.8	86.3	85.7	85.2	84.7	84.1	83.6	13.8
14.0	88.1	87.5	87.0	86.4	85.9	85.4	84.8	84.3	83.8	14.0
14.2	88.2	87.7	87.1	86.6	86.1	85.5	85.0	84.5	84.0	14.2
14.4	88.3	87.8	87.3	86.7	86.2	85.7	85.2	84.7	84.2	14.4
14.6	88.5	88.0	87.4	86.9	86.4	85.9	85.4	84.9	84.4	14.6
14.8	88.6	88.1	87.6	87.1	86.5	86.0	85.5	85.1	84.6	14.8
15.0	88.8	88.2	87.7	87.2	86.7	86.2	85.7	85.2	84.7	15.0
15.2	88.9	88.4	87.9	87.4	86.9	86.4	85.9	85.4	84.9	15.2
15.4	89.0	88.5	88.0	87.5	87.0	86.5	86.0	85.6	85.1	15.4
15.6	89.1	88.6	88.1	87.6	87.2	86.7	86.2	85.7	85.2	15.6
15.8	89.3	88.8	88.3	87.8	87.3	86.8	86.3	85.9	85.4	15.8
16.0	89.4	88.9	88.4	87.9	87.4	87.0	86.5	86.0	85.6	16.0
16.2	89.5	89.0	88.5	88.0	87.6	87.1	86.6	86.2	85.7	16.2
16.4	89.6	89.1	88.6	88.2	87.7	87.2	86.8	86.3	85.9	16.4
16.6	89.7	89.2	88.8	88.3	87.8	87.4	86.9	86.5	86.0	16.6
16.8	89.8	89.4	88.9	88.4	88.0	87.5	87.0	86.6	86.2	16.8
17.0	89.9	89.5	89.0	88.5	88.1	87.6	87.2	86.7	86.3	17.0

TABLE FOR THE DETERMINATION OF COEFFICIENTS OF PURITY.—*Continued.*

Per Cent Sucrose.	Per Cent of Non-Sucrose = Degree Brix minus Per Cent Sucrose.									Per Cent Sucrose.
	2.8	2.9	3.0	3.1	3.2	3.3	3.4	3.5	3.6	
8.0	74.1	73.4	72.7	72.1	71.4	70.8	70.2	69.6	69.0	8.0
8.2	74.5	73.9	73.2	72.6	71.9	71.3	70.7	70.1	69.5	8.2
8.4	75.0	74.3	73.7	73.0	72.4	71.8	71.2	70.6	70.0	8.4
8.6	75.4	74.8	74.1	73.5	72.9	72.3	71.7	71.1	70.5	8.6
8.8	75.9	75.2	74.6	73.9	73.3	72.7	72.1	71.5	71.0	8.8
9.0	76.3	75.6	75.0	74.4	73.8	73.2	72.6	72.0	71.4	9.0
9.2	76.7	76.0	75.4	74.8	74.2	73.6	73.0	72.4	71.9	9.2
9.4	77.0	76.4	75.8	75.2	74.6	74.0	73.4	72.9	72.3	9.4
9.6	77.4	76.8	76.2	75.6	75.0	74.4	73.8	73.3	72.7	9.6
9.8	77.8	77.2	76.6	76.0	75.4	74.8	74.2	73.7	73.1	9.8
10.0	78.1	77.5	76.9	76.3	75.8	75.2	74.6	74.1	73.5	10.0
10.2	78.5	77.9	77.3	76.7	76.1	75.6	75.0	74.5	73.9	10.2
10.4	78.8	78.2	77.6	77.0	76.5	75.9	75.4	74.8	74.3	10.4
10.6	79.1	78.5	77.9	77.4	76.8	76.3	75.7	75.2	74.6	10.6
10.8	79.4	78.8	78.3	77.7	77.1	76.6	76.1	75.5	75.0	10.8
11.0	79.7	79.1	78.6	78.0	77.5	76.9	76.4	75.9	75.3	11.0
11.2	80.0	79.4	78.9	78.3	77.8	77.2	76.7	76.2	75.7	11.2
11.4	80.3	79.7	79.2	78.6	78.1	77.6	77.0	76.5	76.0	11.4
11.6	80.6	80.0	79.4	78.9	78.4	77.9	77.3	76.8	76.3	11.6
11.8	80.8	80.3	79.7	79.2	78.7	78.1	77.6	77.1	76.6	11.8
12.0	81.1	80.5	80.0	79.5	78.9	78.4	77.9	77.4	76.9	12.0
12.2	81.3	80.8	80.3	79.7	79.2	78.7	78.2	77.7	77.2	12.2
12.4	81.6	81.0	80.5	80.0	79.5	79.0	78.5	78.0	77.5	12.4
12.6	81.8	81.3	80.8	80.3	79.7	79.2	78.8	78.3	77.8	12.6
12.8	82.1	81.5	81.0	80.5	80.0	79.5	79.0	78.5	78.0	12.8
13.0	82.3	81.8	81.2	80.7	80.2	79.8	79.3	78.8	78.3	13.0
13.2	82.5	82.0	81.5	81.0	80.5	80.0	79.5	79.0	78.6	13.2
13.4	82.7	82.2	81.7	81.2	80.7	80.2	79.8	79.3	78.8	13.4
13.6	82.9	82.4	81.9	81.4	81.0	80.5	80.0	79.5	79.1	13.6
13.8	83.1	82.6	82.1	81.7	81.2	80.7	80.2	79.8	79.3	13.8
14.0	83.3	82.8	82.3	81.9	81.4	80.9	80.5	80.0	79.5	14.0
14.2	83.5	83.0	82.5	82.1	81.6	81.1	80.7	80.2	79.8	14.2
14.4	83.7	83.2	82.7	82.3	81.8	81.4	80.9	80.4	80.0	14.4
14.6	83.9	83.4	82.9	82.5	82.0	81.6	81.1	80.7	80.2	14.6
14.8	84.1	83.6	83.1	82.7	82.2	81.8	81.3	80.9	80.4	14.8
15.0	84.3	83.8	83.3	82.9	82.4	82.0	81.5	81.1	80.6	15.0
15.2	84.4	84.0	83.5	83.1	82.6	82.2	81.7	81.3	80.8	15.2
15.4	84.6	84.2	83.7	83.2	82.8	82.4	81.9	81.5	81.0	15.4
15.6	84.8	84.3	83.9	83.4	83.0	82.5	82.1	81.7	81.2	15.6
15.8	84.9	84.5	84.0	83.6	83.2	82.7	82.3	81.9	81.4	15.8
16.0	85.1	84.7	84.2	83.8	83.3	82.9	82.5	82.0	81.6	16.0
16.2	85.3	84.8	84.4	83.9	83.5	83.1	82.7	82.2	81.8	16.2
16.4	85.4	84.9	84.5	84.1	83.7	83.2	82.8	82.4	82.0	16.4
16.6	85.6	85.1	84.7	84.3	83.8	83.4	83.0	82.6	82.2	16.6
16.8	85.7	85.3	84.8	84.4	84.0	83.6	83.2	82.8	82.4	16.8
17.0	85.9	85.4	85.0	84.6	84.2	83.7	83.3	82.9	82.5	17.0

TABLE FOR THE DETERMINATION OF COEFFICIENTS OF PURITY.—*Continued.*

Per Cent Sucrose.	Per Cent of Non-Sucrose = Degree Brix minus Per Cent Sucrose.									Per Cent Sucrose.
	3.7	3.8	3.9	4.0	4.1	4.2	4.3	4.4	4.5	
8.0	68.4	67.8	67.2	66.7	66.1	65.6	65.0	64.5	64.0	8.0
8.2	68.9	68.3	67.8	67.2	66.7	66.1	65.6	65.1	64.6	8.2
8.4	69.4	68.8	68.3	67.7	67.2	66.7	66.1	65.6	65.1	8.4
8.6	69.9	69.3	68.8	68.3	67.7	67.2	66.7	66.2	65.6	8.6
8.8	70.4	69.8	69.3	68.8	68.2	67.7	67.2	66.7	66.2	8.8
9.0	70.9	70.3	69.8	69.2	68.7	68.2	67.7	67.2	66.7	9.0
9.2	71.3	70.8	70.2	69.7	69.2	68.7	68.1	67.6	67.2	9.2
9.4	71.8	71.2	70.7	70.1	69.6	69.1	68.6	68.1	67.6	9.4
9.6	72.2	71.6	71.1	70.6	70.1	69.6	69.1	68.6	68.1	9.6
9.8	72.6	72.1	71.5	71.0	70.5	70.0	69.5	69.0	68.5	9.8
10.0	73.0	72.5	71.9	71.4	70.9	70.4	69.9	69.4	69.0	10.0
10.2	73.4	72.9	72.3	71.8	71.3	70.8	70.3	69.9	69.4	10.2
10.4	73.8	73.2	72.7	72.2	71.7	71.2	70.7	70.3	69.8	10.4
10.6	74.1	73.6	73.1	72.6	72.1	71.6	71.1	70.7	70.2	10.6
10.8	74.5	74.0	73.5	73.0	72.5	72.0	71.5	71.1	70.6	10.8
11.0	74.8	74.3	73.8	73.3	72.8	72.4	71.9	71.4	71.0	11.0
11.2	75.2	74.7	74.2	73.7	73.2	72.7	72.3	71.8	71.3	11.2
11.4	75.5	75.0	74.5	74.0	73.5	73.1	72.6	72.2	71.7	11.4
11.6	75.8	75.3	74.8	74.4	73.9	73.4	73.0	72.5	72.0	11.6
11.8	76.1	75.6	75.2	74.7	74.2	73.8	73.3	72.8	72.4	11.8
12.0	76.4	75.9	75.5	75.0	74.5	74.1	73.6	73.2	72.7	12.0
12.2	76.7	76.2	75.8	75.3	74.8	74.4	73.9	73.5	73.1	12.2
12.4	77.0	76.5	76.1	75.6	75.2	74.7	74.3	73.8	73.4	12.4
12.6	77.3	76.8	76.4	75.9	75.4	75.0	74.6	74.1	73.7	12.6
12.8	77.6	77.1	76.6	76.2	75.7	75.3	74.9	74.4	74.0	12.8
13.0	77.8	77.4	76.9	76.5	76.0	75.6	75.1	74.7	74.3	13.0
13.2	78.1	77.6	77.2	76.7	76.3	75.9	75.4	75.0	74.6	13.2
13.4	78.4	77.9	77.5	77.0	76.6	76.1	75.7	75.3	74.9	13.4
13.6	78.6	78.2	77.7	77.3	76.8	76.4	76.0	75.6	75.1	13.6
13.8	78.9	78.4	78.0	77.5	77.1	76.7	76.2	75.8	75.4	13.8
14.0	79.1	78.7	78.2	77.8	77.3	76.9	76.5	76.1	75.7	14.0
14.2	79.3	78.9	78.5	78.0	77.6	77.2	76.8	76.3	75.9	14.2
14.4	79.6	79.1	78.7	78.3	77.8	77.4	77.0	76.6	76.2	14.4
14.6	79.8	79.3	78.9	78.5	78.1	77.6	77.2	76.8	76.4	14.6
14.8	80.0	79.6	79.1	78.7	78.3	77.9	77.5	77.1	76.7	14.8
15.0	80.2	79.8	79.4	78.9	78.5	78.1	77.7	77.3	76.9	15.0
15.2	80.4	80.0	79.6	79.2	78.8	78.4	77.9	77.6	77.2	15.2
15.4	80.6	80.2	79.8	79.4	79.0	78.6	78.2	77.8	77.4	15.4
15.6	80.8	80.4	80.0	79.6	79.2	78.8	78.4	78.0	77.6	15.6
15.8	81.0	80.6	80.2	79.8	79.4	79.0	78.6	78.2	77.8	15.8
16.0	81.2	80.8	80.4	80.0	79.6	79.2	78.8	78.4	78.0	16.0
16.2	81.4	81.0	80.6	80.2	79.8	79.4	79.0	78.6	78.3	16.2
16.4	81.6	81.2	80.8	80.4	80.0	79.6	79.2	78.8	78.5	16.4
16.6	81.8	81.4	81.0	80.6	80.2	79.8	79.4	79.0	78.7	16.6
16.8	82.0	81.6	81.2	80.8	80.4	80.0	79.6	79.2	78.9	16.8
17.0	82.1	81.7	81.3	81.0	80.6	80.2	79.8	79.4	79.1	17.0

266. Value of the Degrees of Polariscopic Scales.

Grams sugar
in 100 cc.

1° scale of Mitscherlich.................. = .750
1° "   " Soleil-Dubosq... .... = .1619
1° "   " Ventzke-Soleil .... = .26048
1° "   " Wild (sugar scale)........... = .10
1° "   " Laurent and Dubosq (Shadow) = .1619

1° scale of Mitscherlich = 4°.635 Soleil-Dubosq = 2°.879 Soleil-Ventzke.

1° scale of Soleil-Dubosq = .215° Mitscherlich = .620° Ventzke-Soleil = 1°.619 Wild.

1° scale of Ventzke = .346° Mitscherlich = 1°.608 Soleil-Dubosq = 2°.648 Wild.

1° scale of Wild (sugar-scale) = .618° Soleil-Dubosq = .384° Soleil-Ventzke = .133° Mitscherlich.

Circular Degrees.

1° Wild (sugar-scale) = .1328 Circular degree D.
J1° Soleil-Dubosq.... = .2167   "   " D.
J1°   "   "   .... = .2450   "   " j.
J1° Soleil-Ventzke.... = .3455   "   " D.
J1°   "   "   .... = .3906   "   " j.

267. Clerget's Constant. Results of Redeterminations. (A. Wohl, Zeit. für Zucker, Aug. 1888.)

Weight of Sucrose.	Concentration of Invert Solution.	Invert Reading.	Constant.
13.024	13.700	− 16.34	142.7
6.512	6.855	− 7.92	142.3
3.256	3.427	− 3.80	140.4

These numbers correspond very nearly with the mean of Landolt's determinations.

BLANK FORMS

FOR PRACTICAL USE

IN

SUGAR-HOUSE WORK.

Dates.	Beets Worked. Tons.	Number of Diffusers.	Beets per Diffuser.	Juice % Beets. Gals. or Litres.

Max. Temp.	Specific Gravity of Juice.	Volume of the Juice. Gals. or Litres.	Weight of the Juice. Pounds.

Dates.	Beets Worked. Tons.	Number of Diffusers.	Beets per Diffuser.	Juice % Beets. Gals. or Litres.

Max. Temp.	Specific Gravity of Juice.	Volume of the Juice. Gals. or Litres.	Weight of the Juice. Pounds.

Dates.	Beets Worked. Tons.	Number of Diffusers.	Beets per Diffuser.	Juice % Beets. Gals. or Litres.

Max. Temp.	Specific Gravity of Juice.	Volume of the Juice. Gals. or Litres.	Weight of the Juice. Pounds.

Dates.	Beets Worked. Tons.	Number of Diffusers.	Beets per Diffuser.	Juice % Beets. Gals. or Litres.

DIFFUSION-JUICE.

Max. Temp.	Specific Gravity of Juice.	Volume of the Juice. Gals. or Litres.	Weight of the Juice. Pounds.

Dates.	Beets Worked. Tons.	Number of Diffusers.	Beets per Diffuser.	Juice % Beets. Gals. or Litres.

Max. Temp.	Specific Gravity of Juice.	Volume of the Juice. Gals. or Litres.	Weight of the Juice. Pounds.

| Dates. | Sucrose % Beets | | |
	Fresh Cossettes.	Diffusion juice.	Diffusion Losses, by Difference.

IN THE COSSETTES, LOSSES, ETC.

Dates	Exhausted Cossettes.	Waste Water.	Total Losses.	Not Determined.

Dates.	SUCROSE % BEETS		
	Fresh Cossettes.	Diffusion juice.	Diffusion Losses, by Difference.

IN THE COSSETTES, LOSSES, ETC.

Dates.	Exhausted Cossettes.	Waste Water.	Total Losses.	Not Determined.

| Dates. | Sucrose % Beets | | |
	Fresh Cossettes.	Diffusion juice.	Diffusion Losses, by Difference.

IN THE COSSETTES, LOSSES, ETC.

Dates.	Exhausted Cossettes.	Waste Water.	Total Losses.	Not Determined.

Dates.	SUCROSE % BEETS		
	Fresh Cossettes.	Diffusion juice.	Diffusion Losses, by Difference.

DIFFUSION.

IN THE COSSETTES, LOSSES, ETC.

Dates.	Exhausted Cossettes.	Waste Water.	Total Losses.	Not Determined.

Dates.	Sucrose % Beets		
	Fresh Cossettes.	Diffusion juice.	Diffusion Losses, by Difference.

IN THE COSSETTES, LOSSES, ETC.

Dates.	Exhausted Cossettes.	Waste Water.	Total Losses.	Not Determined.

Dates.	Sucrose % in the Beets.	DIFFUSION-JUICE.			
		Brix or Baumé.	% Sucrose.	Sucrose % Beets.	% Reducing Sugar.

Dates.	DIFFUSION-JUICE.				
	% Ash.	% Organic Matter not Sugar.	Coefficient of Purity.	Saline Coefficient.	

Dates.	Sucrose % in the Beets.	DIFFUSION-JUICE.			
		Brix or Baumé.	% Sucrose.	Sucrose % Beets.	% Reducing Sugar.

Dates.	DIFFUSION-JUICE.			
	% Ash.	% Organic Matter not Sugar.	Coefficient of Purity.	Saline Coefficient

Dates.	Sucrose % in the Beets.	DIFFUSION-JUICE.			
		Brix or Baumé.	% Sucrose.	Sucrose % Beets.	% Reducing Sugar.

Dates.	DIFFUSION-JUICE.				
	% Ash.	% Organic Matter not Sugar.	Coefficient of Purity.	Saline Coefficient.	

Dates.	Sucrose % in the Beets.	DIFFUSION-JUICE.			
		Brix or Baumé.	% Sucrose.	Sucrose % Beets.	% Reducing Sugar.

Dates.	DIFFUSION-JUICE.				
	% Ash.	% Organic Matter not Sugar.	Coefficient of Purity.	Saline Coefficient.	

Dates.	Sucrose % in the Beets.	DIFFUSION-JUICE.			
		Brix or Baumé.	% Sucrose.	Sucrose % Beets.	% Reducing Sugar.

Dates.	DIFFUSION-JUICE.				
	% Ash.	% Organic Matter not Sugar.	Coefficient of Purity.	Saline Coefficient.	

Dates.	1st Carbonatation.		2d Carbonatation.	
	Lime used, % Beets.	Alkalinity. Grams Lime per Litre.	Lime used, % Beets.	Alkalinity. Grams Lime per Litre.

Dates.	Brix or Baumé.	% Sucrose.	Alkalinity. Grams Lime per Litre.	Coefficient of Purity.	

Dates.	1ST CARBONATATION.		2D CARBONATATION.		Alkalinity after Sulphuring.
	Lime used, % Beets.	Alkalinity. Grams Lime per Litre.	Lime used, % Beets.	Alkalinity. Grams Lime per Litre.	

Dates.	Brix or Baumé.	% Sucrose.	Alkalinity. Grams Lime per Litre.	Coefficient of Purity.	

Dates.	1ST CARBONATATION.		2D CARBONATATION.		Alkalinity after Sulphuring.
	Lime used, % Beets.	Alkalinity. Grams Lime per Litre.	Lime used, % Beets.	Alkalinity. Grams Lime per Litre.	

SIRUPS.

Dates.	Brix or Baumé.	% Sucrose.	Alkalinity. Grams Lime per Litre.	Coefficient of Purity.	

Dates.	1st Carbonatation.		2d Carbonatation.		Alkalinity after Sulphuring.
	Lime used, % Beets.	Alkalinity. Grams Lime per Litre.	Lime used, % Beets.	Alkalinity. Grams Lime per Litre.	

Dates.	Brix or Baumé.	% Sucrose.	Alkalinity. Grams Lime per Litre.	Coefficient of Purity.	

Dates.	1st Carbonatation.		2d Carbonatation.		Alkalinity after Sulphuring.
	Lime used, % Beets.	Alkalinity, Grams Lime per Litre.	Lime used, % Beets.	Alkalinity, Grams Lime per Litre.	

Dates.	Brix or Baumé.	% Sucrose.	Alkalinity. Grams Lime per Litre.	Coefficient of Purity.	

Dates.	Apparent Brix or Baumé.	% Total Solids by Drying.	% Sucrose (Direct).	% Sucrose (Clerget).	% Raffinose.

% Organic Matter not Sucrose.	Apparent Coefficient of Purity.	True Coefficient of Purity.	Saline Coefficient.

Dates.	Apparent Brix or Baumé.	% Total Solids by Drying.	% Sucrose (Direct).	% Sucrose (Clerget).	% Raffinose.

% Ash.	% Reducing Sugars.	% Organic Matter not Sucrose.	Apparent Coefficient of Purity.	True Coefficient of Purity.	Saline Coefficient.

Dates.	Apparent Brix or Baumé.	% Total Solids by Drying.	% Sucrose (Direct).	% Sucrose (Clerget).	% Raffinose.

MASSECUITES.

% Ash.	% Reducing Sugars.	% Organic Matter not Sucrose.	Apparent Coefficient of Purity.	True Coefficient of Purity.	Saline Coefficient.

Dates.	Apparent Brix or Baumé.	% Total Solids by Drying.	% Sucrose (Direct).	% Sucrose (Clerget).	% Raffinose.

MASSECUITES.

% Ash.	% Reducing Sugars.	% Organic Matter not Sucrose.	Apparent Coefficient of Purity.	True Coefficient of Purity.	Saline Coefficient.

Dates.	Apparent Brix or Baumé.	% Total Solids by Drying.	% Sucrose (Direct).	% Sucrose (Clerget).	% Raffinose.

% Ash.	% Reducing Sugars.	% Organic Matter not Sucrose.	Apparent Coefficient of Purity.	True Coefficient of Purity.	Saline Coefficient.

Dates.	Apparent Brix or Baumé.	% Total Solids by Drying.	% Sucrose (Direct).	% Sucrose (Clerget).	% Raffinose.

% Ash.	% Reducing Sugars.	% Organic Matter not Sucrose.	Apparent Coefficient of Purity.	True Coefficient of Purity.	Saline Coefficient.

Dates.	Apparent Brix or Baumé.	% Total Solids by Drying.	% Sucrose (Direct).	% Sucrose (Clerget).	% Raffinose.

MASSECUITES.

% Ash.	% Reducing Sugars.	% Organic Matter not Sucrose.	Apparent Coefficient of Purity.	True Coefficient of Purity.	Saline Coefficient.

Dates.	Apparent Brix or Baumé.	% Total Solids by Drying.	% Sucrose (Direct).	% Sucrose (Clerget).	% Raffinose.

% Ash.	% Reducing Sugars.	% Organic Matter not Sucrose.	Apparent Coefficient of Purity.	True Coefficient of Purity.	Saline Coefficient.

Dates.	Apparent Brix or Baumé.	% Total Solids by Drying.	% Sucrose (Direct).	% Sucrose (Clerget).	% Raffinose.

Ash.	% Reducing Sugars.	% Organic Matter not Sucrose.	Apparent Coefficient of Purity.	True Coefficient of Purity.	Saline Coefficient.

Dates.	Apparent Brix or Baumé.	% Total Solids by Drying.	% Sucrose (Direct).	% Sucrose (Clerget).

% Ash.	% Reducing Sugars.	% Organic Matter not Sucrose.	Apparent Coefficient of Purity.	True Coefficient of Purity.	Saline Coefficient.

SEASON OF.... ....

Dates.	Apparent Brix or Baumé.	% Total Solids by Drying.	% Sucrose (Direct).	% Sucr (Clerg

% Organic Matter not Sucrose.	Apparent Coefficient of Purity.	True Coefficient of Purity.	Saline Coefficient.

Apparent Brix or Baumé.	% Total Solids by Drying.	% Sucrose (Direct).	% Sucrose (Clerget).	% Raffinose.

% Ash.	% Reducing Sugars.	% Organic Matter not Sucrose.	Apparent Coefficient of Purity.	True Coefficient of Purity.	Saline Coefficient.

Dates.	Apparent Brix or Baumé.	% Total Solids by Drying.	% Sucrose (Direct).	% Sucrose (Clerget).	% Raffinose.

% Ash.	% Reducing Sugars.	% Organic Matter not Sucrose.	Apparent Coefficient of Purity.	True Coefficient of Purity.	Saline Coefficient.

Dates.	Apparent Brix or Baumé.	% Total Solids by Drying.	% Sucrose (Direct).	% Sucrose (Clerget).	% Raffinose.

% Ash.	% Reducing Sugars.	% Organic Matter not Sucrose.	Apparent Coefficient of Purity.	True Coefficient of Purity.	Saline Coefficient.

Dates.	Apparent Brix or Baumé.	% Total Solids by Drying.	% Sucrose (Direct).	% Sucrose (Clerget).	% Raffinose.

% Ash.	% Reducing Sugars.	% Organic Matter not Sucrose.	Apparent Coefficient of Purity.	True Coefficient of Purity.	Saline Coefficient.

Dates.	Apparent Brix or Baumé.	% Total Solids by Drying.	% Sucrose (Direct).	% Sucrose (Clerget).	% Raffinose.

Reducing Sugars.	% Organic Matter not Sucrose.	Apparent Coefficient of Purity.	True Coefficient of Purity.	Saline Coefficient.

Dates.	Apparent Brix or Baumé.	% Total Solids by Drying.	% Sucrose (Direct).	% Sucrose (Clerget).	% Raffinose.

Ash.	% Reducing Sugars.	% Organic Matter not Sucrose.	Apparent Coefficient of Purity.	True Coefficient of Purity.	Saline Coefficient.

Dates.	Apparent Brix or Baumé.	% Total Solids by Drying.	% Sucrose (Direct).	% Sucrose (Clerget).	% Raffinose.

MOLASSES.

% Ash.	% Reducing Sugars.	% Organic Matter not Sucrose.	Apparent Coefficient of Purity.	True Coefficient of Purity.	Saline Coefficient.

Dates.	Apparent Brix or Baumé.	% Total Solids by Drying.	% Sucrose (Direct).	% Sucrose (Clerget).	% Raffinose.

% Ash.	% Reducing Sugars.	% Organic Matter not Sucrose.	Apparent Coefficient of Purity.	True Coefficient of Purity.	Saline Coefficient.

Dates.	Apparent Brix or Baumé.	% Total Solids by Drying.	% Sucrose (Direct).	% Sucrose (Clerget).

% Ash.	% Reducing Sugars.	% Organic Matter not Sucrose.	Apparent Coefficient of Purity.	True Coefficient of Purity.	Saline Coefficient.

Lot Nos.	Polarizations.	Pounds of Sugar.	Lot Nos.	Polarizations.	Pounds of Sugar.

Lot Nos.	Polarizations.	Pounds of Sugar.	Lot Nos.	Polarizations.	Pounds of Sugar.

Lot Nos.	Polar- izations.	Pounds of Sugar.	Lot Nos.	Polar- izations.	Pounds of Sugar.

Lot Nos.	Polar-izations	Pounds of Sugar.	Lot Nos.	Polar-izations.	Pounds of Sugar.

Lot Nos.	Polar-izations.	Pounds of Sugar.	Lot Nos.	Polar-izations.	Pounds of Sugar.

Lot Nos.	Polar- izations.	Pounds of Sugar.	Lot Nos.	Polar- izations.	Pounds of Sugar.

Lot Nos.	Polarizations.	Pounds of Sugar.	Lot Nos.	Polarizations.	Pounds of Sugar.

Lot Nos.	Polar- izations.	Pounds of Sugar.	Lot Nos.	Polar- izations.	Pounds of Sugar.

Lot Nos.	Polar- izations	Pounds of Sugar.	Lot Nos.	Polar- izations.	Pounds of Sugar.

Pounds of Sugar.	Lot Nos.	Polar- izations.	Pounds of Sugar.

Lot Nos.	Polar-izations.	Pounds of Sugar.	Lot Nos.	Polar-izations.

Lot Nos.	Polar-izations.	Pounds of Sugar.	Lot Nos.	Polar-izations.	Pounds of Sugar.

Lot Nos.	Polar-izations.	Pounds of Sugar.	Lot Nos.	Polar-izations.	Pounds of Sugar.

Lot Nos.	Polarizations.	Pounds of Sugar.	Lot Nos.	Polarizations.	Pounds of Sugar.

Lot Nos.	Polar- izations.	Pounds of Sugar.	Lot Nos.	Polar- izations.	Pounds of Sugar.

Lot Nos.	Polar- izations.	Pounds of Sugar.	Lot Nos.	Polar- izations.	Pounds of Sugar.

Lot Nos.	Polar-izations.	Pounds of Sugar.	Lot Nos.	Polar-izations.	Pounds of Sugar.

Lot Nos.	Polarizations	Pounds of Sugar.	Lot Nos.	Polarizations	Pounds of Sugar.

Lot Nos.	Polar- izations.	Pounds of Sugar.	Lot Nos.	Polar- izations.	Pounds of Sugar.

Lot Nos.	Polar- izations.	Pounds of Sugar.	Lot Nos.	Polar- izations.	Pounds of Sugar.

Lot Nos.	Polar-izations.	Pounds of Sugar.	Lot Nos.	Polar-izations.	Pounds of Sugar.

Lot Nos.	Polarizations.	Pounds of Sugar.	Lot Nos.	Polarizations.	Pounds of Sugar.

Lot Nos.	Polar-izations.	Pounds of Sugar.	Lot Nos.	Polar-izations.	Pounds of Sugar.

Lot Nos.	Polarizations.	Pounds of Sugar.	Lot Nos.	Polarizations.	Pounds of Sugar.

Lot Nos.	Polar-izations.	Pounds of Sugar.	Lot Nos.	Polar-izations.	Pounds of Sugar.

Lot Nos.	Polar- izations.	Pounds of Sugar.	Lot Nos.	Polar- izations.	Pounds of Sugar.

Lot Nos.	Polar-izations.	Pounds of Sugar.	Dates.	Total Sugar to Date. Pounds.

Lot Nos.	Polar- izations.	Pounds of Sugar.	Dates.	Total Sugar to Date. Pounds.

Lot Nos.	Polar- izations.	Pounds of Sugar.	Dates.	Total Sugar to Date. Pounds.

Lot Nos.	Polar-izations.	Pounds of Sugar.	Dates.	Total Sugar to Date. Pounds

Lot Nos.	Polar-izations.	Pounds of Sugar.	Dates.	Total Sugar to Date. Pounds.

Lot Nos.	Polar- izations.	Pounds of Sugar.	Dates.	Total Sugar to Date. Pounds.

Lot Nos.	Polar-izations.	Pounds of Sugar.	Dates.	Total Sugar to Date. Pounds.

Lot Nos.	Polarizations.	Pounds of Sugar.	Dates.	Total Sugar to Date. Pounds.

Dates.	Carbonic Acid, CO_2. %	Oxygen, O. %	Carbonic Oxide, CO. %	Nitrogen, N. (By difference.) %	

Dates.	Carbonic Acid, CO_2. %	Oxygen, O. %	Carbonic Oxide, CO. %	Nitrogen, N. (By difference.) %	

Dates.	Carbonic Acid, CO_2. %	Oxygen, O. %	Carbonic Oxide, CO. %	Nitrogen, N. (By difference.) %	

418

Dates.	Carbonic Acid, CO_2. %	Oxygen, O. %	Carbonic Oxide, CO. %	Nitrogen, N. (By difference.) %	

Dates.	Carbonic Acid, CO_2. %	Oxygen, O. %	Carbonic Oxide, CO. %	Nitrogen, N. (By difference.) %	

Dates.	Carbonic Acid, CO_2. %	Oxygen, O. %	Carbonic Oxide, CO. %	Nitrogen, N. (By difference.) %	

Dates.	Carbonic Acid, CO_2. %	Oxygen, O. %	Carbonic Oxide, CO. %	Nitrogen, N. (By difference.) %	

Dates.	Carbonic Acid, CO_2. %	Oxygen, O. %	Carbonic Oxide, CO. %	Nitrogen, N. (By difference.) %	

Dates.	Carbonic Acid, CO_2. %	Oxygen, O. %	Carbonic Oxide, CO. %	Nitrogen, N. (By difference.) %	

Dates.	Carbonic Acid, CO_2. %	Oxygen, O. %	Carbonic Oxide, CO. %	Nitrogen, N. (By difference.) %	

Dates.	Carbonic Acid, CO_2. %	Oxygen, O. %	Carbonic Oxide, CO. %	Nitrogen, N. (By difference.) %	

Dates.	Carbonic Acid, CO_2. %	Oxygen, O. %	Carbonic Oxide, CO. %	Nitrogen, N. (By difference.) %	

SUMMARY

OF

YIELD AND LOSSES.

CROP OF............................SUMMARY OF YIELD AND LOSSES.

Period No............   Commencing....................   Ending....................

SUGAR AND MOLASSES.	Grade.	Total weight, Pounds.	Per ton of Beets, Pounds.	Per cent Beets.	Sucrose per ton of Beets, Pounds.	Sucrose per cent Beets.	Sucrose per cent, total Sucrose.
1st Sugar							
2d Sugar............							
3d Sugar............							
4th Sugar...........							
Molasses, gallons ..							
Molasses, pounds ..							
Total............							

430

Beets worked........................	Tons................................			
Sucrose in the beets....................	Pounds.....................			
Juice extracted......................	"	....		
Sucrose in the juice...................	"	....		
First massecuite, total weight	"	....		
Sucrose accounted for in the sugars and molasses	"	....		
Sucrose accounted for in the sugars and molasses	Per cent beets....................			
Sucrose to be accounted for in losses in manufacture.....	"	"	"	....
Sucrose lost in the exhausted cossettes...............................	"	"	"	....
Sucrose lost in the waste waters......	"	"	"	....
" " by inversion in the diffusion-battery.............	"	"	"	....
Sucrose lost in the diffusion, by difference......	"	"	"	....
Sucrose lost in the concentration to sirup...............................	" .	"	"	....
Sucrose lost in the concentration, etc., from sirup to first massecuite.......	"	"	"	....
Sucrose lost in the concentration, etc., from sirup to molasses	"	"	"	....
Sucrose lost in overflows and wastage.	"	"	"	....
" " " the filter press cake...	"	"	"	....
" " " the evaporation........	"	"	"	....
Other losses, sucrose.	"	"	"	....
Total sucrose accounted for in the losses.......................	"	"	"	....
Total sucrose accounted for in the products and losses.	"	"	"	....
" " unaccounted for......	"	"	"	....

431

CROP OF............................SUMMARY OF YIELD AND LOSSES.

Period No..............   Commencing....................   Ending....................

Sugar and Molasses.	Grade.	Total weight, Pounds.	Per ton of Beets, Pounds.	Per cent Beets.	Sucrose per ton of Beets, Pounds.	Sucrose per cent Beets.	Sucrose per cent, total Sucrose.
1st Sugar							
2d Sugar...........							
3d Sugar...........							
4th Sugar..........							
Molasses, gallons..							
Molasses, pounds..							
Total............							

432

Beets worked........	Tons...............................		
Sucrose in the beets....................	Pounds....................		
Juice extracted.........................	"		
Sucrose in the juice...................	"		
First massecuite, total weight	"		
Sucrose accounted for in the sugars and molasses	"		
Sucrose accounted for in the sugars and molasses...	Per cent beets....................		
Sucrose to be accounted for in losses in manufacture.....	"	"	" ....
Sucrose lost in the exhausted cossettes.................................	"	"	" ....
Sucrose lost in the waste waters......	"	"	" ....
" " by inversion in the diffusion-battery............	"	"	" ....
Sucrose lost in the diffusion, by difference........	"	"	" ....
Sucrose lost in the concentration to sirup..................................	"	"	" ....
Sucrose lost in the concentration, etc., from sirup to first massecuite.......	"	"	" ....
Sucrose lost in the concentration, etc., from sirup to molasses..............	"	"	" ....
Sucrose lost in overflows and wastage.	"	"	" ....
" " " the filter press cake...	"	"	" ....
" " " the evaporation........	"	"	" ....
Other losses, sucrose.	"	"	" ....
Total sucrose accounted for in the losses................................	"	"	" ....
Total sucrose accounted for in the products and losses.	"	"	" ....
" " unaccounted for......	"	"	" ....

433

CROP OF.................................SUMMARY OF YIELD AND LOSSES.

Period No...............   Commencing.................   Ending...............

Sugar and Molasses.	Grade.	Total weight, Pounds.	Per ton of Beets, Pounds.	Per cent Beets.	Sucrose per ton of Beets, Pounds.	Sucrose per cent Beets.	Sucrose per cent, total Sucrose.
1st Sugar							
2d Sugar.........							
3d Sugar.........							
4th Sugar.........							
Molasses, gallons..							
Molasses, pounds..							
Total.........							

434

Beets worked............................ Tons.............................

Sucrose in the beets.................... Pounds.....................

Juice extracted......................... " ....

Sucrose in the juice.................... " ....

First massecuite, total weight......... " ....

Sucrose accounted for in the sugars
   and molasses.......................... " ....

Sucrose accounted for in the sugars
   and molasses... .... Per cent beets....................

Sucrose to be accounted for in losses
   in manufacture...... ... . " " " ....

Sucrose lost in the exhausted cos-
   settes................................ " " " ....

Sucrose lost in the waste waters...... " " " ....

   " " by inversion in the diffu-
      sion-battery............. " " " ....

Sucrose lost in the diffusion, by differ-
   ence........ .... " " " ....

Sucrose lost in the concentration to
   sirup.......................... .... " " " ....

Sucrose lost in the concentration, etc.,
   from sirup to first massecuite... ... " " " ....

Sucrose lost in the concentration, etc.,
   from sirup to molasses............. " " " ....

Sucrose lost in overflows and wastage. " " " ....

   " " " the filter press cake... " " " ....

   " " " the evaporation......... " " " ....

Other losses, sucrose. .... " " " ....

Total sucrose accounted for in the
   losses................................ " " " ....

Total sucrose accounted for in the
         products and losses. " " " ....

   " " unaccounted for..... " " " ....

CROP OF................................SUMMARY OF YIELD AND LOSSES.

Period No............... Commencing..................... Ending.....................

SUGAR AND MOLASSES.	Grade.	Total weight, Pounds.	Per ton of Beets, Pounds.	Per cent Beets.	Sucrose per ton of Beets, Pounds.	Sucrose per cent Beets.	Sucrose per cent, total Sucrose.
1st Sugar							
2d Sugar...........							
3d Sugar...........							
4th Sugar...........							
Molasses, gallons..							
Molasses, pounds..							
Total...........							

Beets worked........................	Tons...........................	
Sucrose in the beets....................	Pounds.....................	
Juice extracted........................	"	
Sucrose in the juice....................	"	
First massecuite, total weight.........	"	
Sucrose accounted for in the sugars and molasses	"	
Sucrose accounted for in the sugars and molasses	Per cent beets....................	
Sucrose to be accounted for in losses in manufacture.....................	" " "	
Sucrose lost in the exhausted cossettes................................	" " "	
Sucrose lost in the waste waters......	" " "	
" " by inversion in the diffusion-battery............	" " "	
Sucrose lost in the diffusion, by difference...........................	" " "	
Sucrose lost in the concentration to sirup.............................	" " "	
Sucrose lost in the concentration, etc., from sirup to first massecuite.......	" " "	
Sucrose lost in the concentration, etc., from sirup to molasses	" " "	
Sucrose lost in overflows and wastage.	" " "	
" " " the filter press cake...	" " "	
" " " the evaporation........	" " "	
Other losses, sucrose.	" " "	
Total sucrose accounted for in the losses	" " "	
Total sucrose accounted for in the products and losses.	" " "	
" " unaccounted for......	" " "	

CROP OF................................SUMMARY OF YIELD AND LOSSES.

Period No................ Commencing................ Ending................

SUGAR AND MOLASSES.	Grade.	Total weight. Pounds.	Per ton of Beets, Pounds.	Per cent Beets.	Sucrose per ton of Beets. Pounds.	Sucrose per cent Beets.	Sucrose per cent, total Sucrose.
1st Sugar..........							
2d Sugar...........							
3d Sugar...........							
4th Sugar..........							
Molasses, gallons..							
Molasses, pounds ..							
Total.........							

438

Beets worked .... Tons........................... ...

Sucrose in the beets.................... Pounds.................... .....

Juice extracted....................... " ....

Sucrose in the juice.................... " ....

First massecuite, total weight .... " ....

Sucrose accounted for in the sugars

    and molasses .... " ....

Sucrose accounted for in the sugars

    and molasses... .... Per cent beets....................

Sucrose to be accounted for in losses

    in manufacture..... ... . " " " ....

Sucrose lost in the exhausted cos-

    settes............................... " " " ....

Sucrose lost in the waste waters...... " " " ....

    " " by inversion in the diffu-

        sion-battery........... " " " ....

Sucrose lost in the diffusion, by differ-

    ence................ .... " " " ....

Sucrose lost in the concentration to

    sirup............................... " " " ....

Sucrose lost in the concentration, etc.,

    from sirup to first massecuite....... " " " ....

Sucrose lost in the concentration, etc.,

    from sirup to molasses............. " " " ....

Sucrose lost in overflows and wastage. " " " ....

    " " " the filter press cake... " " " ....

    " " " the evaporation........ " " " ....

Other losses, sucrose. .... " " " ....

Total sucrose accounted for in the

    losses............................... " " " ....

Total sucrose accounted for in the

             products and losses. " " " ....

    " " unaccounted for...... " " " ....

CROP OF................SUMMARY OF YIELD AND LOSSES.

Period No.............. Commencing................. Ending.................

SUGAR AND MOLASSES.	Grade.	Total weight. Pounds.	Per ton of Beets. Pounds.	Per cent Beets.	Sucrose per ton of Beets. Pounds.	Sucrose per cent Beets.	Sucrose per cent, total Sucrose.
1st Sugar							
2d Sugar..........							
3d Sugar..........							
4th Sugar.........							
Molasses, gallons..							
Molasses, pounds..							
Total.........							

Beets worked.......................... Tons...............................

Sucrose in the beets.................. Pounds.................. ......

Juice extracted....................... " ....

Sucrose in the juice.................. " ....

First massecuite, total weight........ " ....

Sucrose accounted for in the sugars

   and molasses .... " ....

Sucrose accounted for in the sugars

   and molasses... .... Per cent beets...................

Sucrose to be accounted for in losses

   in manufacture..... ... . " " " ....

Sucrose lost in the exhausted cos-

   settes............................... " " " ....

Sucrose lost in the waste waters...... " " " ....

   " " by inversion in the diffu-

          sion-battery............ " " " ....

Sucrose lost in the diffusion, by differ-

   ence................ .... " " " ....

Sucrose lost in the concentration to

   sirup........................ .... " " " ....

Sucrose lost in the concentration, etc.,

   from sirup to first massecuite....... " " " ....

Sucrose lost in the concentration, etc.,

   from sirup to molasses............. " " " ....

Sucrose lost in overflows and wastage. " " " ....

   " " " the filter press-cake... " " " ....

   " " " the evaporation........ " " " ....

Other losses, sucrose. .... " " " ....

Total sucrose accounted for in the

   losses............................... " " " ....

Total sucrose accounted for in the

           products and losses. " " " ....

   " " unaccounted for...... " " " ....

CROP OF.................................SUMMARY OF YIELD AND LOSSES.

Period No................ Commencing...................... Ending....................

SUGAR AND MOLASSES.	Grade.	Total weight. Pounds.	Per ton of Beets. Pounds.	Per cent Beets.	Sucrose per ton of Beets. Pounds.	Sucrose per cent Beets.	Sucrose per cent, total Sucrose.
1st Sugar							
2d Sugar.............							
3d Sugar.............							
4th Sugar...........							
Molasses, gallons..							
Molasses, pounds..							
Total.............							

442

Beets worked............................ Tons.............................

Sucrose in the beets.................... Pounds...........................

Juice extracted........................ " ....

Sucrose in the juice................... " ....

First massecuite, total weight........ " ....

Sucrose accounted for in the sugars
    and molasses............ .... " ....

Sucrose accounted for in the sugars
    and molasses .. .. Per cent beets...................

Sucrose to be accounted for in losses
    in manufacture..... ... . " " " ....

Sucrose lost in the exhausted cos-
    settes.................................. " " " ....

Sucrose lost in the waste waters...... " " " ....

" " by inversion in the diffu-
    sion-battery............ " " " ....

Sucrose lost in the diffusion, by differ-
    ence............. .... " " " ....

Sucrose lost in the concentration to
    sirup.............................. ...... " " " ....

Sucrose lost in the concentration, etc.,
    from sirup to first massecuite....... " " " ....

Sucrose lost in the concentration, etc.,
    from sirup to molasses.............. " " " ....

Sucrose lost in overflows and wastage. " " " ....

" " " the filter press cake... " " " ....

" " " the evaporation........ " " " ....

Other losses, sucrose. .... " " " ....

Total sucrose accounted for in the
    losses......................... .... " " " ....

Total sucrose accounted for in the
    products and losses. " " " ....

" " unaccounted for...... " " " ....

443

CROP OF............................SUMMARY OF YIELD AND LOSSES.

Period No..............   Commencing........................   Ending........................

SUGAR AND MOLASSES.	Grade.	Total weight. Pounds.	Per ton of Beets. Pounds.	Per cent Beets.	Sucrose per ton of Beets. Pounds.	Sucrose per cent Beets.	Sucrose per cent, total Sucrose.
1st Sugar							
2d Sugar							
3d Sugar............							
4th Sugar............							

Beets worked .... Tons...............................

Sucrose in the beets.................... Pounds........................ .....

Juice extracted................. .... " ....

Sucrose in the juice.................... " ....

First massecuite, total weight .... " ....

Sucrose accounted for in the sugars
    and molasses .... " ....

Sucrose accounted for in the sugars
    and molasses .. .. Per cent beets...................

Sucrose to be accounted for in losses
    in manufacture..... ... . " " " ....

Sucrose lost in the exhausted cos-
    settes............................ " " " ....

Sucrose lost in the waste waters...... " " " ....

    " " by inversion in the diffu-
        sion-battery............ " " " ....

Sucrose lost in the diffusion, by differ-
    ence .... " " " ....

Sucrose lost in the concentration to
    sirup............................ " " " ....

Sucrose lost in the concentration, etc.,
    from sirup to first massecuite... ... " " " ....

Sucrose lost in the concentration, etc.,
    from sirup to molasses .... " " " ....

Sucrose lost in overflows and wastage. " " " ....

    " " " the filter press cake... " " " ....

    " " " the evaporation........ " " " ....

Other losses, sucrose .... " " " ....

Total sucrose accounted for in the
    losses .... " " " ....

Total sucrose accounted for in the
           products and losses. " " " ....

    " " unaccounted for...... " " " ....

CROP OF.................SUMMARY OF YIELD AND LOSSES.

Period No................ Commencing................ Ending................

SUGAR AND MOLASSES.	Grade.	Total weight, Pounds.	Per ton of Beets. Pounds.	Per cent Beets.	Sucrose per ton of Beets. Pounds.	Sucrose per cent Beets.	Sucrose per cent, total Sucrose.
1st Sugar							
2d Sugar							
3d Sugar							
4th Sugar							
Molasses, gallons..							
Molasses, pounds..							
Total..........							

446

Beets worked .... Tons..............................

Sucrose in the beets.................... Pounds............................

Juice extracted........................ " ....

Sucrose in the juice................... " ....

First massecuite, total weight .... " ....

Sucrose accounted for in the sugars
    and molasses .... " ....

Sucrose accounted for in the sugars
    and molasses ... . Per cent beets....................

Sucrose to be accounted for in losses
    in manufacture..... ... . " " " ....

Sucrose lost in the exhausted cos-
    settes................................ " " " ....

Sucrose lost in the waste waters...... " " " ....

    " " by inversion in the diffu-
        sion-battery............ " " " ....

Sucrose lost in the diffusion, by differ-
    ence........ .... " " " ....

Sucrose lost in the concentration to
    sirup.............................. " " " ....

Sucrose lost in the concentration, etc.,
    from sirup to first massecuite....... " " " ....

Sucrose lost in the concentration, etc.,
    from sirup to molasses..... .... " " " ....

Sucrose lost in overflows and wastage. " " " ....

    " " " the filter press-cake... " " " ....

    " " " the evaporation........ " " " ....

Other losses, sucrose. .... " " " ....

Total sucrose accounted for in the
    losses................................ " " " ....

Total sucrose accounted for in the
              products and losses. " " " ....

    " " unaccounted for...... " " " ....

447

CROP OF.................SUMMARY OF YIELD AND LOSSES.

Period No.............. Commencing............... Ending...............

SUGAR AND MOLASSES.	Grade.	Total weight. Pounds.	Per ton of Beets. Pounds.	Per cent Beets.	Sucrose per ton of Beets. Pounds.	Sucrose per cent Beets.	Sucrose per cent, total Sucrose.
1st Sugar...........							
2d Sugar...........							
3d Sugar...........							
4th Sugar...........							
Molasses, gallons..							
Molasses, pounds..							
Total...........							

448

Beets worked .... Tons ....

Sucrose in the beets .... Pounds ....

Juice extracted .... " ....

Sucrose in the juice .... " ....

First massecuite, total weight .... " ....

Sucrose accounted for in the sugars
and molasses .... " ....

Sucrose accounted for in the sugars
and molasses ... . Per cent beets...................

Sucrose to be accounted for in losses
in manufacture..... ... . "   "   " ....

Sucrose lost in the exhausted cos-
settes................................ "   "   " ....

Sucrose lost in the waste waters...... "   "   " ....

"   '   "   by inversion in the diffu-
sion battery............ "   "   " ....

Sucrose lost in the diffusion, by differ-
ence........ .... "   "   " ....

Sucrose lost in the concentration to
sirup.................... .... "   "   " ....

Sucrose lost in the concentration, etc.,
from sirup to first massecuite... ... "   "   " ....

Sucrose lost in the concentration, etc.,
from sirup to molasses.... .... "   "   " ....

Sucrose lost in overflows and wastage. "   "   " ....

"   "   " the filter press cake... "   "   " ....

"   "   " the evaporation........ "   "   " ....

Other losses, sucrose. .... "   "   " ....

Total sucrose accounted for in the
losses................................ "   "   " ....

Total sucrose accounted for in the
products and losses. "   "   " ....

"   "   unaccounted for...... "   "   " ....

449

CROP OF.................SUMMARY OF YIELD AND LOSSES.

Period No.................Commencing.................Ending.................

SUGAR AND MOLASSES.	Grade.	Total weight. Pounds.	Per ton of Beets. Pounds.	Per cent Beets.	Sucrose per ton of Beets. Pounds.	Sucrose per cent Beets.	Sucrose per cent, total Sucrose.
1st Sugar							
2d Sugar							
3d Sugar							
4th Sugar							
Molasses, gallons..							
Molasses, pounds..							
Total							

Beets worked........ .... Tons........................,...........

Sucrose in the beets.................... Pounds.................... ......

Juice extracted................. .... " ....

Sucrose in the juice.................... " ....

First massecuite, total weight .... " ....

Sucrose accounted for in the sugars

    and molasses .... " ....

Sucrose accounted for in the sugars

    and molasses... .... Per cent beets....................

Sucrose to be accounted for in losses

    in manufacture..... ... . " " " ....

Sucrose lost in the exhausted cos-

    settes............................. " " " ....

Sucrose lost in the waste waters...... " " " ....

    " " by inversion in the diffu-

              sion-battery........... " " " ....

Sucrose lost in the diffusion, by differ-

    ence............ .... " " " ....

Sucrose lost in the concentration to

    sirup................................ " " " ....

Sucrose lost in the concentration, etc.,

    from sirup to first massecuite....... " " " ....

Sucrose lost in the concentration, etc.,

    from sirup to molasses.............. " " " ....

Sucrose lost in overflows and wastage. " " " ....

    " " " the filter press-cake... " " " ....

    " " " the evaporation........ " " " ....

Other losses, sucrose. .... " " " ....

Total sucrose accounted for in the

    losses .... " " " ....

Total sucrose accounted for in the

                 products and losses. " " " ....

    " " unaccounted for...... " " " ....

451

CROP OF.................SUMMARY OF YIELD AND LOSSES.

Period No............   Commencing.................   Ending.............

SUGAR AND MOLASSES.	Grade.	Total weight, Pounds.	Per ton of Beets. Pounds.	Per cent Beets.	Sucrose per ton of Beets. Pounds.	Sucrose per cent Beets.	Sucrose per cent, total Sucrose.
1st Sugar							
2d Sugar...........							
3d Sugar...........							
4th Sugar...........							
Molasses, gallons..							
Molasses, pounds..							
Total...........							

452

Beets worked .... Tons. ....

Sucrose in the beets.................... Pounds..................... ......

Juice extracted.................. ....   " ....

Sucrose in the juice....................   " ....

First massecuite, total weight ....   " ....

Sucrose accounted for in the sugars

  and molasses ....   " ....

Sucrose accounted for in the sugars

  and molasses .. .. Per cent beets....................

Sucrose to be accounted for in losses

  in manufacture..... ... .   "   "   " ....

Sucrose lost in the exhausted cos-

  settes................................   "   "   " ....

Sucrose lost in the waste waters......   "   "   " ....

    "   " by inversion in the diffu-

        sion-battery............   "   "   " ....

Sucrose lost in the diffusion, by differ-

  ence....... ....   "   "   " ....

Sucrose lost in the concentration to

  sirup................................   "   "   " ....

Sucrose lost in the concentration, etc.,

  from sirup to first massecuite.......   "   "   " ....

Sucrose lost in the concentration, etc.,

  from sirup to molasses..... ....   "   "   " ....

Sucrose lost in overflows and wastage.   "   "   " ....

    "   "   " the filter press cake...   "   "   " ....

    "   "   " the evaporation........   "   "   " ....

Other losses, sucrose. ....   "   "   " ....

Total sucrose accounted for in the

  losses................................   "   "   " ....

Total sucrose accounted for in the

           products and losses.   "   "   " ....

    "   "   unaccounted for......   "   "   " ....

453

CROP OF................................SUMMARY OF YIELD AND LOSSES.

Period No............... Commencing............... Ending...............

SUGAR AND MOLASSES.	Grade.	Total weight. Pounds.	Per ton of Beets. Pounds.	Per cent Beets.	Sucrose per ton of Beets. Pounds.	Sucrose per cent Beets.	Sucrose per cent, total Sucrose.
1st Sugar							
2d Sugar.........							
3d Sugar.........							
4th Sugar.........							
Molasses, gallons..							
Molasses, pounds..							
Total...........							

454

Beets worked............................ Tons..............................

Sucrose in the beets.................... Pounds..................... ......

Juice extracted................. .... " ....

Sucrose in the juice.................... " ....

First massecuite, total weight .... " ....

Sucrose accounted for in the sugars

   and molasses .... " ....

Sucrose accounted for in the sugars

   and molasses... .... Per cent beets....................

Sucrose to be accounted for in losses

   in manufacture..... ... . "   "   "   ....

Sucrose lost in the exhausted cos-

   settes................................ "   "   "   ....

Sucrose lost in the waste waters...... "   "   "   ....

   "   " by inversion in the diffu-

       sion-battery............ "   "   "   ....

Sucrose lost in the diffusion, by differ-

   ence....... .... "   "   "   ....

Sucrose lost in the concentration to

   sirup................................. "   "   "   ....

Sucrose lost in the concentration, etc.,

   from sirup to first massecuite....... "   "   "   ....

Sucrose lost in the concentration, etc.,

   from sirup to molasses.... .... "   "   "   ....

Sucrose lost in overflows and wastage. "   "   "   ....

   "   " " the filter press-cake... "   "   "   ....

   "   " " the evaporation........ "   "   "   ....

Other losses, sucrose. .... "   "   "   ....

Total sucrose accounted for in the

   losses............................... "   "   "   ....

Total sucrose accounted for in the

          products and losses. "   "   "   ....

   "   "   unaccounted for...... "   "   "   ....

CROP OF...............SUMMARY OF YIELD AND LOSSES.

Period No................ Commencing................ Ending................

SUGAR AND MOLASSES.	Grade.	Total weight. Pounds.	Per ton of Beets. Pounds.	Per cent Beets.	Sucrose per ton of Beets. Pounds.	Sucrose per cent Beets.	Sucrose per cent, total Sucrose.
1st Sugar							
2d Sugar							
3d Sugar							
4th Sugar							
Molasses, gallons							
Molasses, pounds							
Total							

Beets worked	Tons.		....
Sucrose in the beets....................	Pounds		....
Juice extracted..................	"		....
Sucrose in the juice....................	"		....
First massecuite, total weight	"		....
Sucrose accounted for in the sugars and molasses	"		....
Sucrose accounted for in the sugars and molasses	Per cent beets....................		
Sucrose to be accounted for in losses in manufacture.....	"	" "	....
Sucrose lost in the exhausted cossettes.............................	"	" "	....
Sucrose lost in the waste waters......	"	" "	....
" " by inversion in the diffusion-battery............	"	" "	....
Sucrose lost in the diffusion, by difference........	"	" "	....
Sucrose lost in the concentration to sirup..........................	"	" "	....
Sucrose lost in the concentration, etc., from sirup to first massecuite... ...	"	" "	....
Sucrose lost in the concentration, etc., from sirup to molasses	"	" "	....
Sucrose lost in overflows and wastage.	"	" "	....
" " " the filter press cake...	"	" "	....
" " " the evaporation........	"	" "	....
Other losses, sucrose.	"	" "	....
Total sucrose accounted for in the losses.........	"	" "	....
Total sucrose accounted for in the products and losses.	"	" "	....
" " unaccounted for......	"	" "	....

CROP OF..................................SUMMARY OF YIELD AND LOSSES.

Period No................. Commencing.................. Ending..................

SUGAR AND MOLASSES.	Grade.	Total weight. Pounds.	Per ton of Beets. Pounds.	Per cent Beets.	Sucrose per ton of Beets. Pounds.	Sucrose per cent Beets.	Sucrose per cent, total Sucrose.
1st Sugar							
2d Sugar............							
3d Sugar............							
4th Sugar..........							
Molasses, gallons...							
Molasses, pounds...							

458

Beets worked .... Tons..............................

Sucrose in the beets................... Pounds...................... .....

Juice extracted.................. ....   "   ....

Sucrose in the juice...................   "   ....

First massecuite, total weight ....   "   ....

Sucrose accounted for in the sugars

    and molasses ....   "   ....

Sucrose accounted for in the sugars

    and molasses... .... Per cent beets...................

Sucrose to be accounted for in losses

    in manufacture..... ... .   "   "   "   ....

Sucrose lost in the exhausted cos-

    settes...............................   "   "   "   ....

Sucrose lost in the waste waters......   "   "   "   ....

    "   "   by inversion in the diffu-

        sion-battery............   "   "   "   ....

Sucrose lost in the diffusion, by differ-

    ence............... ....   "   "   "   ....

Sucrose lost in the concentration to

    sirup........................... ....   "   "   "   ....

Sucrose lost in the concentration, etc.,

    from sirup to first massecuite... ...   "   "   "   ....

Sucrose lost in the concentration, etc.,

    from sirup to molasses.............   "   "   "   ....

Sucrose lost in overflows and wastage.   "   "   "   ....

    "   "   " the filter press-cake...   "   "   "   ....

    "   "   " the evaporation........   "   "   "   ....

Other losses, sucrose. ....   "   "   "   ....

Total sucrose accounted for in the

    losses ....   "   "   "   ....

Total sucrose accounted for in the

           products and losses.   "   "   "   ....

    "   "   unaccounted for......   "   "   "   ....

459

CROP OF............................SUMMARY OF YIELD AND LOSSES.

Period No................ Commencing.................... Ending........................

SUGAR AND MOLASSES.	Grade.	Total weight. Pounds.	Per ton of Beets. Pounds.	Per cent Beets.	Sucrose per ton of Beets. Pounds.	Sucrose per cent Beets.	Sucrose per cent, total Sucrose.
1st Sugar							
2d Sugar............							
3d Sugar............							
4th Sugar...........							
Molasses, gallons...							
Molasses, pounds...							
Total............							

460

Beets worked .... Tons............................ ....

Sucrose in the beets.................... Pounds...................... ......

Juice extracted.......................... " ....

Sucrose in the juice.................... " ....

First massecuite, total weight .... " ....

Sucrose accounted for in the sugars

   and molasses .... " ....

Sucrose accounted for in the sugars

   and molasses... .... Per cent beets....................

Sucrose to be accounted for in losses

   in manufacture..... ... . " " " ....

Sucrose lost in the exhausted cos-

   settes................................ " " " ....

Sucrose lost in the waste waters...... " " " ....

   " " by inversion in the diffu-

      sion-battery............ " " " ....

Sucrose lost in the diffusion, by differ-

   ence....... .... " " " ....

Sucrose lost in the concentration to

   sirup.............................. " " " ....

Sucrose lost in the concentration, etc.,

   from sirup to first massecuite....... " " " ....

Sucrose lost in the concentration, etc.,

   from sirup to molasses.............. " " " ....

Sucrose lost in overflows and wastage. " " " ....

   " " " the filter press cake... " " " ....

   " " " the evaporation........ " " " ....

Other losses, sucrose. .... " " " ....

Total sucrose accounted for in the

   losses............. .... " " " ....

Total sucrose accounted for in the

          products and losses. " " " ....

   " " unaccounted for...... " " " ....

461

INDEX.

A.

PAGE

Q.

R.

LIST OF TABLES AND FORMULÆ.

473

INVERT-SUGAR AND INVERSION.

MISCELLANEOUS TABLES AND FORMULÆ.

REAGENTS.

SOLUBILITIES.

STRENGTH OF VARIOUS SOLUTIONS, ETC.

THERMAL DATA.

TABLES FOR CALCULATING SUCROSE, REDUCING SUGAR AND PURITY.

TOTAL SOLIDS.

WATER ANALYSIS.

ADVERTISEMENTS.

Vilmorin=Andrieux & Co.
SEEDSMEN.

4, QUAI DE LA MÉGISSERIE, 4
Paris, France.

THE firm of VILMORIN-ANDRIEUX & Co. has by methodic and scientific selection, during a period of over forty years, produced a beet of perfect form and containing a juice of the greatest richness and purity. This is the well-known "**Vilmorin's Improved White Beet.**" The keeping qualities of this beet are unsurpassed. Late autumnal rains cause less deterioration in this than in most other varieties, and it is better adapted to black virgin soil.

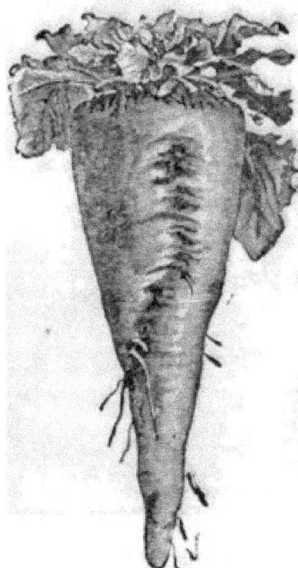

Messrs. VILMORIN - ANDRIEUX & Co. also produce the seed of the **Klein Wanzlebener** and of the **French Very Rich** beet, taking every precaution to select only the highest grade of seed.

B. Improved Vilmorin.

For full information, including circulars and catalogues, address the firm or

AUG. RHOTERT, 26 Barclay St., New York,
AGENT FOR THE UNITED STATES AND CANADA.

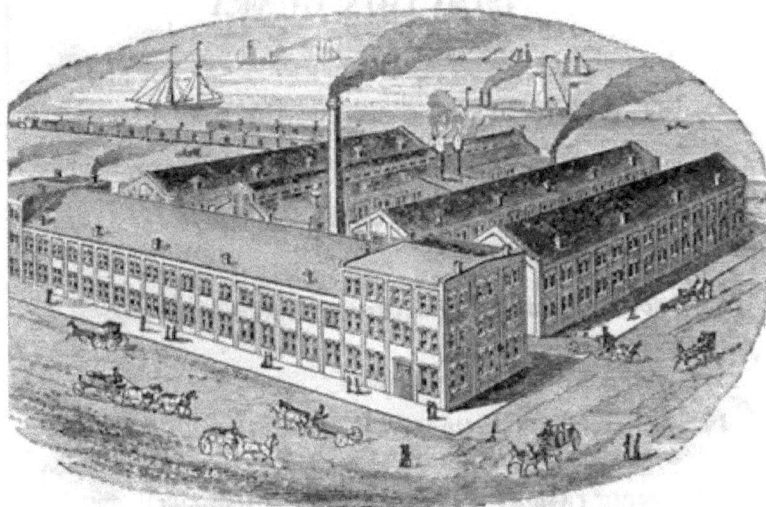

www.ingramcontent.com/pod-product-compliance
Lightning Source LLC
Chambersburg PA
CBHW020900210326
41598CB00018B/1727